J. W. Niemantsverdriet
Spectroscopy in Catalysis

1807–2007 Knowledge for Generations

Each generation has its unique needs and aspirations. When Charles Wiley first opened his small printing shop in lower Manhattan in 1807, it was a generation of boundless potential searching for an identity. And we were there, helping to define a new American literary tradition. Over half a century later, in the midst of the Second Industrial Revolution, it was a generation focused on building the future. Once again, we were there, supplying the critical scientific, technical, and engineering knowledge that helped frame the world. Throughout the 20th Century, and into the new millennium, nations began to reach out beyond their own borders and a new international community was born. Wiley was there, expanding its operations around the world to enable a global exchange of ideas, opinions, and know-how.

For 200 years, Wiley has been an integral part of each generation's journey, enabling the flow of information and understanding necessary to meet their needs and fulfill their aspirations. Today, bold new technologies are changing the way we live and learn. Wiley will be there, providing you the must-have knowledge you need to imagine new worlds, new possibilities, and new opportunities.

Generations come and go, but you can always count on Wiley to provide you the knowledge you need, when and where you need it!

William J. Pesce
President and Chief Executive Officer

Peter Booth Wiley
Chairman of the Board

J. W. Niemantsverdriet

Spectroscopy in Catalysis

An Introduction

Third, Completely Revised and Enlarged Edition

WILEY-VCH Verlag GmbH & Co. KGaA

The Author

Prof. Dr. J. W. Niemantsverdriet
Eindhoven University of Technology
Schuit Institute of Catalysis
Den Dolech 2
5612 AZ Eindhoven
The Netherlands

All books published by Wiley-VCH are carefully produced. Nevertheless, authors, editors, and publisher do not warrant the information contained in these books, including this book, to be free of errors. Readers are advised to keep in mind that statements, data, illustrations, procedural details or other items may inadvertently be inaccurate.

Library of Congress Card No.: applied for

British Library Cataloguing-in-Publication Data
A catalogue record for this book is available from the British Library

Bibliographic information published by the Deutsche Nationalbibliothek
Die Deutsche Nationalbibliothek lists this publication in the Deutsche Nationalbibliografie; detailed bibliographic data are available in the Internet at http://dnb.d-nb.de

© 2007 WILEY-VCH Verlag GmbH & Co. KGaA, Weinheim

All rights reserved (including those of translation into other languages). No part of this book may be reproduced in any form – by photoprinting, microfilm, or any other means – nor transmitted or translated into a machine language without written permission from the publishers. Registered names, trademarks, etc. used in this book, even when not specifically marked as such, are not to be considered unprotected by law.

Printed in the Federal Republic of Germany
Printed on acid-free paper

Typesetting Asco Typesetters, Hong Kong
Printing Strauss GmbH, Mörlenbach
Bookbinding Litges & Dopf GmbH, Heppenheim
Wiley Bicentennial Logo Richard J. Pacifico

ISBN 978-3-527-31651-9

To Marianne, Hanneke, Annemieke, Karin and Peter

Contents

Preface *XIII*

List of Acronyms *XVII*

1 **Introduction** *1*
1.1 Heterogeneous Catalysis *1*
1.2 The Aim of Catalyst Characterization *4*
1.3 Spectroscopic Techniques *5*
1.4 Research Strategies *7*
References *9*

2 **Temperature-Programmed Techniques** *11*
2.1 Introduction *11*
2.2 Temperature-Programmed Reduction *13*
2.2.1 Thermodynamics of Reduction *13*
2.2.2 Reduction Mechanisms *15*
2.2.3 Applications *18*
2.3 Temperature-Programmed Sulfidation *21*
2.4 Temperature-Programmed Reaction Spectroscopy *22*
2.5 Temperature-Programmed Desorption *23*
2.5.1 TPD Analysis *29*
2.5.2 Desorption in the Transition State Theory *31*
2.6 Temperature-Programmed Reaction Spectroscopy in UHV *35*
References *37*

3 **Photoemission and Auger Spectroscopy** *39*
3.1 Introduction *39*
3.2 X-Ray Photoelectron Spectroscopy (XPS) *41*
3.2.1 XPS Intensities and Sample Composition *44*
3.2.2 XPS Binding Energies and Oxidation States *46*
3.2.3 Shake Up, Shake Off, Multiplet Splitting and Plasmon Excitations *50*
3.2.4 Experimental Aspects of XPS *51*

Spectroscopy in Catalysis: An Introduction, Third Edition
J. W. Niemantsverdriet
Copyright © 2007 WILEY-VCH Verlag GmbH & Co. KGaA, Weinheim
ISBN: 978-3-527-31651-9

3.2.5	Charging and Sample Damage	*52*
3.2.6	Dispersion of Supported Particles from XPS	*54*
3.2.7	Angle-Dependent XPS	*59*
3.2.8	*In-Situ* and Real Time XPS Studies	*63*
3.3	Ultraviolet Photoelectron Spectroscopy (UPS)	*65*
3.3.1	Photoemission of Adsorbed Xenon	*71*
3.4	Auger Electron Spectroscopy	*74*
3.4.1	Energy of Auger Peaks	*75*
3.4.2	Intensity of Auger Peaks	*77*
3.4.3	Application of AES in Catalytic Surface Science	*78*
3.4.4	Scanning Auger Spectroscopy	*80*
3.4.5	Depth-Sensitive Information from AES	*80*
	References 81	

4 The Ion Spectroscopies *85*

4.1	Introduction	*85*
4.2	Secondary Ion Mass Spectrometry (SIMS)	*86*
4.2.1	Theory of SIMS	*88*
4.2.2	Electron and Photon Emission under Ion Bombardment	*90*
4.2.3	Energy Distribution of Secondary Ions	*91*
4.2.4	The Ionization Probability	*92*
4.2.5	Emission of Molecular Clusters	*94*
4.2.6	Conditions for Static SIMS	*94*
4.2.7	Charging of Insulating Samples	*95*
4.2.8	Applications on Catalysts	*95*
4.2.9	Model Catalysts	*99*
4.2.10	Single Crystal Studies	*101*
4.2.11	Concluding Remarks	*105*
4.3	Secondary Neutral Mass Spectrometry (SNMS)	*105*
4.4	Ion Scattering: The Collision Process	*106*
4.5	Rutherford Backscattering Spectrometry (RBS)	*108*
4.6	Low-Energy Ion Scattering (LEIS)	*112*
4.6.1	Neutralization	*113*
4.6.2	Applications of LEIS in Catalysis	*114*
	References 117	

5 Mössbauer Spectroscopy *121*

5.1	Introduction	*121*
5.2	The Mössbauer Effect	*122*
5.3	Mössbauer Spectroscopy	*126*
5.3.1	Isomer Shift	*128*
5.3.2	Electric Quadrupole Splitting	*129*
5.3.3	Magnetic Hyperfine Splitting	*131*
5.3.4	Intensity	*132*

5.4	Mössbauer Spectroscopy in Catalyst Characterization	*134*
5.4.1	*In-Situ* Mössbauer Spectroscopy at Cryogenic Temperatures	*137*
5.4.2	Particle Size Determination	*139*
5.4.3	Kinetics of Solid-State Reactions from Single Velocity Experiments	*140*
5.4.4	*In-Situ* Mössbauer Spectroscopy Under Reaction Conditions	*141*
5.4.5	Mössbauer Spectroscopy of Elements Other Than Iron	*143*
5.5	Conclusion	*145*
	References	*145*

6	**Diffraction and Extended X-Ray Absorption Fine Structure (EXAFS)**	*147*
6.1	Introduction	*147*
6.2	X-Ray Diffraction	*148*
6.2.1	*In-Situ* XRD: Kinetics of Solid-State Reactions	*152*
6.2.2	Concluding Remarks	*154*
6.3	Low-Energy Electron Diffraction (LEED)	*155*
6.4	X-Ray Absorption Fine Structure (XAFS)	*159*
6.4.1	EXAFS	*160*
6.4.2	Quick EXAFS for Time-Resolved Studies	*170*
6.4.3	X-Ray Absorption Near Edge Spectroscopy	*172*
	References	*175*

7	**Microscopy and Imaging**	*179*
7.1	Introduction	*179*
7.2	Electron Microscopy	*180*
7.2.1	Transmission Electron Microscopy	*182*
7.2.2	Scanning Electron Microscopy	*184*
7.2.3	Scanning Transmission Electron Microscopy	*186*
7.2.4	Element Analysis in the Electron Microscope	*190*
7.3	Field Emission Microscopy and Ion Microscopy	*193*
7.3.1	Theory of FEM and FIM	*193*
7.4	Scanning Probe Microscopy: AFM and STM	*197*
7.4.1	AFM and SFM	*198*
7.4.1.1	Contact Mode AFM	*199*
7.4.1.2	Non-Contact Mode AFM	*200*
7.4.1.3	Tapping Mode AFM	*200*
7.4.2	AFM Equipment	*200*
7.4.3	Scanning Tunneling Microscopy (STM)	*205*
7.4.4	Applications of STM in Catalytic Surface Science	*208*
7.5	Other Imaging Techniques	*211*
7.5.1	Low-Energy Electron Microscopy and Photoemission Electron Microscopy	*212*
	References	*214*

8	**Vibrational Spectroscopy** *217*	
8.1	Introduction *217*	
8.2	Theory of Molecular Vibrations *218*	
8.3	Infrared Spectroscopy *224*	
8.3.1	Equipment *226*	
8.3.2	Applications of Infrared Spectroscopy *226*	
8.3.3	Transmission Infrared Spectroscopy *227*	
8.3.4	Diffuse Reflectance Infrared Fourier Transform Spectroscopy (DRIFTS) *230*	
8.3.5	Attenuated Total Reflection *233*	
8.3.6	Reflection Absorption Infrared Spectroscopy (RAIRS) *234*	
8.4	Sum-Frequency Generation *235*	
8.5	Raman Spectroscopy *238*	
8.5.1	Applications of Raman Spectroscopy *240*	
8.6	Electron Energy Loss Spectroscopy (EELS) *243*	
8.7	Concluding Remarks *247*	
	References *248*	
9	**Case Studies in Catalyst Characterization** *251*	
9.1	Introduction *251*	
9.2	Supported Rhodium Catalysts *251*	
9.2.1	Preparation of Alumina-Supported Rhodium Model Catalysts *252*	
9.2.2	Reduction of Supported Rhodium Catalysts *254*	
9.2.3	Structure of Supported Rhodium Catalysts *257*	
9.2.4	Disintegration of Rhodium Particles Under CO *261*	
9.2.5	Concluding Remarks *264*	
9.3	Alkali Promoters on Metal Surfaces *264*	
9.4	Cobalt–Molybdenum Sulfide Hydrodesulfurization Catalysts *272*	
9.4.1	Sulfidation of Oxidic Catalysts *272*	
9.4.2	Structure of Sulfided Catalysts *276*	
9.5	Chromium Polymerization Catalysts *284*	
9.6	Concluding Remarks *292*	
	References *293*	
Appendix	**Metal Surfaces and Chemisorption** *297*	
A.1	Introduction *297*	
A.2	Theory of Metal Surfaces *297*	
A.2.1	Surface Crystallography *297*	
A.2.2	Surface Free Energy *301*	
A.2.3	Lattice Vibrations *302*	
A.2.4	Electronic Structure of Metal Surfaces *305*	
A.2.5	Work Function *309*	
A.3	Chemisorption on Metals *311*	

A.3.1 Adsorption of Molecules on Jellium *315*
A.3.2 Adsorption on Metals with d-Electrons *317*
A.3.3 Concluding Remarks *319*
References 319

Index *321*

Preface

Spectroscopy in Catalysis is an introduction to the most important analytical techniques that are nowadays used in catalysis and in catalytic surface chemistry. The aim of the book is to give the reader a feeling for the type of information that characterization techniques provide about questions concerning catalysts or catalytic phenomena, in routine or more advanced applications.

The title *Spectroscopy in Catalysis* is attractively compact, but not quite precise. The book also introduces microscopy, diffraction and temperature-programmed reaction methods, as these are important tools in the characterization of catalysts. As to applications, I have limited myself to supported metals, oxides, sulfides and metal single crystals. Zeolites, as well as techniques such as nuclear magnetic resonance and electron spin resonance, have been left out – mainly because I have little personal experience with these subjects. Catalysis would not be what it is without surface science. Hence, techniques that are applicable to study the surfaces of single crystals or metal foils used to model catalytic surfaces, have been included.

The book has been written as an introductory text rather than as an exhaustive review. It is meant for students at the start of their Ph.D. projects, and also for anyone else who needs a concise introduction to catalyst characterization. Each chapter describes the physical background and principles of a technique, a few recent applications to illustrate the type of information that can be obtained, and an evaluation of possibilities and limitations. Chapter 9 contains case studies which highlight a few important catalyst systems and illustrates the power of combining techniques. The Appendix, which incorporates the surface theory of metals and details of chemical bonding at surfaces, is included to provide a better insight into the results of photoemission, vibrational spectroscopy, and thermal desorption.

Finally, an important starting point is that reading the book should be enjoyable. Therefore, the book contains many illustrations, as few theoretical formulas as possible, and no mathematical derivations. I hope that the book will be useful and that it conveys some of the enthusiasm I feel for research in catalysis.

About the Third Edition

The present version of the book represents a completely revised update of the first edition as it appeared in 1993, and the second from 2000. Significant new developments in, for example, electron and scanning probe microscopy, synchrotron techniques and vibrational techniques called for revision and additions to the respective chapters. However, the other chapters have also been updated with recent examples, and references to relevant new literature. Many figures from the first two editions have subsequently been improved to make them more informative.

Since its publication, I have used the book as an accompanying text in courses on catalyst characterization, both at the Eindhoven University of Technology, the Netherlands Institute for Catalysis Research, NIOK, as well as in several short courses all over the world. It has been very rewarding to learn that several colleagues in catalysis have also adopted the book for their courses. I will be very grateful for comments and corrections.

Acknowledgments

When I obtained my Masters Degree in experimental physics at the Free University in Amsterdam in 1978, I was totally unaware that as interesting an area as catalysis, with so many challenges for the physicist, existed. I am particularly grateful to Adri van der Kraan and Nick Delgass who introduced me, via the Mössbauer effect in iron catalysts, to the field of catalysis. My Ph.D. advisors at Delft, Adri van der Kraan, Jan van Loef and Vladimir Ponec (Leiden), together with Roel Prins from Eindhoven, stimulated and helped me to pursue a career in catalysis, which now extends back almost 25 years.

Throughout all of these years, Eindhoven has been a stimulating environment. Working with Rutger van Santen, Diek Koningsberger, Roel Prins, Dick van Langeveld, Jan van Hooff(†), Ben Nieuwenhuys, Jaap Schouten, Gert-Jan Kramer, San de Beer, Theo Beelen, Barbara Mojet, Tonek Jansen, Rob van Veen, Bruce Anderson, Dieter Vogt, Armando Borgna and Peter Thüne, and outside our own department Hidde Brongersma, Kees Flipse, Guy Marin, Piet Lemstra, Joachim Loos, Martien de Voigt, Leo van IJzendoorn and Peter Hilbers, has been a great experience. It is a pleasure to mention pleasant and fruitful collaborations with Nick Delgass, Steen Mørup, Ib Chorkendorff, Klaus Wandelt, Fabio Ribeiro, Bruce Gates, Josep Ricart, Peter van Berge, Jan van de Loosdrecht and Gabor Somorjai, who have taught me a lot over the years. I gratefully acknowledge the generous financial support of "Huygens" and "Pionier" fellowships from the Netherlands Organization for Scientific Research (NWO).

This book contains many examples taken from a group of people with whom I worked at the Eindhoven University of Technology. I thank Paco Ample, Adelaida Andoni, Sander van Bavel, Herman Borg, Tracy Bromfield, Srilakshmi Chilukoti, Leon Coulier, Dani Curulla, Andre Engelen, Lyn Eshelman, Wouter van Gennip,

Ashriti Govender, Leo van Gruijthuijsen, Pieter Gunter, Martijn van Hardeveld, Marco Hopstaken, Maarten Jansen, Ton Janssens, Arthur de Jong, Ramesh Kanaparthi, Emiel van Kimmenade, Gurram Kishan, Eero Kontturi, Hanna Korpik, Ralf Linke, Joachim Loos, Stefan Louwers, Toon Meijers, Prabashini and Denzil Moodley, Hannie Muijsers, Davy Nieskens, Leon van den Oetelaar, Joost Reijerse, Abdool Saib, Arno Sanders, Freek Scheijen, Elina Siokou, Peter Thüne, Rik van Veen, Tiny Verhoeven, Thomas Weber, Han Wei, and many others for the very pleasant and fruitful collaboration we have had.

Finally, I wish to thank my wife Marianne, our daughters Hanneke, Annemieke, Karin and son Peter for the many weekends and evenings they patiently allowed me to work on this book and their updated versions. I dedicate this book to them.

Nuenen, April 2007   *Hans Niemantsverdriet*

List of Acronyms

ADF	annular dark field
AES	Auger electron spectrsocopy
AFM	atomic force microscopy
ATR	attenuated total reflection
bcc	body-centered cubic
BET	Brunauer–Emmett–Teller equation
BIS	Bremstrahlung isochromat spectroscopy
DOS	density of states
DRIFTS	diffuse reflection infrared Fourier transform spectrsocopy
EDX	energy dispersive X-ray analysis
EELS	electron energy loss spectroscopy
EMSI	ellipsometry microscopy for surface imaging
ESCA	electron spectroscopy for chemical analysis
ESEM	environmental scanning electron microscopy
EXAFS	extended X-ray absorption fine structure
fcc	face-centerd cubic
FEM	field emission microscopy
FIM	field ion microscopy
FT	Fourier transform
FTIR	Fourier transform infra red
HAADF	high-angle annular dark field
hcp	hexagonally closed packed
HREELS	high resolution electron energy loss spectroscopy
INS	inelastic neutron scattering
IRAS	Infrared reflection absorption spectroscopy
IRES	Infrared emission spectroscopy
IRI	infrared imaging
ISS	ion scattering spectroscopy
LEED	low energy electron diffraction
LEIS	low energy ion scattering
MAES	metastable atom excitation spectroscopy
MAS	Mossbauer absorption spectroscopy
MES	Mossbauer emission spectroscopy

Spectroscopy in Catalysis: An Introduction, Third Edition
J. W. Niemantsverdriet
Copyright © 2007 WILEY-VCH Verlag GmbH & Co. KGaA, Weinheim
ISBN: 978-3-527-31651-9

MIES	metastable ion excitation spectroscopy
NEXAFS	NEXAFS
PAX	Photoemission of adsorbed xenon
PEEM	photoelectron emission microscopy
PFDMS	pulsed field desorption mass spectrometry
QEXAFS	Quick EXAFS
RAIRS	reflection absorption infrared spectroscopy
RAM	reflection anisotropy microscopy
RBS	Rutherford backscattering spectroscopy
SAM	scanning Auger microscopy
SAXS	small angle X-ray scattering
SEM	scanning electron microscopy
SFG	sum frequency generation
SFM	scanning force microscopy
SIMS	secondary ion mass spectrometry
SNMS	SNMS
SPM	scanning probe microscopy
SSIMS	static secondary ion mass spectrometry
STEM	scanning transmission electron microscopy
STM	scanning tunneling microscopy
STS	scanning tunneling spectroscopy
TDS, see TPD	thermal desorption spectroscopy, see TPD
TEM	transmission electron microscopy
TPD	temperature programmed desorption
TPO	temperature programmed oxidation
TPR	temperature programmed reduction
TPRS	temperature programmed reaction spectroscopy
TPS	temperature programmed sulfidation
TPSIMS	temperature programmed secondary ion mass spectrometry
UPS	ultra violet photoelectron spectroscopy
XAFS	X-ray absorption spectroscopy
XANES	X-ray absorption near edge spectroscopy
XPS	X-ray photoelectron spectroscopy
XRD	X-ray diffraction
XRF	X-ray fluorescence

1
Introduction

Keywords

Heterogeneous catalysis
The aim of catalyst characterization
Spectroscopic techniques
Research strategies

1.1
Heterogeneous Catalysis

Today, catalysis plays a prominent role in our society, and the majority of all chemicals and fuels produced within the chemical industry has been in contact with one or more catalysts. In fact, catalysis has become indispensable in environmental pollution control, with selective catalytic routes replacing stoichiometric processes that generate waste problems. One clear example is the way in which a three-way catalyst leads to an effective reduction in the pollution from car engines, and catalytic processes to clean industrial exhaust gases have now been developed and installed. In short, whilst catalysis is vitally important for our economies of today, it will become increasingly important in the future.

A heterogeneous catalytic reaction begins with the adsorption of the reacting gases onto the surface of the catalyst, where intramolecular bonds are broken or weakened (exactly how this happens on metals in terms of simplified molecular orbital theory is explained in the Appendix). Next, the adsorbed species react on the surface, often in several consecutive steps. Finally, the products desorb from the surface into the gas phase, thereby regenerating the active sites on the surface for the following catalytic cycle. The function of the catalyst is to provide an energetically favorable pathway for the desired reaction, in which the activation barriers of all intermediate steps are low compared to the activation energy of the gas phase reaction. The sequence for the catalytic oxidation of carbon monoxide is illustrated schematically in Figure 1.1. Here, the key role of the catalyst is to dissociate the O_2 molecule, which occurs at the surface of many metals. For some introductory texts on the theory of catalysis, see [1–10].

Spectroscopy in Catalysis: An Introduction, Third Edition
J. W. Niemantsverdriet
Copyright © 2007 WILEY-VCH Verlag GmbH & Co. KGaA, Weinheim
ISBN: 978-3-527-31651-9

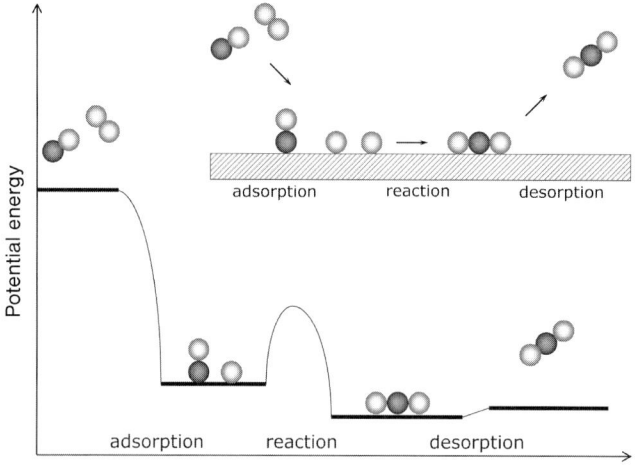

Fig. 1.1 Schematic representation of a well-known catalytic reaction, the oxidation of carbon monoxide on noble metal catalysts: $CO + \frac{1}{2}O_2 \rightarrow CO_2$. The catalytic cycle begins with the associative adsorption of CO and the dissociative adsorption of O_2 onto the surface. As adsorption is always exothermic, the potential energy decreases. Next, CO and O combine to form an adsorbed CO_2 molecule, which represents the rate-determining step in the catalytic sequence. The adsorbed CO_2 molecule desorbs almost instantaneously, thereby liberating adsorption sites that are available for the following reaction cycle. This regeneration of sites distinguishes catalytic from stoichiometric reactions.

As catalysis proceeds at the surface, a catalyst should preferably consist of small particles with a high fraction of surface atoms. This is often achieved by dispersing particles on porous supports such as silica, alumina, titania or carbon (see Fig. 1.2). Unsupported catalysts are also in use; examples of these include the iron catalysts for ammonia synthesis and CO hydrogenation (the Fischer–Tropsch synthesis), or the mixed metal oxide catalysts used in the production of acrylonitrile from propylene and ammonia.

The important properties of small particles are defined in Figure 1.3. For unsupported catalysts and for support materials, it is necessary to know how much material is exposed to the gas phase. This property is expressed by the specific area, in units of $m^2\ g^{-1}$. Typical supports such as silica and alumina have specific areas on the order of 200 to 300 $m^2\ g^{-1}$, while active carbons may have specific areas of up to 1000 $m^2\ g^{-1}$, or more. Unsupported catalysts have much lower surface areas, typically in the range of 1 to 50 $m^2\ g^{-1}$.

Surface areas are determined by *physisorption*. The most common procedure to determine surface area is to measure how much N_2 is adsorbed onto a certain amount of material. The uptake is measured at a constant low temperature (i.e., 80 K) as a function of N_2 pressure, and is usually very well described by the Brunauer–Emmett–Teller (BET) isotherm. After determining the number of N_2

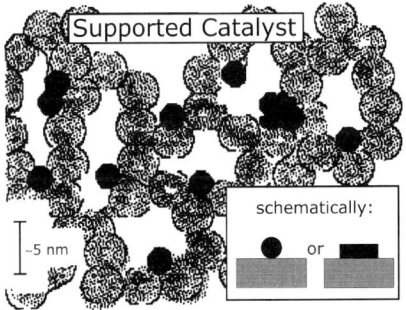

Fig. 1.2 An impression of a silica-supported catalyst. The inset shows the usual schematic representations.

molecules that form a monolayer on the support, one obtains the total area by setting the area of a single N_2 molecule to 0.16 nm^2 [11].

For particles on a support, dispersion is an important property, and is straightforwardly defined as the fraction of atoms in a particle located at the surface. The dispersion is usually determined by chemisorption of a gas that adsorbs only on the supported particles, and not on the support. For metals, hydrogen or CO are the obvious choices. It is necessary to adopt a certain stoichiometry for the number of H-atoms or CO molecules that can be accommodated per surface atom in the particle (this is usually taken as one CO or H-atom per metal atom) [11].

Supported catalysts can be prepared in several ways [12], but the simplest is that of *impregnation*. A support has a characteristic pore volume (e.g., 0.5 mL g^{-1}); hence, adding this volume of a solution containing the appropriate amount of a convenient catalyst precursor (e.g., nickel nitrate in water to prepare a supported nickel catalyst) to the support will simply fill all the pores. However, by allowing the system to dry, and then heating it in air to decompose any undesired salts, the supported material will be converted to the oxidic form. Reduction in hydrogen then converts the oxidic precursor – at least partially – into a supported metal catalyst.

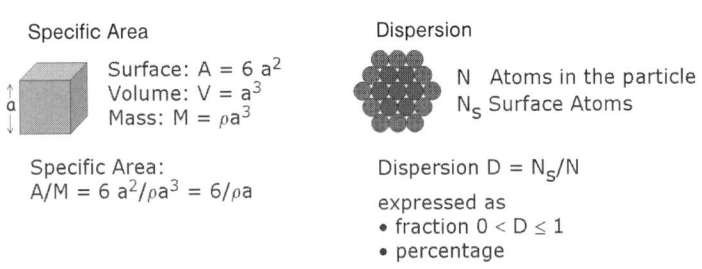

Fig. 1.3 Specific area and dispersion are important characteristic properties of a supported catalyst.

Catalysts may be metals, oxides, zeolites, sulfides, carbides, organometallic complexes, and enzymes [1–10]. The principal properties of a catalyst are its activity, its selectivity, and its stability. *Chemical promoters* may be added to optimize the quality of a catalyst, while *structural promoters* improve the mechanical properties and stabilize the particles against sintering. As a result, catalysts may be quite complex. The state of the catalytic surface often also depends on the conditions under which it is used, and in this respect spectroscopy, microscopy, diffraction and reaction techniques represent methods by which the appearance of the active catalyst can be investigated.

1.2
The Aim of Catalyst Characterization

The catalytic properties of a surface are determined by its composition and structure on the *atomic scale*. Hence, it is not sufficient to know that a surface consists of a metal and a promoter – perhaps iron and potassium – but it is essential to know the exact structure of the iron surface, including any defects and steps, as well as the exact location of the promoter atoms. Thus, from a fundamental point of view, the ultimate goal of catalyst characterization should be to examine the surface atom by atom, and under reaction conditions. The well-defined surfaces of single crystals offer the best perspectives for atom-by-atom characterization, although occasionally atomic scale information can also be obtained from real catalysts under *in-situ* conditions (for some examples, see Chapter 9). The many aspects that we need to study in order to properly understand supported catalysts on a fundamental level are shown schematically in Figure 1.4 [13].

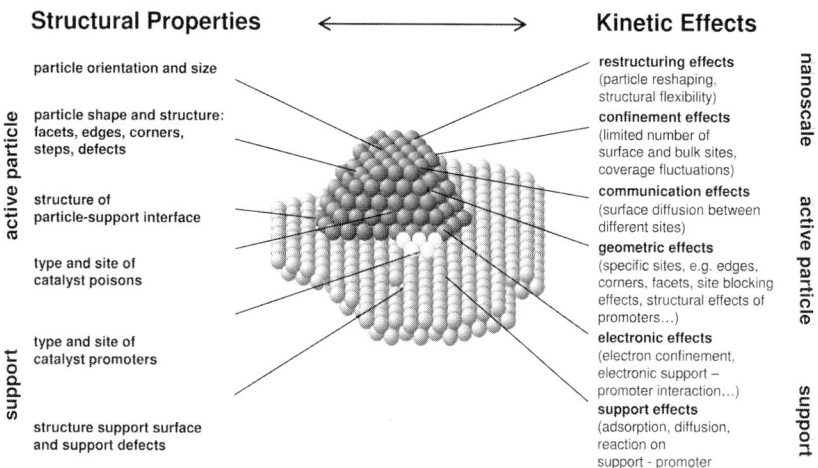

Fig. 1.4 Scheme of the many aspects of supported catalysts that call for characterization on the molecular level. (Adapted from [13]).

The industrial view on catalyst characterization is different, however. Here, the emphasis is placed mainly on developing an active, selective, stable and mechanically robust catalyst. In order to accomplish this, tools are needed which identify those structural properties that discriminate efficient from less-efficient catalysts, and all spectroscopic information that helps to achieve this is welcome. The establishment of empirical relationships between the factors that govern catalyst composition, particle size and shape and pore dimensions on the one hand, and catalytic performance on the other hand, are extremely useful in the process of catalyst development. However, such relationships may not provide much fundamental insight into how the catalyst operates in molecular detail.

Van Santen [14] identifies three levels of research in catalysis:

- The *macroscopic* level is the world of reaction engineering, test reactors and catalyst beds. Questions concerning the catalyst deal with such aspects as activity per unit volume, mechanical strength, and whether it should be used in the form of extrudates, spheres, or loose powders.

- The *mesoscopic* level comprises kinetic studies, activity per unit surface area, and the relationship between the composition and structure of a catalyst versus its catalytic behavior. Much of the characterization studies belong to this category.

- The *microscopic* level is that of fundamental studies, and deals with the details of adsorption on surfaces, reaction mechanisms, theoretical modeling, and surface science.

In simplifying, one could say that catalyst characterization in industrial research deals with the *materials science of catalysts* on a more or less mesoscopic scale, whereas the ultimate goal of fundamental catalytic research is to characterize the surface of a catalyst at the microscopic level – that is, on the *atomic* scale.

Catalyst characterization is a lively and highly relevant discipline in catalysis, with the literature revealing a clear desire to work with defined materials. For example, about 80% of the 143 oral reports at the 11th International Congress on Catalysis [15] contained at least some results on the catalyst(s) obtained by characterization techniques, whereas only 20% of these reports dealt with catalytic reactions over uncharacterized catalysts. Another remarkable fact obtained from these statistics is that about 10% of the reports included the results of theoretical calculations. Clearly, the modern trend is to approach catalysis from many different viewpoints, using a combination of sophisticated experimental and theoretical tools.

1.3
Spectroscopic Techniques

There are many ways to obtain information on the physico-chemical properties of materials. Figure 1.5 presents a scheme from which almost all techniques can be

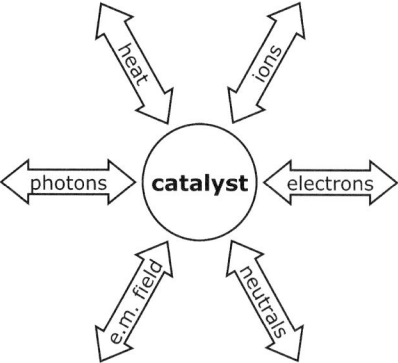

Fig. 1.5 Diagram from which most characterization techniques can be derived. The circle represents the sample under study, the inward arrow represents an excitation, and the outward arrow indicates how the information should be extracted.

derived. Spectroscopies are based on some type of excitation, represented by the in-going arrow in Figure 1.5, to which the catalyst responds, as symbolized by the outgoing arrow. For example, one can irradiate a catalyst with X-rays and study how the X-rays are diffracted (X-ray diffraction, XRD), or one can study the energy distribution of electrons that are emitted from the catalyst due to the photoelectric effect (X-ray photoelectron spectroscopy, XPS). One can also heat up a spent catalyst and examine which temperatures reaction intermediates and products desorb from the surface (temperature programmed desorption, TPD).

Characterization techniques become surface sensitive if the particles or radiation to be detected are derived from the outer layers of the sample. Low-energy electrons, ions and neutrals can only travel over distances of between one and ten interatomic spacings in the solid state, which implies that such particles leaving a catalyst may reveal surface-specific information. The inherent disadvantage of the small mean free path is that measurements must be carried out *in vacuo*, which conflicts with the wish to investigate catalysts under reaction conditions.

Photons that are scattered, absorbed or emitted by a catalyst form a versatile source of information. Figure 1.6 shows the electromagnetic spectrum, along with a number of techniques involving photons. In addition to the common sources of photons (lamps, lasers, helium discharge and X-ray sources) available for laboratory use, synchrotrons offer a broad spectrum of highly intense, polarized light. Electromagnetic radiation penetrates solids significantly, although if the solid responds by emitting electrons – as in the photoelectric effect – one obtains nevertheless surface-specific information.

In this book, we describe some the most often-used techniques in catalyst characterization, and also provide some statistics showing how often certain methods have been described in the most important catalysis journals since 2002 (see Fig. 1.7) [16]. In the following chapters we will highlight those methods that have been particularly useful in studies of metal, oxide and sulfide catalysts, and re-

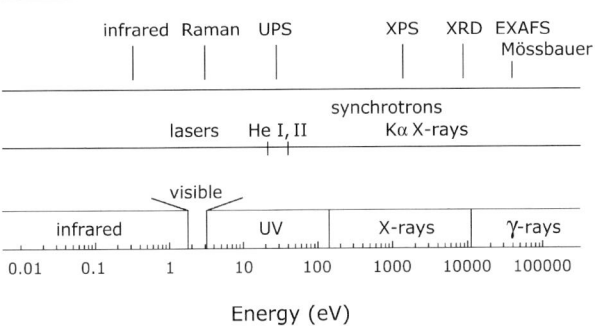

Fig. 1.6 The electromagnetic spectrum, along with common photon sources and a number of characterization techniques based on photons.

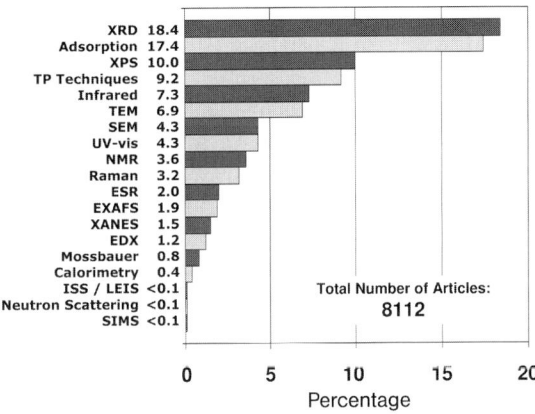

Fig. 1.7 Percentage of articles in *Applied Catalysis A* and *B*, *Catalysis Letters* and *Journal of Catalysis* between 2002 and 2006 (end October) that cite results obtained with various characterization techniques. Typical surface-science methods have been excluded. (Reproduced from [16]).

ISS: ion-scattering spectroscopy; LEIS: low-energy ion scattering; SEM: scanning electron microscopy; SIMS: Secondary ion mass spectrometry; TEM: transmission electron microscopy; TP: temperature-programmed; XANES: X-ray absorption near edge spectroscopy.

lated model systems. Zeolites and techniques such as nuclear magnetic resonance (NMR) fall beyond the scope of this book. Several books covering the subject of catalyst characterization are included among the references [17–21].

1.4
Research Strategies

Bearing in mind that the aim of characterization in fundamental catalysis is to obtain information concerning the active surface under reaction conditions in molecular detail, one might consider that surface science would be the discipline to offer the best opportunities. However, many of the tools do not work on technical catalysts, and consequently one must resort to model systems. For example, one can model the catalytic surface with that of a single crystal [5], and by using the appropriate combination of surface spectroscopies the desired characterization on the atomic scale is certainly possible in favorable cases. The disadvantage here, however, is that although one may be able to study the catalytic properties of such samples under realistic conditions (pressures of 1 atm or higher), most of the characterization is by necessity carried out in ultra-high vacuum, and not under reaction conditions.

The other approach is to study real catalysts by using *in-situ* techniques such as infrared and Mössbauer spectroscopy, extended X-ray absorption fine structure (EXAFS) and XRD, either under reaction conditions, or – as occurs more often – under a controlled environment after quenching the reaction. These *in-situ* techniques, however, are usually not sufficiently specific to yield the desired atom-by-atom characterization of the surface, and often they determine the overall properties of the particles. The situation is represented schematically in Figure 1.8.

The dilemma is thus that investigations of real catalysts, when conducted under relevant conditions by *in-situ* techniques, provide little information on the surface of the catalyst, and that the techniques which are surface-sensitive can often only be applied on model surfaces under vacuum. Model catalysts consisting of particles on flat, conducting supports often present an alternative with respect to surface analysis [22–26], as several examples in this book show. Also, *in-situ* tools that work on single crystals or model surfaces can in some cases be used to assess the significance of results obtained in vacuum for application in the real world [27]. Bridging the gap between ultra-high vacuum and high pressure – as well as the gap between the surface of a single crystal and a that of a real catalyst – are important issues in catalysis.

	Real catalyst	**Single crystal**
Reaction conditions	XRD, TP techniques Infrared and Raman EXAFS, XANES, AFM Mossbauer, ESR, NMR	Infrared TP techniques STM, AFM
Vacuum	XPS, SIMS, SNMS LEIS, RBS, TEM, SEM	All surface science techniques

Fig. 1.8 Possibilities for spectroscopic research in catalysis (for abbreviations, see Fig. 1.7). AFM: Atomic force microscopy; ESR: Electron spin resonance; RBS: Rutherford backscattering; SNMS: secondary neutral mass spectrometry.

Another point that concerns the relevance of spectroscopic research in catalysis is the following. Both catalysis and spectroscopy are disciplines that demand considerable expertise. For instance, the state of a catalyst often depends critically on the method of preparation, its pretreatment, or its environment. It is therefore essential to investigate a catalyst under carefully chosen, relevant conditions, and after the proper treatment. Catalytic scientists recognize these requirements precisely.

Spectroscopy, on the other hand, is by no means simple, and quick and easy experiments barely exist for catalyst characterization. The correct interpretation of spectra requires experience based on practice, together with a sound theoretical background in spectroscopy, in physical chemistry, and often also in solid-state physics. Intensive cooperation between spectroscopists and experts in catalysis is the best way to warrant meaningful and correctly interpreted results.

It is good to realize that, although many techniques undoubtedly provide valuable results on catalysts, the most useful information is almost always derived from a combination of several characterization techniques. The case studies in Chapter 9 present some examples where such an approach has been remarkably successful.

In conclusion, the investigation of catalytic problems by using a combination of spectroscopic techniques, applied under conditions which resemble as closely as possible those of the reaction in which the catalyst operates, according to an integrated approach in which experts in catalysis and spectroscopy work in close harmony, offers the best perspectives for successful research into catalysts.

References

1 M. Bowker, *The Basis and Applications of Heterogeneous Catalysis*. Oxford University Press, Oxford, 1998.
2 J.M. Thomas and W.J. Thomas, *Principles and Practice of Heterogeneous Catalysis*. VCH, Weinheim, 1997.
3 I. Chorkendorff and J.W. Niemantsverdriet, *Concepts of Modern Catalysis and Kinetics*. Wiley-VCH, Weinheim, 2003.
4 R.J. Farrauto and C.H. Bartholomew, *Fundamentals of Industrial Catalytic Processes*. Blackie Academic and Professional, Chapman & Hall, London, 1997.
5 G.A. Somorjai, *Introduction to Surface Chemistry and Catalysis*. Wiley, New York, 1994.
6 R.A. van Santen, P.N.W.M. van Leeuwen, J.A. Moulijn, and B.A. Averill (Eds.), *Catalysis, an Integrated Approach*. Elsevier, Amsterdam, 1999.
7 B.C. Gates, *Catalytic Chemistry*. Wiley, New York, 1992.
8 I.M. Campbell, *Catalysis at Surfaces*. Chapman & Hall, London, 1988.
9 H.H. Kung, *Transition Metal Oxides: Surface Chemistry and Catalysis*. Elsevier, Amsterdam, 1989.
10 R.A. van Santen and J.W. Niemantsverdriet, *Chemical Kinetics and Catalysis*. Plenum, New York, 1995.
11 J.A. Anderson and M. Fernandez Garcia (Eds.), *Supported Metals in Catalysis*. Imperial College Press, London, 2005.
12 G. Ertl, H. Knözinger and J. Weitkamp (Eds.), *Handbook of Heterogeneous Catalysis*. Wiley-VCH, Weinheim, 1997, Vol. 1, p. 49.
13 J. Libuda and H.-J. Freund, *Surface Sci. Rep.* **57** (2005) 157.
14 R.A. van Santen, *Theoretical Heterogeneous Catalysis*. World Scientific Singapore, 1991.

15 J.W. Hightower, W.N. Delgass, E. Iglesia, and A.T. Bell (Eds.), *Proceedings, 11th International Congress on Catalysis, Baltimore, 1996*. Elsevier, Amsterdam, 1996.
16 J.P. Dormans, R.J. Lancee, and J.W. Niemantsverdriet, *Statistics on Catalyst Characterization, Internal Report*. Eindhoven University of Technology, 2006.
17 B. Imelik and J.C. Vedrine (Eds.), *Catalyst Characterization: Physical Techniques for Solid Materials*. Plenum Press, London, 1994.
18 I.E. Wachs and L.E. Fitzpatrick, *Characterization of Catalytic Materials*. Butterworth-Heinemann, London, 1992.
19 J.L.G. Fierro (Ed.), *Spectroscopic Characterization of Heterogeneous Catalysts*. Elsevier, Amsterdam, 1990.
20 J.F. Haw (Ed.), *In-Situ Spectroscopy in Heterogeneous Catalysis*. Wiley-VCH, Weinheim, 2002.
21 B.M. Weckhuysen (Ed.), *In-Situ Spectroscopy of Catalysts*. American Scientific Publishers, 2004.
22 D.W. Goodman, *J. Catal.* **216** (2003) 213.
23 H.-J. Freund, M. Bäumer, and H. Kuhlenbeck, *Adv. Catal.* **45** (2000) 333.
24 P.L.J. Gunter, J.W. Niemantsverdriet, F.H. Ribeiro, and G.A. Somorjai, *Catal. Rev. Sci. Eng.* **39** (1997) 77.
25 C.T. Campbell, *Surface Sci. Rep.* **30** (1997) 227.
26 C.R. Henry, *Surface Sci. Rep.* **31** (1998) 231.
27 G.A. Somorjai, *CaTTech* **3** (1999) 84.

2
Temperature-Programmed Techniques

Keywords

Temperature-programmed reduction (TPR)
Temperature-programmed oxidation (TPO)
Temperature-programmed sulfidation (TPS)
Temperature-programmed desorption (TPD)
Temperature-programmed reaction spectroscopy (TPRS)

2.1
Introduction

Temperature-programmed (TP) reaction methods form a class of techniques in which a chemical reaction is monitored while the temperature increases linearly in time [1, 2]. Whilst several forms of these techniques are currently in use, they are all applicable to real catalysts and single crystals, and have the advantage that they are experimentally simple and inexpensive in comparison to many other spectroscopies. Although interpretation on a qualitative basis is rather straightforward, obtaining reaction parameters such as activation energies or pre-exponential factors from TP methods, is a complicated matter.

The instrumentation for TP investigations is relatively simple; the set-ups for TP reduction (TPR) and TP oxidation (TPO) studies of catalysts are shown in Figure 2.1. The reactor, charged with catalyst, is controlled by a processor, which heats the reactor at a rate of typically 0.1 to 20 °C min^{-1}. A thermal conductivity detector measures the hydrogen or oxygen content of the gas mixture before and after reaction. For TPR, one uses a mixture of typically 5% H_2 in Ar, or for TPO 5% O_2 in He, to optimize the thermal conductivity difference between reactant and carrier gas. With this type of apparatus, a TPR (TPO) spectrum is a plot of the hydrogen (oxygen) consumption of a catalyst as a function of temperature.

Temperature-programmed sulfidation or TP reaction spectroscopy (TPRS) usually deal with more than one reactant or product gas. In these cases, a

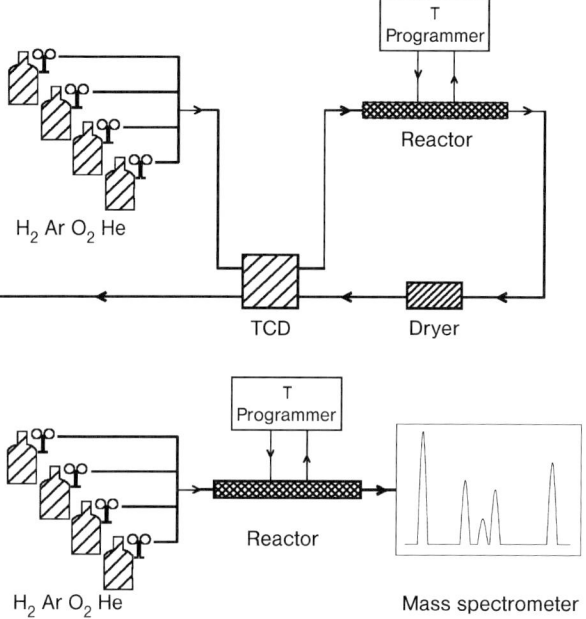

Fig. 2.1 Experimental set-ups for temperature-programmed (TP) reduction, oxidation and desorption. The reactor is inside the oven, the temperature of which can be increased linearly in time. Gas consumption by the catalyst is derived from the change in thermal conductivity of the gas mixture; it is essential to remove traces of water, etc., because these would affect the thermal conductivity measurement. The lower part of the figure shows a TP apparatus equipped with a mass spectrometer.

thermal conductivity detector (TCD) is inadequate and a mass spectrometer is required in order to detect all of the reaction products. With such equipment one obtains a much more complete picture of the reaction process, because one measures simultaneously the consumption of reactants and the formation of products.

In this chapter, we discuss TPR and reduction theory in some detail, and show how TPR provides insight into the mechanism of reduction processes. Next, we present examples of TPO, TP sulfidation (TPS) and TPRS applied on supported catalysts. In the final section we describe how thermal desorption spectroscopy reveals adsorption energies of adsorbates from well-defined surfaces in vacuum. A short treatment of the transition state theory of reaction rates is included to provide the reader with a feeling for what a pre-exponential factor of desorption tells about a desorption mechanism. The chapter is completed with an example of TPRS applied in ultra-high vacuum (UHV), in order to illustrate how this method assists in unraveling complex reaction mechanisms.

2.2
Temperature-Programmed Reduction

2.2.1
Thermodynamics of Reduction

Reduction is an inevitable step in the preparation of metallic catalysts. It is often also a critical step, because if it is not performed correctly the catalyst may sinter or may not reach its optimum state of reduction. The reduction of a metal oxide MO_n by H_2 is described by the equation

$$MO_n + nH_2 \rightarrow M + nH_2O \tag{2-1}$$

Thermodynamics predicts under which conditions a catalyst can be reduced: As with every reaction, the reduction will proceed when the change in Gibbs free energy, ΔG, has a negative value. Equation (2-2) shows how ΔG depends on pressures and temperature:

$$\Delta G = \Delta G^\circ + nRT \ln\left(\frac{p_{H_2O}}{p_{H_2}}\right) \tag{2-2}$$

where:
ΔG is the change in Gibbs free energy for the reduction;
ΔG° is the same under standard conditions (see e.g. [3]);
n is the stoichiometric coefficient of reaction (2-1);
R is the gas constant;
T is the temperature;
p is the partial pressure.

If one reduces the catalyst under flowing hydrogen, the reaction product – water – is removed effectively and the second term in Eq. (2-2) is therefore always negative. For many oxides, such as those of cobalt, nickel and the noble metals, ΔG° is already negative and reduction is thermodynamically feasible. All one has to do is find a temperature where the kinetics is rapid enough to achieve complete reduction. Oxides such as Fe_2O_3 and SnO_2, however, have a positive ΔG° [3]. Now, the second term in Eq. (2-2) determines whether ΔG is negative, or not. In order to see if reduction is thermodynamically permitted, Eq. (2-2) is written in the form

$$\Delta G = nRT \ln\left[\left(\frac{p_{H_2O}}{p_{H_2}}\right) \bigg/ \left(\frac{p_{H_2O}}{p_{H_2}}\right)_{eq}\right] \tag{2-3}$$

with the symbols as defined above and the subscript "eq" for the equilibrium ratio. Thus, ΔG is negative when the ratio $p(H_2O)/p(H_2)$ is smaller than the

Table 2.1 Thermodynamic data for the reduction of metal oxides in hydrogen at 400 °C [4].

Metal	Oxide	$(p(H_2O)/p(H_2))_{eq}$
Ti	TiO_2	$4 \cdot 10^{-16}$
	TiO	$2 \cdot 10^{-19}$
V	V_2O_5	$6 \cdot 10^{-4}$
	VO	$2 \cdot 10^{-11}$
Cr	Cr_2O_3	$3 \cdot 10^{-9}$
Mn	MnO_2	10
	MnO	$2 \cdot 10^{-10}$
Fe	Fe_2O_3	0.7
	FeO	0.1
Co	CoO	50
Ni	NiO	500
Cu	CuO	$2 \cdot 10^8$
	Cu_2O	$2 \cdot 10^6$
Mo	MoO_3	40
	MoO_2	0.02
Ru	RuO_2	10^{12}
Rh	RhO	10^{13}
Pd	PdO	10^{14}
Ag	Ag_2O	$3 \cdot 10^{17}$
Ir	IrO_2	10^{13}

equilibrium value and the efficiency with which water is removed from the reactor becomes the decisive factor.

Table 2.1 lists equilibrium ratios for the reduction of selected metal oxides [4], while Figure 2.2 provides a complete phase diagram for the reduction of iron oxide at different temperatures [3, 5]. In order to reduce bulk iron oxide to metallic iron at 600 K, the water content of the hydrogen gas above the sample must be below a few percent, which is easily achieved. However, in order to reduce Cr_2O_3, the water content should be as low as a few parts per billion, which is much more difficult to realize. The data in Table 2.1 also illustrate that, in many cases, only partial reduction to a lower oxide may be expected. Reduction of Mn_2O_3 to MnO is thermodynamically allowed at relatively high water contents, but further reduction to manganese is unlikely.

In fact, phase diagrams as shown in Figure 2.2 form indispensable background information for the interpretation of reduction experiments. However, it should be realized that equilibrium data as in Figure 2.2 and Table 2.1 refer to the reduction of bulk compounds. Numbers valid for the reduction of surface phases may be quite different. Also, traces of water present on the surface of catalyst particles or on the support represent a locally high concentration and may cause the surface to be oxidized under conditions which, interpreted macroscopically, would give rise to complete reduction.

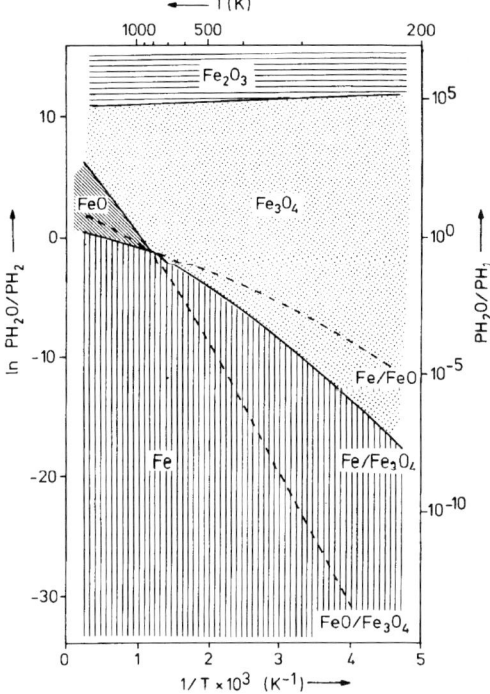

Fig. 2.2 Phase diagram of the iron–iron oxide system in mixtures of H_2 and H_2O. (Reproduced from [5]).

If the reaction, in which the metallic fraction serves as a catalyst, produces water as a byproduct, it may well be that the catalyst converts back to an oxide. One should always be aware that in fundamental catalytic studies, where reactions are usually carried out under differential conditions (i.e., low conversions) the catalyst may be more reduced than is the case under industrial conditions. An example is the behavior of iron in the Fischer–Tropsch reaction, where the industrial iron catalyst at work contains substantial fractions of Fe_3O_4, while fundamental studies report that iron is entirely carbidic and in the zero-valent state when the reaction is run at low conversions [6].

2.2.2
Reduction Mechanisms

Reduction reactions of metal oxides by hydrogen begin with the dissociative adsorption of H_2, which is a much more difficult process on oxides than on metals. Atomic hydrogen is utilized for the actual reduction. Depending on how fast or how slow the dissociative adsorption is with respect to the subsequent reduction reactions, comprising diffusion of atomic hydrogen into the lattice, reaction with

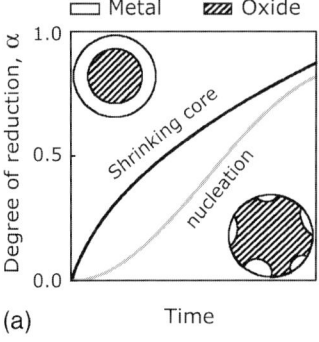 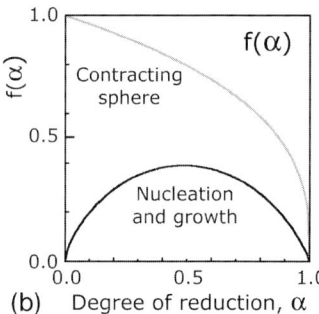

Fig. 2.3 (a) Reduction models. In the shrinking core or contracting sphere model the rate of reduction is initially fast and decreases progressively due to diffusion limitations. The nucleation model applies when the initial reaction of the oxide with molecular hydrogen is difficult. Once metal nuclei are available for the dissociation of hydrogen, reduction proceeds at a higher rate until the system enters the shrinking core regime. (b) The reduction rate depends on the concentration of unreduced sample $(1-\alpha)$ as $f(\alpha)$ (see Eqs. (2-5) and (2-6)].

oxygen and removal of the hydroxyl species formed, two limiting cases are distinguished [1, 7].

If the *initiation step* – the activation of H_2 – is fast (as may be the case on noble metal oxides or highly defective oxide surfaces), the shrinking core or contracting sphere model applies (see Fig. 2.3). The essence of this model is that nuclei of reduced metal atoms form rapidly over the entire surface of the particle and grow into a shell of reduced metal. Further reduction is limited by the transport of lattice oxygen out of the particle. Initially, the extent of reduction increases rapidly, but then slows down as the metal shell grows.

The *nucleation model* represents the other extreme. Here, the dissociation of hydrogen is the slow step. Once a nucleus of reduced metal exists, it acts as a catalyst for further reduction, as it provides a site where H_2 is dissociated. Atomic hydrogen diffuses to adjacent sites on the surface or into the lattice and reduces the oxide. As a result, the nuclei grow in three dimensions until the whole surface is reduced, after which further reduction takes place, as in the shrinking core model. The extent of reduction (see Fig. 2.3a) shows an induction period, but then increases rapidly and slows down again when the reduction enters the shrinking core regime.

In TPR one follows the degree of reduction of the catalyst as a function of time, while the temperature increases at a linear rate. We follow the theoretical treatment of the TPR process as given by Hurst et al. [1] and by Wimmers et al. [8]. Formally, one can write the rate expression for the reduction reaction in Eq. (2-1), under conditions where the reverse reaction from metal to oxide can be ignored, as

$$-\frac{d[MO_n]}{dt} = k_{red}[H_2]^p f([MO_n]) \qquad (2\text{-}4)$$

2.2 Temperature-Programmed Reduction

where:
- $[MO_n]$ is the concentration of metal oxide;
- $[H_2]$ is the concentration of hydrogen gas;
- k_{red} is the rate constant of the reduction reaction;
- p is the order of the reaction in hydrogen gas;
- f is a function which describes the dependence of the rate of reduction on the concentration of metal oxide;
- t is the time.

If we write α for the degree of reduction, assume that the reaction is conducted in excess hydrogen ($p = 0$), use the fact that the temperature increases linearly in time ($dT = \beta\, dt$) and replace k_{red} by the Arrhenius equation, we see the temperature dependence of the reduction process:

$$\frac{d\alpha}{dT} = \frac{\nu}{\beta} e^{-E_{red}/RT} f(1-\alpha) \qquad (2\text{-}5)$$

where:
- α is the fraction of reduced material;
- $f(1-\alpha)$ is a function of the fraction of unreduced material as in Eq. (2-4);
- ν is the pre-exponential factor;
- β is the heating rate, dT/dt;
- E_{red} is the activation energy of the reduction reaction;
- R is the gas constant;
- T is the temperature.

The function $f(1-\alpha)$ depends on the model that describes the reduction process; the simplest choices would be $f(1-\alpha) = (1-\alpha)$ or $(1-\alpha)^q$, as noted in the review of Hurst et al. [1]. More realistic expressions are those for the nucleation and the contracting sphere models of Figure 2.3:

$$\begin{aligned} f(\alpha) &= 3(1-\alpha)^{1/3} \qquad &\text{contracting sphere} \\ f(\alpha) &= (1-\alpha)[-\ln(1-\alpha)]^{2/3} \qquad &\text{nucleation and growth} \end{aligned} \qquad (2\text{-}6)$$

These functions are shown in Figure 2.3b.

The activation energy of the reduction can be estimated from the temperature T_{max} at which the reduction rate is a maximum by using the following equation (valid under the assumption that both $f(1-\alpha)$ and $\alpha(T_{max})$ do not depend on the heating rate [8, 9]):

$$\ln\left(\frac{\beta}{T_{max}^2}\right) = -\frac{E_{red}}{RT_{max}} + \ln\left(\frac{\nu R}{E_{red}}\right) + \text{constant} \qquad (2\text{-}7)$$

The way to use Eq. (2-7) is to record a series of TPR patterns at different heating rates, and plot the left-hand side of (2-7) against $1/T_{max}$. The result should give a straight line with slope $-E_{red}/R$. Furthermore, Eq. (2-5) can be used in combination with Eq. (2-6) to simulate TPR patterns, which can then be compared with the measured patterns. In this way, it may be possible to identify the prevailing reduction mechanism. An example of this is provided later in the chapter.

2.2.3
Applications

The first useful information that TPR provides are the temperatures needed for the complete reduction of a catalyst. Figure 2.4 shows the TPR and TPO patterns of silica-supported iron and rhodium catalysts [10]. Here, there are three things to note:

- The difference in reduction temperature between the noble metal rhodium and the non-noble metal iron, which is in agreement with Table 2.1.

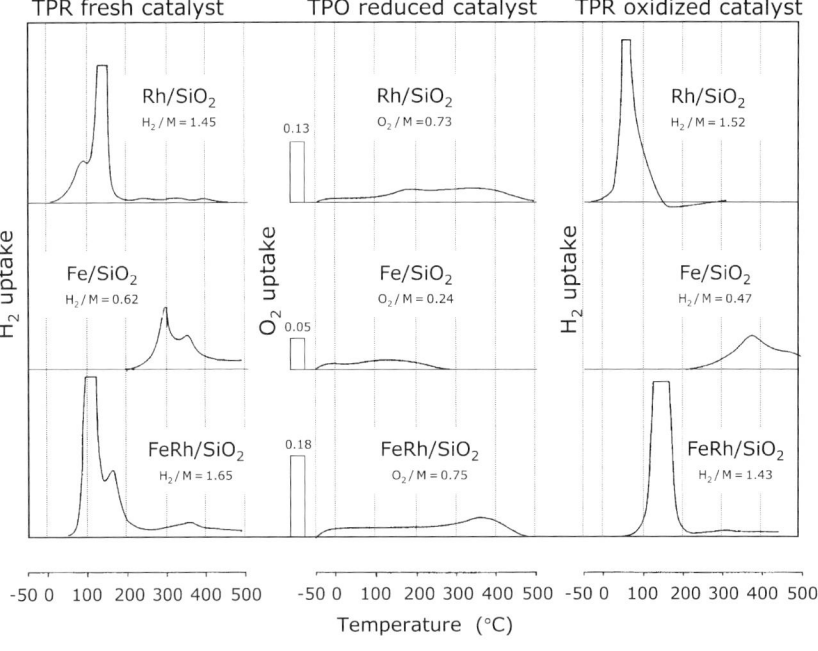

Fig. 2.4 TPR and TPO of silica-supported Rh, Fe, and Fe-Rh catalysts. The left TPR curves have been measured with freshly prepared catalysts; those on the right after a complete oxidation. The latter curves provide a reference for reduction of the purely oxidic systems, and assist in the interpretation of more complex TPR patterns of the freshly prepared samples. (Adapted from [10]).

- The TPR patterns of the freshly prepared catalysts contain two peaks. The assignment of these peaks is possible if we take a reference spectrum of the fully oxidized catalysts (Fig. 2.4, right panel). Comparison suggests that the first peak in the TPR pattern of the fresh Rh/SiO$_2$ catalyst is associated with the reduction of Rh–O bonds and the second with Rh–Cl.

- The area under a TPR or TPO curve represents the total hydrogen consumption, and is commonly expressed in moles of H$_2$ consumed per mole of metal atoms (H$_2$/M). The ratios of almost 1.5 for rhodium in the right panel of Figure 2.4 indicate that rhodium was present as Rh$_2$O$_3$. For iron, the H$_2$/M ratios are significantly lower, indicating that this metal is only partially reduced.

The TPR of supported bimetallic catalysts often reveals whether the two metals are in contact, or not. The TPR pattern of the 1:1 FeRh/SiO$_2$ catalyst in Figure 2.4 shows that the bimetallic combination reduces largely in the same temperature range as the rhodium catalyst does, indicating that rhodium catalyzes the reduction of the less-noble iron. This provides evidence that rhodium and iron are well mixed in the fresh catalyst. The reduction mechanism is as follows: As soon as rhodium becomes metallic it dissociates hydrogen; atomic hydrogen migrates to iron oxide in contact with metallic rhodium and reduces the oxide instantaneously.

In general, TPR measurements are interpreted on a qualitative basis, as in the example discussed above. Attempts to calculate the activation energies of reduction by means of Eq. (2-7) can only be undertaken if the TPR pattern represents a single, well-defined process. This requires, for example, that all catalyst particles are equivalent. In a supported catalyst, all particles should have the same morphology and all atoms of the supported phase should be affected by the support in the same way, otherwise the TPR pattern would represent a combination of different reduction reactions. Such strict conditions are seldom obeyed in supported catalysts, but are more easily met in unsupported particles. As an example we discuss the TPR studies conducted by Wimmers et al. [8] on the reduction of unsupported Fe$_2$O$_3$ particles (diameter ca. 300 nm). Such research is of interest in regards to the synthesis of ammonia and the Fischer–Tropsch process, both of which are carried out over unsupported iron catalysts.

The phase diagram of the iron–iron oxide system in H$_2$O/H$_2$ mixtures (see Fig. 2.2) predicts that the reduction of Fe$_2$O$_3$ to metallic iron is not a direct process but rather goes through Fe$_3$O$_4$ as the intermediate. Water in the gas phase may be critical. As the data in Figure 2.2 indicate, water is expected to retard the final reduction of Fe$_3$O$_4$ to Fe more than the initial conversion of Fe$_2$O$_3$ to Fe$_3$O$_4$. This knowledge helps to interpret the TPR spectra, as will be seen later.

Figure 2.5 illustrates two TPR patterns for the reduction of the Fe$_2$O$_3$ particles: one measured with dry hydrogen as the reductant; and the other with a few percent of water added to the hydrogen. The former spectrum consists of two partially overlapping peaks in which the low-temperature component is attributed to the conversion of Fe$_2$O$_3$ to Fe$_3$O$_4$, while the one at higher temperature is thought to reflect the reduction to metallic iron. The interpretation is confirmed by the

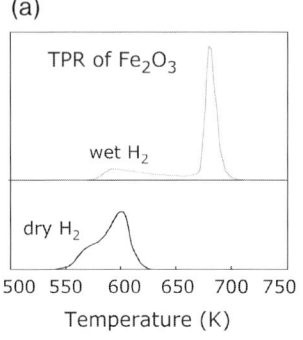
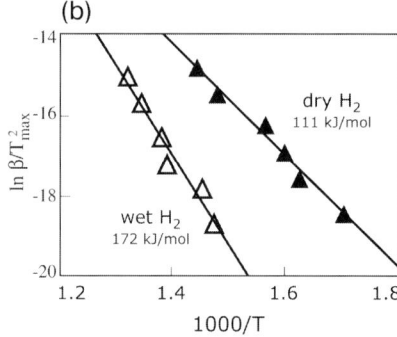

Fig. 2.5 (a) TPR patterns of 300-nm particles of Fe_2O_3 in dry and wet hydrogen indicate that the reduction of Fe_3O_4 to Fe (the high-temperature peak) is more retarded than the initial conversion of Fe_2O_3 to Fe_3O_4 (low-temperature peak), in agreement with the thermodynamic data in Figure 2.2. (b) Arrhenius plots according to Eq. (2-7) for the reduction of small Fe_3O_4 particles to iron in dry and wet hydrogen, along with the activation energies of reduction as determined from the slopes of the plots. (Data from [8]).

TPR curve measured in the presence of water: both peaks shift to higher temperatures, but the one attributed to the reduction of Fe_3O_4 to Fe is significantly more affected, in agreement with our expectation based on the phase diagram.

Activation energies for the reduction can be derived by measuring a series of TPR patterns at different heating rates and using Eq. (2-7). A plot of $\ln(\beta/T_{max}^2)$ versus $1/T_{max}$ (see Fig. 2.5b) yields a straight line corresponding to an activation energy of 111 kJ mol^{-1} for the reduction of Fe_3O_4 to Fe in dry H_2. The addition of water increases the activation energy for this step to 172 kJ mol^{-1}. The activation energy for the initial step – the conversion of Fe_2O_3 to Fe_3O_4 – cannot be determined for the dry reduction because its peak cannot be distinguished from that of the main reduction step. For reduction in wet hydrogen, the initial step has an activation energy of 124 kJ mol^{-1}.

Comparison of the measured peak shape with simulations based on Eqs. (2-5) and (2-6) reveals that a nucleation and growth model describes the reduction of Fe_3O_4 to Fe best. Thus, the formation of metallic iron nuclei at the surface of the particles is the difficult step. Once these nuclei have formed, they provide the site where molecular hydrogen dissociates to yield atomic hydrogen, which takes care of further reduction. The studies of Wimmers and co-workers [8] show nicely that TPR allows for detailed conclusions on reduction mechanisms, albeit in favorable cases only.

To summarize, TPR is a highly useful technique which provides a quick characterization of metallic catalysts. It also provides information on the phases present after impregnation, and on the eventual degree of reduction. For bimetallic catalysts, TPR patterns often indicate if the two components are mixed, or not. In favorable cases where the catalyst particles are uniform, TPR yields activation energies for the reduction, as well as information on the mechanism of reduction.

2.3
Temperature-Programmed Sulfidation

Catalysts used for the hydrodesulfurization (HDS) and hydrodenitrogenation (HDN) of heavy oil fractions are largely based on alumina-supported molybdenum or tungsten, to which cobalt or nickel is added as a promoter [11]. As the catalysts are active in the sulfided state, activation is carried out by treating the oxidic catalyst precursor in a mixture of H_2S and H_2 (or by exposing the catalyst to the sulfur-containing feed). The function of hydrogen is to prevent the decomposition of the relatively unstable H_2S to elemental sulfur, which otherwise would accumulate on the surface of the catalyst. Moulijn and co-workers have studied the sulfidation mechanism of several catalyst systems with a variation on TPR/TPO, which they referred to as TPS. As an illustration of these investigations, we discuss the TPS study conducted by Moulijn et al. on MoO_3/Al_2O_3 catalysts [12].

Moulijn and colleagues used a TPS apparatus equipped with a mass spectrometer such that the consumption or production of the gases H_2S, H_2 and H_2O could be detected separately. In the TPS patterns of MoO_3/Al_2O_3 (Fig. 2.6), the presence of negative peaks mean that the corresponding gas is consumed, whereas positive peaks mean that it is produced. Below 400 K, H_2S is consumed

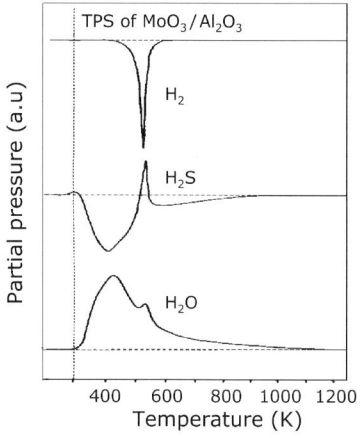

Fig. 2.6 Temperature-programmed sulfidation (TPS) of MoO_3/Al_2O_3 catalysts in a mixture of H_2S and H_2, showing the consumption of these gases and the production of H_2O as a function of temperature. Note that H_2S evolves from the catalyst at ca. 500 K, which is attributed to the hydrogenation of elementary sulfur. (Reproduced from [12]).

and H_2O produced, while no uptake of H_2 is detected. This suggests that the prevailing reaction at low temperatures is the exchange of sulfur for oxygen:

$$MoO_3 + H_2S \rightarrow MoO_2S + H_2O \tag{2-8}$$

At about 500 K, the catalyst consumes H_2 in a sharp peak, while simultaneously H_2S and some additional H_2O are produced. Arnoldy et al. [12] assigned the uptake of hydrogen and the evolution of H_2S to the hydrogenation of excess sulfur formed via the decomposition of the oxysulfide species in Eq. (2-8) at low temperatures:

$$MoO_2S \rightarrow MoO_2 + S \tag{2-9}$$

$$S + H_2 \rightarrow H_2S \tag{2-10}$$

At higher temperatures, the catalyst continues to exchange oxygen for sulfur until all molybdenum is present as MoS_2:

$$MoO_2 + 2H_2S \rightarrow MoS_2 + 2H_2O \tag{2-11}$$

Although Eqs. (2-8) to (2-11) explain the results satisfactorily, one needs to be aware that TP studies detect only those reactions in the catalyst that are accompanied by a net production or consumption of gases. Suppose, for instance, that Eq. (2-11) is the result of two consecutive steps:

$$MoO_2 + 2H_2 \rightarrow Mo + 2H_2O \tag{2-12}$$

$$Mo + 2H_2S \rightarrow MoS_2 + 2H_2 \tag{2-13}$$

If Eq. (2-13) follows Eq. (2-12) instantaneously, the effect will not be noticeable in the H_2 signal [12]. Despite these limitations, it is concluded that TPS with mass spectrometric detection is a highly useful technique for studying the sulfidation of hydrotreating catalysts. We return to the sulfidation of molybdenum oxides in Chapter 3 (photoemission), Chapter 4 (ion spectroscopy), and also in a case study on hydrodesulfurization catalysts (Chapter 9).

2.4
Temperature-Programmed Reaction Spectroscopy

Important information on reaction mechanisms and on the influence of promoters can be deduced from temperature-programmed reactions [2]. Figure 2.7 illustrates how the reactivity of adsorbed surface species on a real catalyst can be measured with TPRS. In this figure, the reactivity of adsorbed CO towards H_2 on a reduced Rh catalyst is compared with that of CO on a vanadium-promoted Rh catalyst [13]. The reaction sequence, in a simplified form, is thought to be as follows:

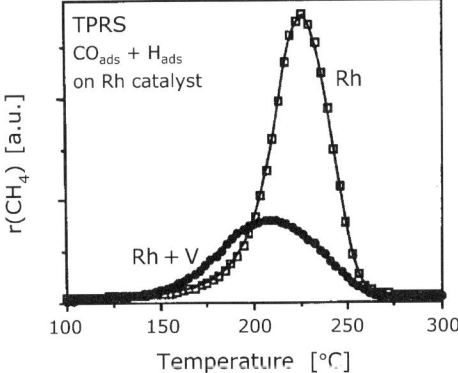

Fig. 2.7 The surface reaction between adsorbed carbon monoxide and hydrogen to methane over rhodium catalysts occurs at lower temperatures in the presence of a vanadium oxide promoter, which is known to enhance the rate of CO dissociation. (Reproduced from [13]).

$$CO_{ads} + {}^* \rightarrow C_{ads} + O_{ads}$$
$$H_2 + 2^* \rightarrow 2H_{ads}$$
$$O_{ads} + 2H_{ads} \rightarrow H_2O + 3^*$$
$$C_{ads} + 4H_{ads} \rightarrow CH_4 + 5^*$$
(2-14)

where * denotes an empty site at the surface. Due to the promoting effect of vanadium, the chemisorbed CO molecules react at a somewhat lower temperature to methane than on the unpromoted catalyst. As the rate-limiting step in this reaction is the dissociation of CO into C and O atoms, the TPSR results imply that vanadium enhances the rate of the CO dissociation. This explains the higher activity of vanadium-promoted Rh catalysts in the synthesis gas reactions [13].

2.5
Temperature-Programmed Desorption

Temperature-programmed desorption (TPD) – which is also referred to as thermal desorption spectroscopy (TDS) – can be used on technical catalysts, but is particularly useful in surface science, where the desorption of gases from single crystals and polycrystalline foils into vacuum is examined [2]. Because TPD offers interesting opportunities to interpret desorption in terms of reaction kinetic theories (such as the transition state formalism), we will discuss TPD in somewhat more detail than would be justified from the point of view of practical catalyst characterization alone.

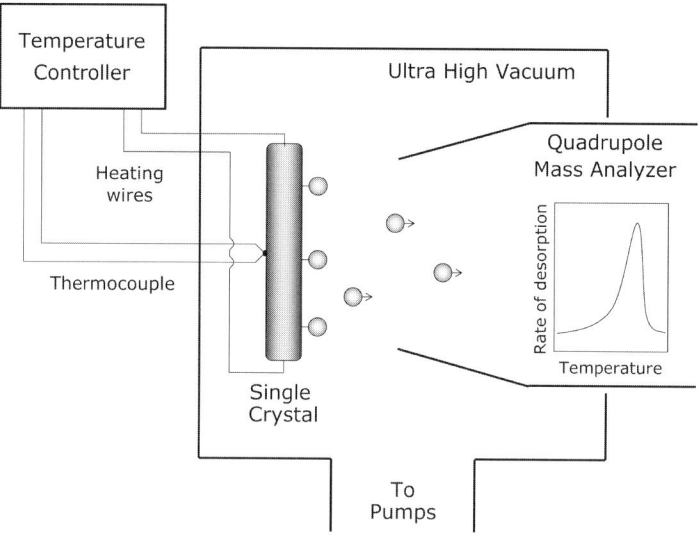

Fig. 2.8 Experimental set-up for temperature-programmed desorption studies in ultra-high vacuum. The heat dissipated in the tantalum wires resistively heats the crystal; the temperature is measured by a thermocouple which is spot-welded to the back of the crystal. Desorption of gases is followed using mass spectrometry.

Figure 2.8 shows a schematic set-up for TPD. The crystal, mounted on a manipulator in a UHV chamber is heated resistively via thin tantalum or tungsten wires. A thermocouple which is spot-welded to the back of the crystal is used to monitor the temperature. This type of sample responds much more rapidly to heating than a catalyst in a reactor. Hence, the heating rates used in TPD can be much higher than those used for TPR, and are typically between 0.1 and 25 K per second. The concentration of desorbing species is usually measured with a quadrupole mass spectrometer, but can also be determined with an ionization manometer or by monitoring the work function of the sample. Optimized geometries for TPD measurements have been described by Feulner and Menzel [14].

Pumping capacity is an important consideration in TPD. The pumping speed should be sufficiently high to prevent readsorption of the desorbed species back onto the surface. The effect is that spectra broaden towards higher desorption temperatures and should not be underestimated. For example, studies of H_2 desorption in UHV systems pumped by turbo molecular pumps with their unfavorable compression factors for light gases will certainly suffer from readsorption features (actually, turbo–molecular drag combination pumps are better suited to studies involving hydrogen). For heavier gases, the situation becomes less critical.

If the pumping speed is infinitely high, readsorption may be ignored, and the relative rate of desorption – defined as the change in adsorbate coverage per unit of time – is given by [15]:

2.5 Temperature-Programmed Desorption

$$r = -\frac{d\theta}{dt} = k_{des}\theta^n = v(\theta)\theta^n \exp\left(-\frac{E_{des}(\theta)}{RT}\right)$$

$$T = T_o + \beta t$$

(2-15)

where:
- r is the rate of desorption;
- θ is the coverage in monolayers;
- t is the time;
- k_{des} is the reaction rate constant for desorption;
- n is the order of desorption;
- v is the pre-exponential factor of desorption;
- E_{des} is the activation energy of desorption;
- R is the gas constant;
- T is the temperature;
- T_o is the temperature at which the experiment starts;
- β is the heating rate, equal to dT/dt.

Attractive or repulsive interactions between the adsorbate molecules make the desorption parameters E_{des} and v dependent on coverage [16].

The TPD spectra of three different adsorbate systems, corresponding to zeroth-, first- and second-order kinetics, are shown in Figure 2.9. Each trace corresponds to a different initial adsorbate coverage, as indicated in the figure. The simplest case in TPD corresponds to first-order desorption kinetics, represented by the CO/Rh(111) series in Figure 2.9 [17, 18]. For CO coverages up to 0.5 monolayer (ML), the CO molecules do not interact on Rh(111) and the desorption traces all fall in the same temperature range, all with the same peak maximum temperature. Hence, the rate of desorption is proportional to the surface concentration of CO. Above 0.5 ML, CO starts to populate additional sites (from vibrational spectroscopy studies we know that in addition to on-top sites also threefold hollow sites are occupied; see Fig. 8.15), and a faster reaction channel for desorption opens up, as seen by the development of a shoulder at lower temperatures [18].

The right-hand panel of Figure 2.9 corresponds to the recombinative desorption of N-atoms on Rh(100). As two adsorbed N-atoms are involved, the rate of desorption depends on the coverage squared; this leads to TPD peaks which shift to lower temperatures with increasing coverage [19]. The presence of lateral interactions is difficult to recognize without detailed analysis of the kinetics (this point is discussed later).

The Ag/Ru(001) series (Fig. 2.9, left panel) corresponds to zeroth-order desorption [20]. To appreciate what this means, we need to explain this system in more detail. Adsorbate systems of low-melting metals such as Cu, Ag and Au, adsorbed on high-melting metals such as Ru and W, are highly suited to fundamental desorption studies, for two reasons: (1) readsorption on the sample surface does not occur; and (2) the rate of desorption is measured without any systematic error due to background pressures, as would be the case with gases such as H_2, CO,

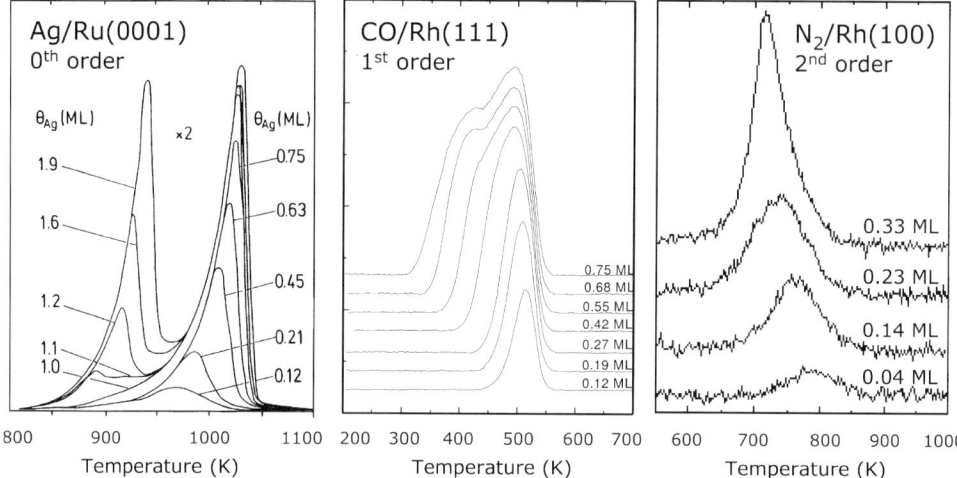

Fig. 2.9 Examples of temperature-programmed desorption following zeroth-, first- and second-order kinetics. Each curve corresponds to a different initial coverage of the adsorbate. Ag/Ru(001): Silver forms islands on the ruthenium substrate. Desorption of Ag from the edges of these islands gives rise to zeroth-order kinetics; note the exponential increase of the low-temperature sides of the peak, as expected from Eq. (2-15). Desorption from the second layer of silver occurs at lower temperatures, indicating that Ag–Ag bonds are weaker than Ag–Ru bonds [20]. CO/Rh(111): At coverages up to 0.5 monolayer (ML) CO desorbs in a single peak, indicating that all CO molecules bind in a similar configuration to the surface. Hence, the rate of desorption is proportional to the CO coverage. At higher coverages, an additional desorption peak appears, indicative of a different adsorption geometry [17]. N_2/Rh(100): Two N-atoms have to react to form N_2, which desorbs instantaneously upon formation. Hence, the rate of desorption depends on the coverage squared, implying that peaks shift to lower temperature with increasing coverage [19].

and N_2. Metallic overlayers have attracted attention as models for bimetallic surfaces in reactivity studies [21, 22]. The data on Ag/Ru(001) were used to illustrate that TPD provides information on:

- adsorbate coverage (quantitatively);
- the adsorption energy, which in general equals the activation energy of desorption;
- lateral interactions between the adsorbates, through the coverage dependence of the adsorption energy; and
- the pre-exponential factor of desorption, which in turn reflects the desorption mechanism (when interpreted in the transition state theory of reaction rates).

TPD is an excellent technique for determining surface coverages. As shown in Figure 2.9, the desorption of silver occurs in two temperature regions. At small coverages, Ag desorbs between 950 and 1050 K, while at higher coverages it desorbs first between 850 and 950 K, followed by desorption from the high-temperature peak. We assign the latter to Ag desorption from the first monolayer on Ru(001). The low-temperature peak is due to the desorption of Ag from thicker layers. When one has such a clear calibration point for the saturation of the first layer, quantitation is straightforward: all coverages as indicated in Figure 2.9 were determined from the area under the TPD peaks and calibrated against the saturation intensity of the first-layer peak at 1020 K. Thus, determining surface concentrations is a highly useful application of TPD. With a good calibration point one obtains absolute coverages.

TPD also provides information on the strength of the bond between adsorbate and substrate. An important check is obtained from the desorption of Ag from a thick layer: here, the activation energy of desorption should be equal to the heat of vaporization of Ag (254 kJ mol^{-1}). Of course, the more interesting information is in the adsorption energies of Ag on Ru, but this requires much effort, as will be shown later.

The analysis of TPD spectra is a time-consuming matter, if it is to be performed correctly! On examining Eq. (2-15), the task is to evaluate activation energies, prefactors and orders of desorption from a series of spectra, as in Figure 2.9. The problem is rather complex because both E_{des} and v (and also sometimes n) depend on coverage, θ. Thus, we need to determine E_{des} and v for every coverage separately. Figure 2.10 illustrates the procedure: We choose a coverage θ' (e.g.,

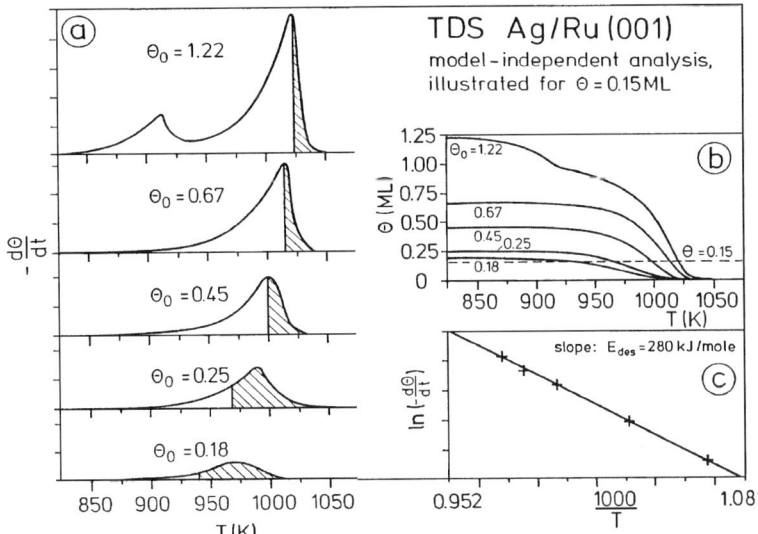

Fig. 2.10 The so-called "complete analysis" of TPD data is based on the rigorous application of Eq. (2-15) for the rate of desorption. (a) The spectra are integrated to determine points on the spectra corresponding to a fixed coverage, in this example 0.15 of a monolayer (ML) (b). This procedure gives a pair of (r, T) values for every desorption trace, from which an Arrhenius plot is made (c). The slope yields the activation energy, the intercept equals $\ln v + n \ln \theta'$, each corresponding to a coverage of 0.15 ML.

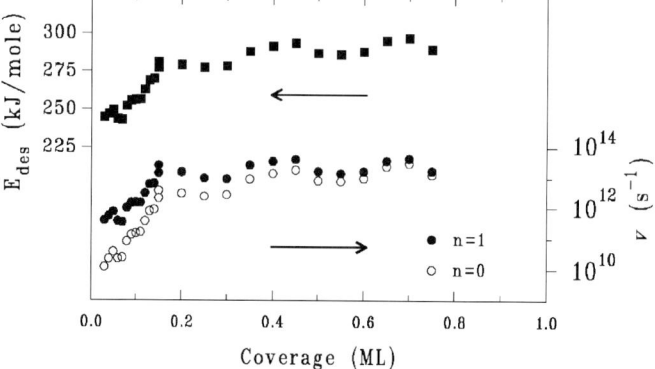

Fig. 2.11 Activation energies and prefactors for the desorption of Ag from Ru(001), as determined with the complete analysis. The desorption parameters become essentially constant for coverages above 0.15 monolayer (ML), indicative of zero-order kinetics. This suggests that Ag atoms desorb from the edges of relatively large two-dimensional islands. (Data from [20]).

0.15) and find the points corresponding to $\theta' = 0.15$ on all TPD curves. This gives a pair of r, T values from every curve with initial coverage θ_o larger than $\theta' = 0.15$. An Arrhenius plot of all $\ln r$ versus $1/T$ values for this particular coverage yields $E_{des}(\theta')$ from the slope of the plot. The prefactor follows from the intercept $n \ln \theta' + \ln v(\theta')$, when we know the order of desorption. Procedures exist to derive the order from TPD spectra, but for coverages above 0.1 ML the term $n \ln \theta'$ is much smaller than $\ln v(\theta')$, and we hardly make an error if we ignore it. This method, based on the rigorous application of Eq. (2-15), is called "the complete analysis", and was first proposed by King [15]. The results for the Ag/Ru(001) system are shown in Figure 2.11 [20].

Lateral interactions between the Ag adsorbate atoms become apparent if we examine the coverage dependence of the adsorption energy. The activation energy of desorption starts at 240 kJ mol^{-1} for single Ag atoms on Ru, but increases rapidly to about 290 kJ mol^{-1} for coverages of 0.15 ML and up. This indicates that Ag atoms attract each other on the Ru(001) surface. This attractive interaction leads to island formation at coverages above 0.15 ML. As a result, the desorption energy becomes virtually constant and the desorption process is zero-order in the silver concentration. Apparently, Ag atoms desorb out of the edge of the islands, where all Ag atoms feel the same interaction with the other Ag atoms. Note that the leading (i.e., low-temperature) side of the desorption curves indeed resembles an exponential function, as is expected from Eq. (2-15) if one takes $n = 0$ and E_{des} and v independent of θ. The difference in adsorption energy between a single Ag atom on Ru(001), of 240 kJ mol^{-1}, and an Ag atom at the edge of an island, of 290 kJ mol^{-1}, may be considered as the two-dimensional heat of "vaporization" of Ag on Ru, and amounts to about 50 kJ mol^{-1}. Thus, the desorption energy depends not only on the strength of the bond between adsorbate and substrate, but also on interactions between the adsorbate atoms. Both contributions can be estimated from TPD.

Information on the desorption mechanism is provided by the pre-exponential factors, if we interpret them in terms of Eyring's transition state theory [23] (see below). According to Figure 2.11, the prefactor for Ag desorption varies between 10^{11} at low and 5×10^{13} at higher coverages, which points to a rather rigid transition state for desorption. An interesting feature illustrated by Figure 2.11 is that $\log \nu$ and E_{act} depend very similarly on the coverage. In fact, the correspondence is so good that one can even write $\log \nu(\theta) = bE_{des}(\theta) + c$, where b and c are constants. This is a manifestation of the so-called "compensation effect", which has been shown to occur in many desorption systems [24], as well in all types of chemical reaction [25]. For discussions of this still-not-entirely understood phenomenon, the reader is referred elsewhere [25–27].

2.5.1
TPD Analysis

The complete desorption analysis described above is rarely used in TPD. Many authors rely on simplified methods, which make use of easily accessible spectral features such as the temperature of the peak maximum, T_{max}, and the peak width at half-maximum intensity, W. We describe these methods briefly and evaluate their merits.

Particularly popular among surface scientists is the Redhead method [28], in which the activation energy of desorption is given by:

$$E_{des} = RT_{max}\left[\ln\left(\frac{\nu T_{max}}{\beta}\right) - 3.46\right] \quad (2\text{-}16)$$

where:
E_{des} is the activation energy of desorption;
R   is the gas constant;
T_{max} is the peak maximum temperature;
ν is the pre-exponential factor;
β is the heating rate, dT/dt.

Equation (2-16) is approximately correct for first-order desorption and for values of ν/β between 10^8 and 10^{13} K^{-1}. It is very often applied to determine E_{des} from a single TPD spectrum. The critical point, however, is that one must choose a value for ν, the general choice being 10^{13} s^{-1}, independent of coverage. As we explain below, this choice is only valid when there is little change in entropy between the molecule in the ground state and the transition state [27, 29, 30]. The Redhead formula should only be used if a reliable value for the prefactor is available!

Another popular method has been developed by Chan, Aris and Weinberg [31]. These authors expressed $E_{des}(\theta)$ and $v(\theta)$ in terms of the peak maximum temperature T_{max} and the peak width, either at half or at three-quarters of the maximum intensity. Their expressions for first-order desorption are:

$$E_{des} = RT_{max}\left[-1 + \sqrt{\left(\frac{W}{T_{max}}\right)^2 + 5.832\frac{T_{max}}{W}}\right] \quad \text{for } n=1 \quad (2\text{-}17)$$

$$v = \frac{E_{des}\beta}{RT_{max}^2}e^{E_{des}/RT_{max}} \quad (2\text{-}18)$$

in which W is the peak width at half the maximum intensity; the other symbols are as defined for Eq. (2-16). Similar expressions exist for the peak width at three-quarters of the maximum intensity and for second-order desorption. It is essential is that the thus-obtained $E_{des}(\theta)$ and $v(\theta)$ values are extrapolated to zero coverage, in order to obtain the desorption parameters of a single molecule adsorbed onto an otherwise empty surface [31].

Among all the approximate analysis methods, there is one that deserves special mention because it is the only one that operates quite well at high coverages, where even the complete analysis inevitably becomes inaccurate. The "leading edge analysis", as described in detail by Küppers [32], acknowledges that $v(\theta)$ may depend on temperature. In order to fix both T and θ, a relatively small temperature interval is selected at the low-temperature, high-coverage side of a TPD spectrum, in which the variations of both T and θ are insignificantly small. An Arrhenius plot of this short interval yields a straight line with slope $-E_{des}(\theta)/R$ and intercept $n \ln \theta + \ln v(T)$. Alternatively, if one expects first-order kinetics, one may plot $\ln(r/\theta)$ against $1/T$, allowing use of a larger part of the desorption curve [17]. The advantage of the leading edge analysis is that it rests on a minimum number of assumptions; the disadvantage is that data of extremely good quality are needed.

The best way to determine which procedures can be used with confidence, and which not, is to test them with a set of simulated spectra [33, 34]. Figure 2.12a shows calculated spectra for which E_{des} and v are known at all coverages, while Figure 2.12b shows the results of the most popular analysis methods for the activation energy. The heavy line represents the input values. The complete method and the leading edge analysis perform well over the entire coverage range. The Chan–Aris–Weinberg equations [see Eq. (2-17)] indeed predict the correct value at zero coverage, but values at higher coverages are without any meaning. The Falconer–Madix method, in which the heating rate is varied [35], and the Redhead formula [Eq. (2-16)] provide imprecise answers.

Lateral interactions between adsorbed species may make the kinetic parameters a function of coverage. In this case, it would be incorrect to rely on integral methods that depend on the properties of the entire TPD curve. Miller et al. [33] and Nieskens et al. [36] showed that doing so may induce artificial compensation effects in the results. Differential methods, which analyze a part of a trace are – in

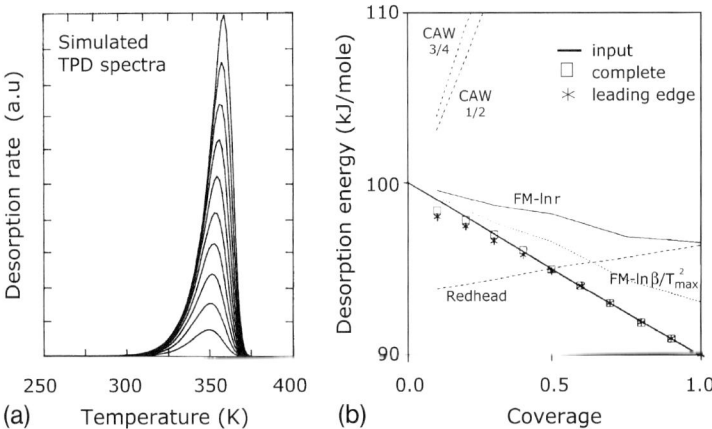

Fig. 2.12 (a) Simulated TPD spectra and (b) results of a number of different analysis procedures for determining the activation energy of desorption. The solid line represents the input for the simulations. Note that only the complete analysis [15] and the leading edge procedure of Habenschaden and Küppers [32] give reliable results. The Chan–Aris–Weinberg (CAW) curves [31] extrapolate to the correct activation energies at zero coverage. FM: Falconer–Madix method [35]. (Reproduced from [34]).

principle – suitable to cope with coverage-dependent kinetic parameters. If one has a series of TPD traces at different initial adsorbate coverage, the leading edge method yields activation energy and pre-exponential factor for each curve separately. Examples of such a rigorous approach are scarce, however. Miller et al. [33] successfully applied this method to CO desorption from Ni(111), while Hopstaken [37] used it to derive coverage-dependent kinetic parameters for a surface reaction – the oxidation of CO – which can be monitored by the desorption of CO_2.

2.5.2
Desorption in the Transition State Theory

The transition state theory of reaction rates [23] provides the link between macroscopic reaction rates and molecular properties of the reactants, such as translational, vibrational, and rotational degrees of freedom [27, 29, 30]. The desorption of a molecule M proceeds as follows:

$$M_{ads} \xrightarrow{K^{\#}} M_{ads}^{\#} \xrightarrow{kT/h} M_{gas} \quad (2\text{-}19)$$

where:
M_{ads} is the adsorbed molecule, the superscript referring to the transition state for desorption;
$K^{\#}$ is the equilibrium constant for the excitation of M_{ads} into the transition state;

k   is Boltzmann's constant;
h   is Planck's constant;
T   is the temperature;
M_{gas} is the desorbed molecule in the gas phase.

The factor kT/h represents the rate constant for the reaction over the activation barrier, in this case from the transition state to the gas phase.

One degree of freedom of the adsorbed molecule serves as the reaction coordinate. For desorption, the reaction coordinate is the vibration of the molecule with respect to the substrate: in the transition state this vibration is highly excited, and the chance that the adsorption bond breaks is given by the factor kT/h. All other degrees of freedom of the excited molecule are in equilibrium with those of the molecule in the ground state, and are accounted for by their partition functions.

The expression for the rate constant of desorption in the transition state theory is:

$$k_{des} = \frac{kT}{h} K^{\#} = \frac{kT}{h} \frac{Q^{\#}}{Q} e^{-\Delta E/RT} \tag{2-20}$$

where:
k_{des} is the rate constant of the desorption;
$Q^{\#}$ is the partition function of $M_{ads}^{\#}$, excluding the reaction coordinate;
Q is the partition function of M_{ads};
ΔE is the adsorption energy (not exactly equal to the activation energy of desorption); and the other symbols as defined after Eq. (2-19).

The partition functions Q contain separate terms for translation, vibration, and rotation:

$$Q = Q_{trans} \cdot Q_{vib} \cdot Q_{rot} \tag{2-21}$$

and

$$Q_{trans} = l\frac{\sqrt{2\pi mkT}}{h}; \quad Q_{vib} = \frac{1}{1 - \exp(-h\nu/kT)}; \quad Q_{rot} = \frac{8\pi^2 IkT}{h^2} \tag{2-22}$$

where:
l is the characteristic linear dimension of an adsorption site;
m is the mass of the molecule M;
ν is the vibrational frequency;
I is the moment of inertia of the molecule M.

Note that Q_{trans} is given per degree of freedom, implying that the total translational partition function for an adsorbed molecule is given by $(Q_{trans})^2$. Also, the total partition functions for vibration and rotation are the products of terms for

Table 2.2 Partition functions for selected molecules[a] and atoms at 500 K.

	Q_{trans}[b]	Q_{rot}	Q_{vib}	Vibration
H_2	33	2.9	1.0	H–H
CO	460	180	1.0	C–O
Cl_2	1200	710	1.3	Cl–Cl
Br_2	2600	2100	1.7	Br–Br
H	16.4	–	1.5	W–H[c]
O	264	–	3	Ni–O[d]
Ag	1750	–	≫3	Ag–Ru[e]

[a] Partition functions of the molecules taken from [39].
[b] Adsorption site of 10^{-15} cm^2.
[c] Calculated for H adsorbed in a twofold position on W(100) with a symmetric stretch frequency of 1200 cm^{-1}, an asymmetric stretch frequency of 960 cm^{-1}, and a wagging mode of 440 cm^{-1}. (Data from [40]).
[d] Estimated for O adsorbed in a fourfold position on Ni(100) with one perpendicular mode (350 cm^{-1}) and a doubly degenerate parallel mode (450 cm^{-1}). (Data from [41]).
[e] Vibrational data for Ag on Ru not available; the partition function will be much higher than for adsorbed O due to low-frequency modes.

each individual vibration and rotation, respectively. Table 2.2 provides values for the partition functions for adsorbed atoms and molecules at 500 K. Vibrational partition functions are usually close to 1, but rotational and translational partition functions have larger values.

We use this knowledge to derive pre-exponential factors from Eq. (2-20) for a few desorption pathways (see Fig. 2.13). The simplest case arises if the partition functions Q and $Q^{\#}$ in Eq. (2-20) are almost equal. This corresponds to a transition state that resembles the ground state of the adsorbed molecule. In order to compare Eq. (2-20) with the Arrhenius expression [Eq. (2-15)], we need to apply the definition of the activation energy:

$$E_{act} = -RT^2 \frac{\partial}{\partial T} \ln k_{des} \tag{2-23}$$

which implies that the activation energy E_{des} for desorption through a tight transition state equals $\Delta E + kT$, whereas the prefactor becomes ekT/h, equal to 1.6×10^{13} s^{-1} at 300 K. We conclude that prefactors on the order of 10^{13} s^{-1} correspond to transition states that resemble the ground state.

Higher pre-exponential factors result if the molecule rotates or moves in the transition state. For example, suppose that CO has free rotation in a plane perpendicular to the surface. As rotational partition functions are usually large (see

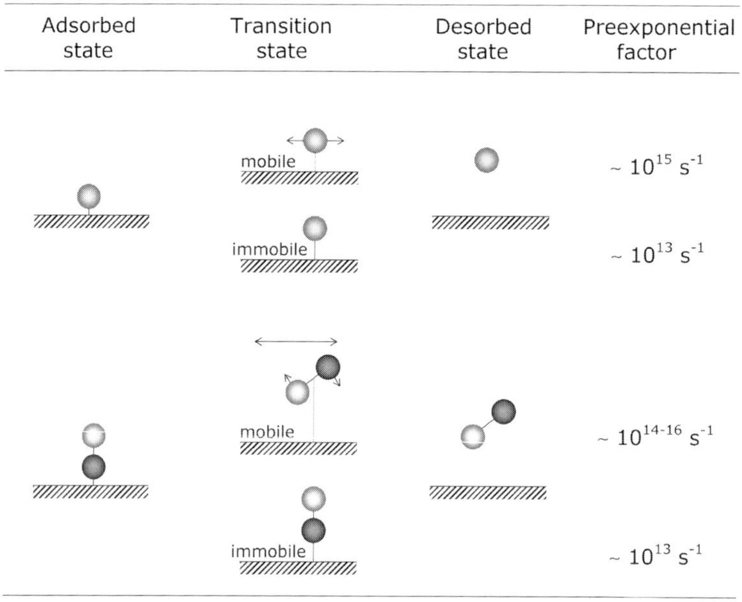

Adsorbed state	Transition state	Desorbed state	Preexponential factor
	mobile		$\sim 10^{15}$ s^{-1}
	immobile		$\sim 10^{13}$ s^{-1}
	mobile		$\sim 10^{14\text{-}16}$ s^{-1}
	immobile		$\sim 10^{13}$ s^{-1}

Fig. 2.13 Microscopic-level diagrams of the desorption of atoms and molecules via mobile and immobile transition states. If the transition state resembles the ground state, we expect a prefactor of desorption on the order of 10^{13} s^{-1}. If the adsorbates are mobile in the transition state, the prefactor goes up by one or two orders of magnitude. In the case of desorbing molecules, free rotation in the transition state increases the prefactor even further. The prefactors are roughly characteristic for atoms such as C, N, and O, and molecules such as N_2, CO, NO, and O_2. See also the partition functions in Table 2.2 and the prefactors for CO desorption in Table 2.3.

Table 2.2), the factor $Q^{\#}/Q$ in Eq. (2-20) becomes much larger than 1, and we expect a prefactor that is two orders of magnitude higher than 10^{13} s^{-1}. If in addition the CO molecule would also have translational freedom in the transition state, the prefactor would again go up by a factor of a few hundred. The experimentally determined prefactors for desorption of CO from Ru(001) and Co(001) are in the range 10^{15} to 10^{16} (Table 2.3), indicating that CO desorbs out of a rather loose transition state.

To summarize, Eqs. (2-20) to (2-22) and Table 2.2 indicate that a high prefactor for desorption is found when an immobile adsorbate desorbs through a mobile or freely rotating transition state. Prefactors on the order of 10^{13} are expected when the transition state resembles the ground state. Prefactors significantly lower than 10^{13} s^{-1} indicate that the transition state is more restricted than the ground state. As pre-exponential factors from TPD spectra reflect the desorption mechanism, some insight into expressions such as Eq. (2-20) and a feeling for what prefactors of desorption mean, certainly help in the correct interpretation of TPD spectra.

In conclusion, TPD of adsorbates on single crystal surfaces measured in UHV systems with sufficiently high pumping speeds, provides information on adsor-

Table 2.3 Activation energies and pre-exponential factors for CO desorption (from [39]).

Substrate	E_{des} [kJ mol^{-1}]	ν [s^{-1}]
Ir(110)	155	10^{13}
Pt(111)	134	10^{14}
Rh(111)	134	10^{14}
Co(001)	117	10^{15}
Ru(001)	159	10^{16}

bate coverage, the adsorption energy, the existence of lateral interactions between the adsorbates, and the pre-exponential factor of desorption, which in turn depends on the desorption mechanism. These analyses of spectra should be conducted with care, as simplified analysis procedures may easily produce erroneous results.

In principle, TPD can also be applied to high-surface area catalysts in a reactor, and this may yield useful qualitative information. Deriving quantitative information from TPD on supported catalysts is also possible, but requires that mass transfer properties such as intraparticle diffusion are properly taken into account. For details of this approach, the reader is referred to an interesting discussion by Kanervo et al. [38].

2.6
Temperature-Programmed Reaction Spectroscopy in UHV

The same set-up as used for TPD can be applied to study reactions between adsorbed molecules. In the experiment shown in Figure 2.14, O-atoms have been co-adsorbed with CO [17]. During temperature programming, CO_2 forms and desorbs instantaneously (see inset of Fig. 2.14). In order to derive the activation energy and pre-exponential factor, we start by assuming that the reaction exhibits first-order kinetics in the surface concentrations of CO and O; hence, we can write the rate as:

$$r = k\theta_O \theta_{CO} = \nu \theta_O \theta_{CO} e^{-E_a/RT} \tag{2-24}$$

As on every point of the TPRS curve the corresponding coverages of CO and O that are left on the surface can be calculated, we can thus divide the rate by the actual coverages at every temperature. Then, if we make the usual Arrhenius plot of $(\ln r/\theta_O \theta_{CO})$ against $1/T$, a straight line evolves (Fig. 2.14), indicating that our assumption of first-order kinetics in the reactants is justified and that the reaction $CO_{ads} + O_{ads}$ can be considered an elementary step that is followed

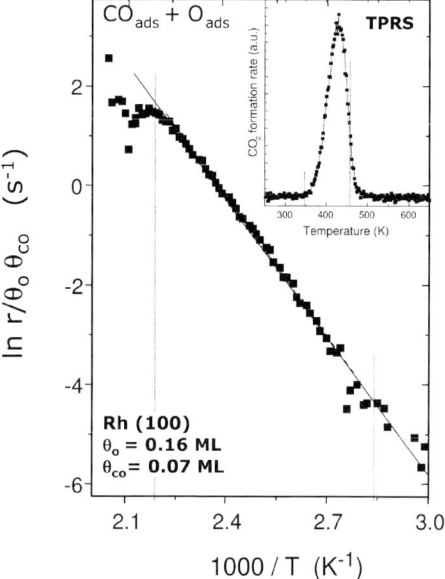

Fig. 2.14 Temperature-programmed reaction between O-atoms and CO adsorbed on Rh(100) along with an Arrhenius plot based on Eq. (2-24). Note the wide temperature range over which the Arrhenius plot forms a straight line. (Adapted from [17]).

by the rapid (and thus kinetically insignificant) desorption of CO_2. The activation energy is calculated as 103 ± 5 kJ mol^{-1}, and the pre-exponential factor as $10^{12.7 \pm 0.7}$ s^{-1} [17]. The latter corresponds to a "normal" transition state for which the prefactor is on the order of $ek_BT/h \approx 10^{13}$ s^{-1}.

We end the chapter with an example of how TPRS reveals information on reaction mechanisms. NO reduction by CO is a crucial reaction in the automotive exhaust catalyst. Figure 2.15 compares the performance of two rhodium single crystal surfaces in this reaction [42]. The experiments are performed by dosing small amounts of CO and NO on the surface at temperatures below 200 K, after which the surface is heated linearly in time, and all relevant gases are monitored with a mass spectrometer. CO and N_2 have equal masses, as well as N_2O and CO_2. Using isotopically labeled CO helps to resolve overlapping signals. In addition, one can also record the mass signals of the C, N, and O atoms (masses 12, 14, and 16) to distinguish between, for example, CO and N_2.

The results in Figure 2.15 reveal that the CO+NO reaction is obviously structure-sensitive. On Rh(100), the first signal that appears is that of CO_2 at 300 K. This implies that at, or somewhere below, this temperature the NO has already dissociated into N and O atoms. In Chapter 4 (when we have a technique that can monitor reactions on the surface) we will see that NO already starts to dissociate at around 200 K on Rh(100). Note that all NO has dissociated, because desorption of NO or any other form of NOx is not observed. When most of the oxygen is consumed, CO desorption competes with CO_2 formation in the temper-

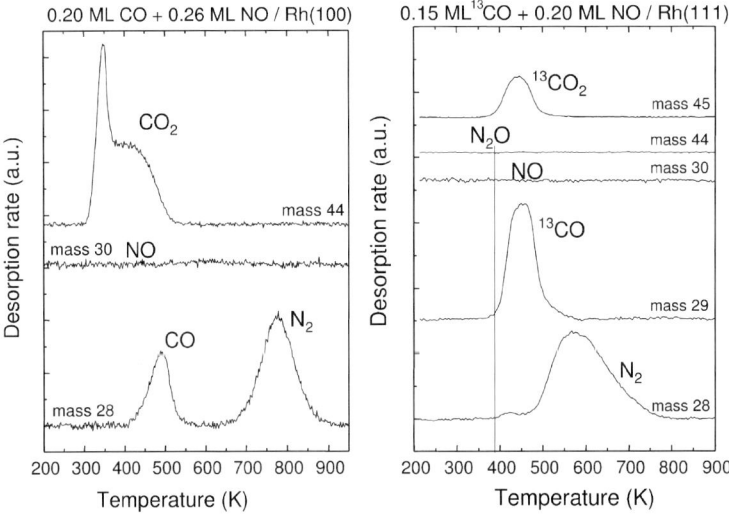

Fig. 2.15 Temperature-programmed reaction between CO and NO on the (100) (left) and (111) (right) surfaces of rhodium. See text for explanation. (Adapted from [42]).

ature range between 400 and 525 K. Above 650 K the N-atoms recombine and desorb as N_2, leaving a Rh(100) surface that is partly covered by O-atoms. Hence, Rh(100) is quite effective in dissociating NO and oxidizing CO to CO_2, but desorption of N_2 is slow, as it occurs at high temperature. Rh(111) on the other hand is much less reactive than Rh(100). Although it dissociates all NO, most of the CO desorbs unreacted and only a small part is oxidized to CO_2. A favorable point here is that N_2 desorbs at significantly lower temperature than on the Rh(100) surface. Note how the signals are nicely separated owing to the use of isotopically labeled CO; without this, the signals of CO and N_2 would have overlapped considerably.

TPRS is a very useful tool for investigating which reactions can take place when several species are present on a surface. If desorption follows instantaneously, its peak can be used to derive an activation energy for the rate-determining step that precedes it.

References

1 N.W. Hurst, S.J. Gentry, A. Jones, and B.D. McNicol, *Catal. Rev. – Sci. Eng.* **24** (1982) 233.
2 J.L. Falconer and J.A. Schwartz, *Catal. Rev. – Sci. Eng.* **25** (1983) 141.
3 I. Barin and O. Knacke, *Thermochemical Properties of Inorganic Substances*. Springer-Verlag, Berlin, 1973, and supplement, 1977.
4 J.R. Anderson, *Structure of Metallic Catalysts*. Academic Press, London, 1975.
5 A.J.H.M. Kock and J.W. Geus, *Progr. Surface Sci.* **20** (1985) 165.

6 M.E. Dry, in: *Catalysis, Science and Technology*, J.R. Anderson and M. Boudart (Eds.), Vol. 1. Springer-Verlag, Berlin, 1981, p. 159.
7 H.H. Kung, *Transition Metal Oxides: Surface Chemistry and Catalysis*. Elsevier, Amsterdam, 1989.
8 O.J. Wimmers, P. Arnoldy and J.A. Moulijn, *J. Phys. Chem.* **90** (1986) 1331.
9 H.E. Kissinger, *Anal. Chem.* **29** (1957) 1702.
10 H.F.J. van't Blik and J.W. Niemantsverdriet, *Appl. Catal.* **10** (1984) 155.
11 B.C. Gates, J.R. Katzer, and G.C.A. Schuit, *Chemistry of Catalytic Processes*. McGraw-Hill, New York, 1979.
12 P. Arnoldy, J.A.M. van den Heijkant, G.D. de Bok, and J.A. Moulijn, *J. Catal.* **92** (1985) 35.
13 T. Koerts, W.J.J. Welters, and R.A. van Santen, *J. Catal.* **134** (1992) 1.
14 P. Feulner and D. Menzel, *J. Vac. Sci. Technol.* **17** (1980) 662.
15 D.A. King, *Surface Sci.* **47** (1975) 384.
16 A. Cassuto and D.A. King, *Surface Sci.* **102** (1981) 388.
17 M.J.P. Hopstaken, W.E. van Gennip, and J.W. Niemantsverdriet, *Surface Sci.* **433–435** (1999) 69.
18 R. Linke, D. Curulla, M.J.P. Hopstaken, and J.W. Niemantsverdriet, *J. Chem. Phys.* **115** (2001) 8209.
19 M.J.P. Hopstaken and J.W. Niemantsverdriet, *J. Phys. Chem. B* **104** (2000) 3058.
20 J.W. Niemantsverdriet, P. Dolle, K. Markert, and K. Wandelt, *J. Vac. Sci. Technol.* **A5** (1987) 875.
21 J.A. Rodriguez and D.W. Goodman, *J. Phys. Chem.* **95** (1991) 4196.
22 C.T. Campbell, *Surface Sci. Rep.* **27** (1997) 1.
23 S. Glasstone, K.J. Laidler, and H. Eyring, *The Theory of Rate Processes*. McGraw-Hill, New York, 1941.
24 J.W. Niemantsverdriet, K. Markert, and K. Wandelt, *Appl. Surface Sci.* **31** (1988) 211.
25 M. Boudart and G. Djega-Mariadassou, *Kinetics of Heterogeneous Catalytic Reactions*. Princeton University Press, Princeton, 1984.
26 A. Clark, *The Theory of Adsorption and Catalysis*. Academic Press, New York, 1970.
27 V.P. Zhdanov, *Elementary Physicochemical Processes on Solid Surfaces*. Plenum, New York, 1991.
28 P.A. Redhead, *Vacuum* **12** (1962) 203.
29 R.A. van Santen and J.W. Niemantsverdriet, *Chemical Kinetics and Catalysis*. Plenum, New York, 1995.
30 I. Chorkendorff and J.W. Niemantsverdriet, *Concepts of Modern Catalysis and Kinetics*. Wiley-VCH, Weinheim, 2003.
31 C.M. Chan, R. Aris, and W.H. Weinberg, *Appl. Surface Sci.* **1** (1978) 360.
32 E. Habenschaden and J. Küppers, *Surface Sci.* **138** (1984) L147.
33 J.B. Miller, H.R. Siddiqui, S.M. Gates, J.N. Russel, Jr., J.T. Yates, Jr., J.C. Trully, and M.J. Cardillo, *J. Chem. Phys.* **87** (1987) 6725.
34 A.M. de Jong and J.W. Niemantsverdriet, *Surface Sci.* **233** (1990) 355.
35 J.L. Falconer and R.J. Madix, *Surface Sci.* **48** (1975) 393.
36 D.L.S. Nieskens, A.P. van Bavel, and J.W. Niemantsverdriet, *Surface Sci.* **546** (2003) 159.
37 M.J.P. Hopstaken and J.W. Niemantsverdriet, *J. Chem. Phys.* **113** (2000) 5457.
38 J.M. Kanervo, T.J. Keskitalo, R.I. Slioor, and A.O.I. Krause, *J. Catal.* **238** (2006) 382.
39 V.P. Zhdanov, J. Pavlicek, and Z. Knor, *Catal. Rev. – Sci. Eng.* **30** (1988) 501.
40 R.F. Willis, *Surface Sci.* **89** (1979) 457.
41 J.M. Szeftel, S. Lehwald, H. Ibach, T.S. Rahman, J.E. Black, and D.C. Mills, *Phys. Rev. Lett.* **51** (1983) 268.
42 M.J.P. Hopstaken and J.W. Niemantsverdriet, *J. Vac. Sci. Technol. A* **18** (2000) 1503.

3
Photoemission and Auger Spectroscopy

Keywords

X-ray photoelectron spectroscopy (XPS)
Ultraviolet photoelectron spectroscopy (UPS)
Auger electron spectroscopy (AES)

3.1
Introduction

Photoemission spectroscopy is based on the photoelectric effect: a sample that is irradiated with light of sufficiently small wavelength emits electrons. The number of photoelectrons depends on the light intensity, and the energy of the electrons on the wavelength of the light.

In hindsight, it was Hertz who, unknowingly, reported the first photoemission experiments in 1887, when he noticed that electrical sparks induced the formation of a second spark in a variety of samples [1]. Hertz correctly recognized that the effect was due to ultraviolet light generated by the first spark, but he did not understand the nature of the induced spark. In fact he could not, because the electron had not yet been discovered! About ten years later, Thomson identified the radiation in Hertz's experiment as being caused by electrons [2]. The phenomenon of photoemission has played an important role in confirming Einstein's famous postulate, published in 1905, that light is quantized in photons of energy, $h\nu$ [3]. We refer to Margaritondo [4] for an interesting review on the early days of the photoelectric effect.

It was a half-century later before the photoelectric effect was applied in spectroscopy. As described by Ertl and Küppers [5], three parallel developments took place:

- During the 1950s, Siegbahn's group worked to improve the energy resolution of electron spectrometers and then to combine it with X-ray sources. This led to a technique called electron spectroscopy for chemical analysis (ESCA), which today is more commonly referred to as X-ray photoelectron spectroscopy (XPS)

Spectroscopy in Catalysis: An Introduction, Third Edition
J. W. Niemantsverdriet
Copyright © 2007 WILEY-VCH Verlag GmbH & Co. KGaA, Weinheim
ISBN: 978-3-527-31651-9

[6]. Commercial instruments of this type have been available since the early 1970s, and in 1981 Siegbahn received the Nobel Prize for his efforts in the field.

- Turner and co-workers [7] later applied the photoelectric effect to gases. By using the sharp ultraviolet line from a helium resonance they were able even to resolve the vibrational fine structure of the electron levels. This topic is beyond the scope of this volume, but the interested reader is referred to some excellent books on the subject [7, 8].

- Spicer [9] measured photoelectron spectra from solids in vacuum irradiated with ultraviolet (UV) light. Unfortunately, the light had to pass through a LiF window, thereby cutting off all photons with energies above 11.6 eV, and consequently Spicer was able to examine only a small part of the valence band. The introduction of differentially pumped windowless UV sources by Eastman and Cashion in 1971 [10] extended the energy range to about 40 eV, and this was the start of laboratory ultraviolet photoelectron spectroscopy (UPS) in the form in which it is still used today. Synchrotrons have subsequently extended the photon energy range further, such that all energies between the UPS and XPS regime are available.

The development of Auger electron spectroscopy (AES) has its own history. In 1925, Pierre Auger discovered that photographic plates exposed to hard X-rays exhibited traces due to electrons [11]. He interpreted them as the result of a relax-

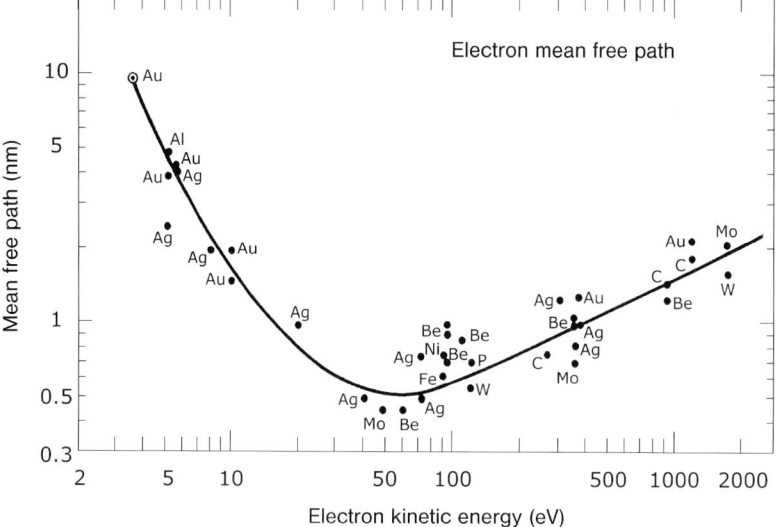

Fig. 3.1 The mean free path of an electron depends on its kinetic energy, and determines how much surface information it carries. Optimum surface sensitivity is obtained with electrons in the 25 to 200 eV range. (Adapted from [16]).

ation process now called "Auger decay". In 1953, Lander suggested that Auger electrons might be used for surface analysis [12], and 15 years later this was accomplished, first with a retarding field analyzer in use in low-energy electron diffraction [13], and a few years later with the more convenient cylindrical mirror analyzer, as proposed by Palmberg [14].

Today, XPS and AES are among the most often applied techniques in the characterization of solid surfaces [15], while UPS is a typical surface science method which is best suited to fundamental studies on single crystals. All three spectroscopies provide surface-sensitive information, however.

A technique becomes surface-sensitive if the radiation or particles to be detected travel no more than a few atomic distances through the solid. The diagram in Figure 3.1 shows that the mean free path, λ, of electrons in elemental solids depends on the kinetic energy, but is limited to less than 1–2 nm for kinetic energies in the range 15 to 1000 eV [16]. Optimum surface sensitivity ($\lambda \approx 0.5$ nm) is achieved with electrons at kinetic energies in the range of 50 to 250 eV, where almost half of the photoelectrons come from the outermost layer.

3.2
X-Ray Photoelectron Spectroscopy (XPS)

Today, XPS is among the most frequently used techniques in catalysis, as it provides information on the elemental composition, the oxidation state of the elements and, in favorable cases, on the dispersion of one phase over another. When working with flat layered samples, depth-selective information is obtained by varying the angle between the sample surface and the analyzer. Several excellent text-books on XPS are available [5, 8, 17–21]. In this section, we first describe briefly the theory behind XPS, followed by details of the instrumentation; finally, we illustrate the type of information that XPS offers with regards to catalysts and model systems.

XPS and UPS are based on the photoelectric effect, whereby an atom absorbs a photon of energy, hν, after which a core or valence electron with binding energy E_b is ejected with kinetic energy (Fig. 3.2):

$$E_k = h\nu - E_b - \varphi \tag{3-1}$$

where:
E_k is the kinetic energy of the photoelectron;
h is Planck's constant;
ν is the frequency of the exciting radiation;
E_b is the binding energy of the photoelectron with respect to the Fermi level of the sample;
φ is the work function of the spectrometer, as explained later in this chapter.

Routinely used X-ray sources are Mg Kα (1253.6 eV) and Al Kα (1486.3 eV). In XPS, one measures the intensity of photoelectrons N(E) as a function of their

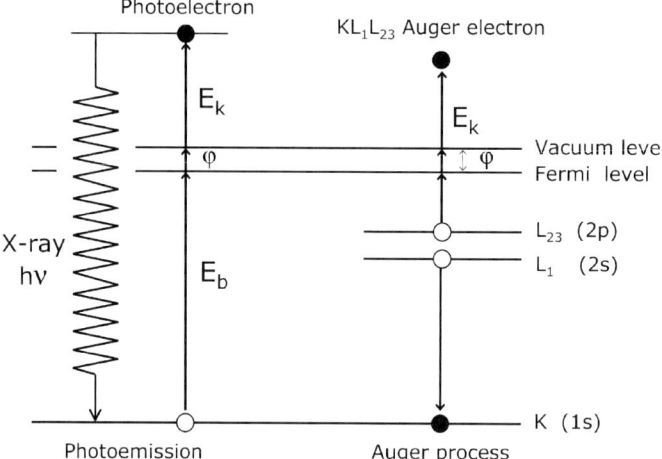

Fig. 3.2 Photoemission and the Auger process. Left: An incident X-ray photon is absorbed and a photoelectron emitted. Measurement of its kinetic energy allows the binding energy of the photoelectron to be calculated. The atom stays behind as an unstable ion with a hole in one of the core levels. Right: The excited ion relaxes by filling the core hole with an electron from a higher shell. The energy released by this transition is taken up by another electron, the Auger electron, which leaves the sample with an element-specific kinetic energy. In Auger spectroscopy a beam of energetic (2–5 keV) electrons creates the initial core holes.

kinetic energy. The XPS spectrum, however, is usually a plot of $N(E)$ versus E_k, or, more often, versus the binding energy E_b. Figure 3.3 shows the XPS spectrum of an alumina-supported rhodium catalyst, prepared by impregnating the support with $RhCl_3$ in water. Peaks due to Rh, Cl, Al, O, and C, due to an always-present contamination by hydrocarbons, are readily assigned if one consults binding energy tables [20, 22].

In addition to the expected photoelectron peaks, the spectrum in Figure 3.3 also contains peaks due to Auger electrons. The latter arise from the de-excitation of the photo ion by an Auger transition (Fig. 3.2). Although Auger electrons have fixed *kinetic* energies, which are independent of the X-ray energy, Auger peaks are nevertheless plotted on the binding energy scale, which has of course no physical significance. The main peak of the O KVV Auger signal in Figure 3.3 has a kinetic energy of about 500 eV, but appears at a binding energy of about 986 eV, because the spectrum was taken with Al Kα X-rays of 1486 eV. Auger peaks can be recognized by recording the spectrum at two different X-ray energies: XPS peaks appear at the same binding energies, while Auger peaks will shift on the binding energy scale. This is the main reason why X-ray sources often contain a dual anode of Mg and Al. By varying the X-ray energy Auger peaks are readily identified, and sometimes an overlap between XPS and Auger peaks can be avoided.

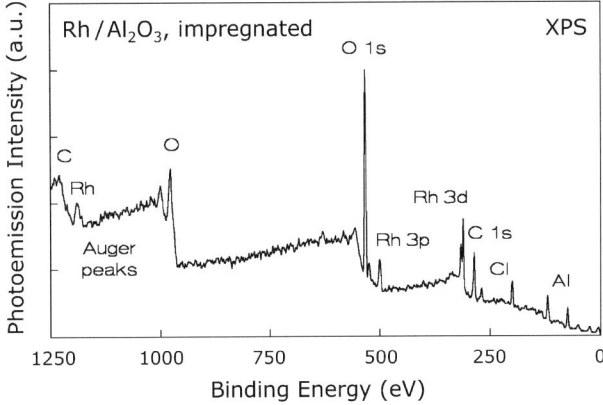

Fig. 3.3 X-ray photoelectron spectroscopy (XPS) spectrum of a Rh/Al$_2$O$_3$ model catalyst prepared by impregnating a thin film of Al$_2$O$_3$ on aluminum with a solution of RhCl$_3$ in water. (Figure courtesy of L.C.A. van den Oetelaar, Eindhoven).

Photoelectron peaks are labeled according to the quantum numbers of the level from which the electron originates. An electron with orbital momentum l $(0, 1, 2, 3, \ldots$ indicated as s, p, d, f, $\ldots)$ and spin momentum s has a total momentum $j = l + s$. As the spin may be either up $\left(s = +\frac{1}{2}\right)$ or down $\left(s = -\frac{1}{2}\right)$, each level with $l \geq 1$ has two sublevels, with an energy difference called the spin-orbit splitting. Thus, the Pt 4f level gives two photoemission peaks, 4f$_{7/2}$ (with $l = 3$ and $j = 3 + \frac{1}{2}$) and 4f$_{5/2}$ ($l = 3$ and $j = 3 - \frac{1}{2}$). The spectroscopic nomenclature is summarized in Table 3.1.

Table 3.1 Spectroscopic notation used in XPS and AES.

n	l	j	X-ray level	Electron level
1	0	1/2	K	1s
2	0	1/2	L$_1$	2s
2	1	1/2	L$_2$	2p$_{1/2}$
2	1	3/2	L$_3$	2p$_{3/2}$
3	0	1/2	M$_1$	3s
3	1	1/2	M$_2$	3p$_{1/2}$
3	1	3/2	M$_3$	3p$_{3/2}$
3	2	3/2	M$_4$	3d$_{3/2}$
3	2	5/2	M$_5$	3d$_{5/2}$
4	3	5/2	N$_6$	4f$_{5/2}$
4	3	7/2	N$_7$	4f$_{7/2}$

Spin-orbit splittings as well as binding energies of a particular electron level increase with increasing atomic number. The intensity ratio of the two peaks of a spin-orbit doublet is determined by the multiplicity of the corresponding levels, equal to 2j + 1. Hence, the intensity ratio of the j = 7/2 and j = 5/2 components of the Pt 4f doublet is 8:6, and that of the 5/2 and 3/2 peaks of the 4d doublet is 6:4, etc. Thus, photoelectron peaks from core levels come in pairs (doublets) except for s levels, which give a single peak (although further splitting of all peaks is possible, as we will see later).

3.2.1
XPS Intensities and Sample Composition

Because a set of binding energies is characteristic for an element, XPS can be used to analyze the composition of samples. Almost all photoelectrons used in XPS have kinetic energies in the range of 0.2 to 1.5 keV. According to the inelastic mean free path data in Figure 3.1, the probing depth of XPS (usually taken as 3λ) varies between 1.5 and 6 nm, depending on the kinetic energy of the photoelectron. For example, the Al 2p peak in the spectrum of the Rh/Al_2O_3 catalyst of Figure 3.3 probes deeper into the sample than the O 1s peak. When determining concentrations, this effect must be accounted for.

Actually, concentrations cannot be calculated without first assuming a structure model. For instance, a metal foil with a thin oxide passivation layer on top will give an intense peak of oxygen in the XPS spectrum, whereas the nominal oxygen concentration for the entire foil is negligible. If the O intensity were taken as representing a homogeneous distribution of O through the sample, then the concentration would become considerable. The general expression for the intensity of an XPS peak is:

$$I = F_X S(E_k) \sigma(E_k) \int_0^\infty n(z) e^{-z/\lambda(E_k)\cos\theta} \, dz \tag{3-2}$$

where:
I is the intensity of the XPS peak (area);
F_X is the X-ray flux on the sample;
$S(E_k)$ is the spectrometer efficiency for detecting the electron at kinetic energy E_k (also called transmission function);
$\sigma(E_k)$ is the cross-section for photoemission;
$n(z)$ is the concentration, in number of atoms per unit volume;
z is the depth below the surface;
$\lambda(E_k, z)$ is the mean free path of the photoelectron at kinetic energy E_k through the material present at depth z;
θ is the take-off angle, that is, the angle between the direction in which the photoelectron is emitted and the surface normal.

In the case of a homogeneous concentration through the sample, the following expression holds for the intensity of each element:

$$I = F_X S(E_k) \sigma(E_k) n \lambda(E_k) \cos \theta \qquad (3\text{-}3)$$

Cross-sections for the elements have been calculated and tabulated by Scofield [23].

Mean free path values are often approximated by a calculating them from a general formula [24], but data which take material properties into account are available also [25]. That this is important is illustrated by the mean free path of Si 2p photoelectrons in SiO_2 (3.7 nm) and in pure silicon (3.2 nm, valid when using Al Kα radiation): the λ-values differ considerably, although the kinetic energies of the electrons are the same. In a recent review, Jablonski and Powell discussed developments in the understanding of electron attenuation lengths [26].

Figure 3.4 illustrates the use of Eq. (3-3) in an example on organoplatinum complexes. The sample for XPS analysis was prepared by allowing a solution of the complexes in dichloromethane to dry on a stainless steel sample stub. The sample should thus be homogeneous, and the use of Eq. (3-3) permitted. The

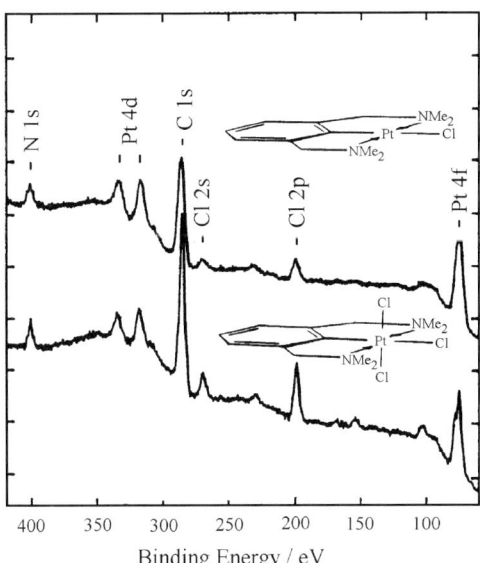

Fig. 3.4 XPS scans between 0 and 450 eV of two organoplatinum complexes showing peaks due to Pt, Cl, N, and C. The C 1s signal not only represents carbon in the compound but also contaminant hydrocarbon fragments, as on any sample. The abbreviation "Me" in the structures represents CH_3. (Figure courtesy of J.C. Muijsers, Eindhoven).

spectrum shows the peaks of all elements expected from the compounds in the binding energy range up to 410 eV: the Pt 4f and 4d doublets (the 4f doublet is unresolved due to the low-energy resolution employed for broad-energy range scans), Cl 2p and Cl 2s, N 1s, and C 1s. However, the C 1s cannot be taken as characteristic for the complex only. All surfaces that have not been cleaned by sputtering or oxidation in the XPS spectrometer contain carbon. The reason is that adsorbed hydrocarbons, originating either from pump oil or from the atmosphere, give the optimum lowering of the surface free energy. Hence, in practice all surfaces are covered by hydrocarbon fragments [16]. The data in Figure 3.4 show immediately that the Cl peaks in the spectrum of the trichloride complex are about three times as intense as in the spectrum of the compound with one Cl. If we apply Eq. (3-3), using Scofield factors, mean free path values and transmission corrections for the elements, we obtain Pt:N:Cl = 1:1.9:4 for the trichloride complex, which is close to the true stoichiometry of 1:2:3 [27].

3.2.2
XPS Binding Energies and Oxidation States

Binding energies are not only element-specific but also contain chemical information, because the energy levels of core electrons depend slightly on the chemical state of the atom. Chemical shifts are typically in the range of 0 to 3 eV. Figure 3.5 shows spectra at high resolution of three possible oxidation states of platinum in a metal foil and in two organometallic complexes [27]. The binding energy in-

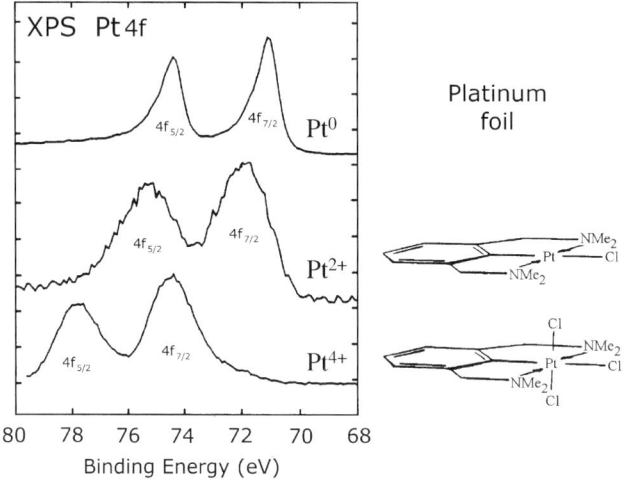

Fig. 3.5 Pt 4f XPS spectra of platinum metal (*top*) and of the two organoplatinum compounds shown in Figure 3.4, showing that the Pt 4f binding energy reflects the oxidation state of platinum. (Adapted from [27]).

Table 3.2 Binding energies of Fe $2p_{3/2}$ electrons in several compounds [20, 22].

Compound	E_b [eV]	Compound	E_b [eV]
Iron metal	706.7	FeBr$_3$	710.0
Fe(CO)$_5$	709.4	FeCl$_3$	711.1
FeO	710.0	FeF$_3$	714.0
Fe$_2$O$_3$	710.7		

creases with the oxidation state of the platinum; the reason for this is that the 74 electrons in the Pt^{4+} ion feel a higher attractive force from the nucleus with a positive charge of 78$^+$, than the 76 electrons in Pt^{2+} or the 78 in the neutral Pt atom. In general, the binding energy increases with increasing oxidation state, and for a fixed oxidation state with the electronegativity of the ligands, as the series FeBr$_3$, FeCl$_3$, FeF$_3$ in Table 3.2 illustrates. Figure 3.6 shows an example of how the C 1s spectrum of a fluorocarbon polymer reveals all the different bonding arrangements of the carbon atoms [28]. However, the rule of thumb that the binding energy of a particular atom goes up with its oxidation state, or with the electronegativity of its neighbors, does not always hold. Alkali metals, for example, present an exception.

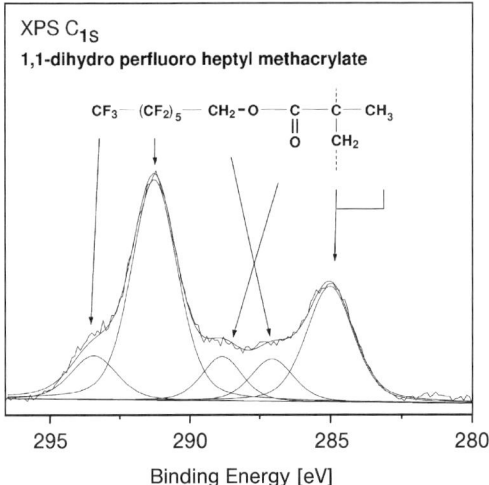

Fig. 3.6 C 1s XPS spectra of a fluorinated methacrylate (1,1-dihydro-perfluoroheptyl methacrylate), showing how the binding energy of carbon depends sensitively on the electronegativity of its neighbors. Fluorine possesses the highest electronegativity, and hence the CF$_3$ moiety gives the highest C 1s binding energy. (Adapted from [28]).

In order to appreciate the meaning of a binding energy, it is necessary to consider final state effects. In practice, we use XPS data as if they were characteristic for the atoms as they are before the photoemission event takes place. We must realize that this is not correct: photoemission data represent a state from which an electron has just left. Thus, we must analyze the event in more detail.

We start with an atom containing N electrons, with a total energy E in the initial state, denoted with the superscript i. The atom absorbs a photon of energy $h\nu$, the absorption event taking less than 10^{-17} s. Some 10^{-14} s later, the atom has emitted the photoelectron with kinetic energy E_k, and is itself in the final state with one electron less and a hole in one of the core levels. The energy balance of the event is

$$E_N^i + h\nu = E_{N-1,l}^f + E_k \qquad (3\text{-}4)$$

where:
- E is the total energy of the atom with N electrons in the initial state, i.e., before the photoemission has taken place;
- $h\nu$ is the energy of the photon;
- $E_{N-1,l}$ is the total energy of the atom with $N-1$ electrons and a hole in core level l, in the final state;
- E_k is the kinetic energy of the photoelectron.

A rearrangement of terms shows that the binding energy with respect to the Fermi level equals:

$$E_b = h\nu - E_k - \varphi = E_{N-1,k}^f - E_N^i - \varphi \qquad (3\text{-}5)$$

where φ is the work function of the spectrometer.

The important point to note is that the $N-1$ remaining electrons in the final state atom, as well as the electrons in neighboring atoms, feel the presence of the core hole. As a result they relax to lower the total energy of the atom by an amount ΔE_{relax}. This relaxation energy, which has both intra- and extra-atomic contributions, is included in the kinetic energy of the photoelectron. Thus, a binding energy is *not* equal to the energy of the orbital from which the photoelectron is emitted; the difference is caused by the reorganization of the remaining electrons when an electron is removed from an inner shell [5, 18, 29].

As a consequence, the binding energy of a photoelectron contains both information on the state of the atom before photoionization (the initial state) and on the core-ionized atom left behind after the emission of an electron (the final state). Fortunately, it is often correct to interpret binding energy shifts as those in Figures 3.5 and 3.6 in terms of initial state effects. The charge potential model [30, 31] elegantly explains the physics behind such binding energy shifts, by means of the formula:

$$E_b^i = kq_i + \sum_j \frac{q_j}{r_{ij}} + E_b^{\text{ref}} \qquad (3\text{-}6)$$

where:
E_b^i is the binding energy of an electron from an atom I;
q_i is the charge on the atom;
k   is a constant;
q_j is the charge on a neighboring atom j;
r_{ij} is the distance between atom i and atom j;
E_b^{ref} is a suitable energy reference.

The first term in Eq. (3-6) indicates that the binding energy increases with increasing positive charge on the atom from which the photoelectron originates. In ionic solids, the second term counteracts the first, because the charge on a neighboring atom will have the opposite sign. Because of its similarity to the lattice potential in ionic solids, the second term is often referred to as the "Madelung sum".

In addition to studying core levels, XPS can also be used to image the valence band. Figure 3.7 shows valence band spectra of Rh and Ag. The step at $E_b = 0$ corresponds to the Fermi level, the highest occupied electron level. The data in Figure 3.7 illustrate that the Fermi level of rhodium lies in the d-band where the

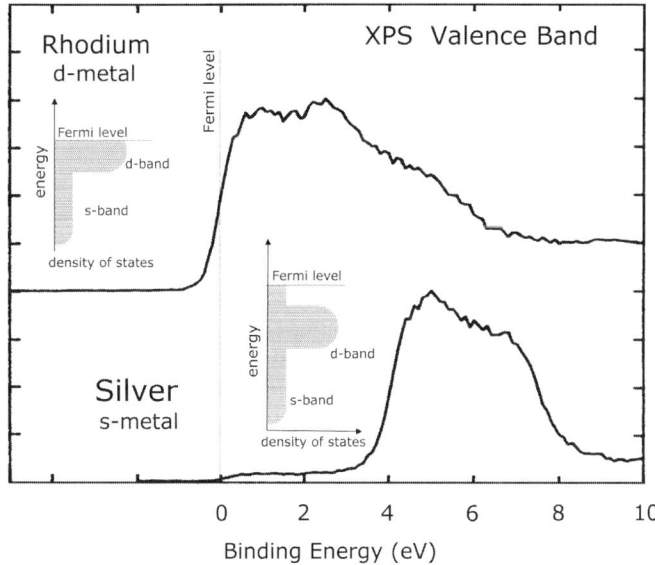

Fig. 3.7 XPS spectra of the valence bands of rhodium and silver along with schematic density of states (see Appendix). The Fermi level, the highest occupied level of a metal, is taken as the zero of the binding energy scale. Rhodium is a d-metal, meaning that the Fermi level lies in the d-band, where the density of states is high. Silver, on the other hand, is an s-metal. The d-band is completely filled and the Fermi level lies in the s-band where the density of states is low: the onset of photoemission at the Fermi level can just be observed.

density of states is high, whereas the Fermi level of silver, with its completely filled d-band, falls in the s-band, where the density of states is low (see also the Appendix).

Although valence band spectra probe those electrons that are involved in chemical bond formation, they are rarely used in studying catalysts. One reason is that all elements have valence electrons, which makes valence band spectra of multicomponent systems difficult to interpret. A second reason is that the mean free path of photoelectrons from the valence band is at its maximum, implying that the chemical effects of for example chemisorption, which are limited to the outer surface layer, can hardly be distinguished from the dominating substrate signal. In this respect UPS (which is discussed later in this chapter) is much more surface-sensitive and therefore better suited to adsorption studies.

At this point a brief comment about energy referencing might be in order. The Fermi levels of conducting samples that are in contact are the same, and hence the Fermi level provides a convenient energy zero. Of course, electrons at the Fermi level are still bound to the metal, and have an ionization potential equal to the work function, φ. However, electrons coming from the Fermi level of the sample are detected with a kinetic energy equal to $(h\nu - \varphi_{sp})$, where φ_{sp} is the work function of the spectrometer, and *not* that of the sample. The reason for this is as follows. Because the spectrometer and the sample are in electrical contact, their Fermi levels are lined up, but the work functions of the two differ. To remove an electron at the Fermi level from the sample costs an amount of energy equal to φ, the work function of the sample. The electron liberated from the sample is now at the potential just outside the sample (which can be considered as the "local vacuum level"), but not yet at the potential of the spectrometer, which in this set-up defines the vacuum level. Thus, while en route to the analyzer the electron is accelerated or decelerated by the work function difference, which is the contact potential between the sample and the spectrometer [20].

3.2.3
Shake Up, Shake Off, Multiplet Splitting and Plasmon Excitations

We have tacitly assumed that the photoemission event occurs sufficiently slowly to ensure that the escaping electron feels the relaxation of the core-ionized atom. This is termed the "adiabatic" limit: all relaxation effects to the energetic ground state of the core-ionized atom are accounted for in the kinetic energy of the photoelectron (but not the decay via Auger or fluorescence processes to a ground state ion, which occurs on a slower time scale). At the other extreme – the "sudden limit" – the photoelectron is emitted immediately after absorption of the photon, and *before* the core ionized atom relaxes. This is often accompanied by shake-up, shake-off and plasmon loss processes, which produce additional peaks in the spectrum.

Shake-up and shake off losses are final state effects, which arise when the photoelectron imparts energy to another electron of the atom. Ultimately, this electron will be in a higher unoccupied state (shake-up) or in an unbound state (shake-off). As a consequence, the photoelectron loses kinetic energy and appears at a higher binding energy in the spectrum. Discrete shake-up losses are prominently present in the spectra of several oxides of nickel, iron, and cobalt, and of many other compounds. Such losses have diagnostic value as the precise loss structure depends on the environment of the atom. In metals, shake-up of electrons in the valence band to empty states at the Fermi level produces a continuous range of energy losses from zero to the energy of the bottom of the band. The result is that the XPS peaks of metals such as Rh and Pt are asymmetrically broadened towards higher binding energies (e.g., see the asymmetric Pt 4f lines in the spectrum of the metal foil in Figure 3.5). As the probability of valence band shake-up depends on the density of states at or just above the Fermi level, the effect is most pronounced in d-metals, but is hardly of any significance for s-metals such as Cu, Ag, and Au, all of which give sharp symmetric peaks. In fact, the narrow Ag 3d lines are often used to test and demonstrate the energy resolution of an XPS instrument.

Multiplet splitting occurs if the initial state atom contains unpaired electrons. Upon photoemission, this electron may interact through its spin moment with the spin of the additional unpaired electron in the core level from which the photoelectron left. Parallel or anti-parallel spins give final states which differ by 1 to 2 eV in energy. A well-known example is the splitting of the N 1s and O 1s peaks (normally single lines) in the spectrum of gas phase NO, due to an unpaired spin in the 2π level of the NO molecule [30]. It should be noted that multiplet splitting is a *final* state effect, whereas the spin-orbit splitting of p, d, and f levels is an *initial* state effect.

3.2.4
Experimental Aspects of XPS

An XPS spectrometer contains an X-ray source – usually Mg Kα (1253.6 eV) or Al Kα (1486.3 eV) – and an analyzer which, in most commercial spectrometers, is hemispherical in design. In the entrance tube, the electrons are retarded or accelerated to a value called the "pass energy", at which they travel through the hemispherical filter. The lower the pass energy, the smaller the number of electrons that reaches the detector, but the more precisely is their energy determined. Behind the energy filter is the actual detector, which consists of an electron multiplier or a channeltron, which amplifies the incoming photoelectrons to measurable currents. Advanced hemispherical analyzers contain up to five multipliers. For further details of these instruments the interested reader should refer to other textbooks [20, 21].

The resolution of XPS is determined by the line width of the X-ray source, the broadening due to the analyzer, and the natural line width of the level under study. These three factors are related as follows:

$$(\Delta E)^2 = (\Delta E_X)^2 + (\Delta E_{an})^2 + (\Delta E_{nat})^2 \qquad (3\text{-}7)$$

where:
ΔE is the width of a photoemission peak at half-maximum;
ΔE_X is the line width of the X-ray source;
ΔE_{an} is the broadening due to the analyzer;
ΔE_{nat} is the natural line width.

The line width of the X-ray source is on the order of 1 eV for Al or Mg Kα sources, but can be reduced to better than about 0.3 eV with the use of a *monochromator*. A monochromator contains a quartz crystal which is positioned at the correct Bragg angle for Al Kα radiation. The monochromator not only narrows this line significantly and focuses it onto the sample, but also cuts out all unwanted X-ray satellites and background radiation. One important advantage of using a monochromator is that heat and secondary electrons generated by the X-ray source cannot reach the sample.

The broadening due to the analyzer depends on the energy at which the electrons travel through the analyzer, and the width of the slits between the energy filter and the actual detector. The analyzer contribution to the line width becomes irrelevant at low pass energies, albeit at the cost of intensity.

The natural line width is determined by Heisenberg's uncertainty relationship:

$$\Delta E_{nat} \cdot \Delta t \approx \frac{h}{2\pi} \qquad (3\text{-}8)$$

where:
ΔE_{nat} is the natural line width;
Δt is the life time of the core-ionized atom;
h   is Planck's constant.

The lifetime of the core-ionized atom is measured from the moment it emits a photoelectron until it decays by Auger processes or X-ray fluorescence. As the number of decay possibilities for an ion with a core hole in a deep level (e.g., the 3s level) is greater than that for an ion with a core hole in a shallow level (e.g., the 3d level), a 3s peak is broader than a 3d peak.

3.2.5
Charging and Sample Damage

An experimental problem in XPS (and also in other electron or ion spectroscopies) is that electrically insulating samples may charge during measurement, because photoelectrons leave the sample. The potential that the sample acquires is determined by the photoelectric current of electrons leaving the sample, the current through the sample holder towards the sample, and the flow of Auger and secondary electrons from the source window onto the sample. Due to the positive charge on the sample, all XPS peaks in the spectrum shift by the same amount to higher binding energies.

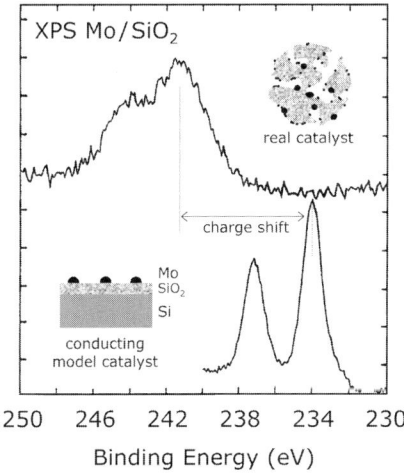

Fig. 3.8 Comparison of the monochromatic Mo 3d XPS spectra of MoO$_3$ in an insulating silica-supported catalyst and in a conducting, thin silica film-supported model catalyst, showing the effect of inhomogeneous charge broadening. (Figure courtesy of H. Korpik, Eindhoven).

Calibration is carried out by using the binding energy of a known compound. In SiO$_2$-supported catalysts, for example, one uses the binding energy of the Si 2p electrons, which should be 103.4 eV. If nothing else is available, it is possible to use the always-present carbon contamination with a C 1s binding energy of 284.6 eV.

In addition to shifting, the peaks may broaden when the sample charges inhomogeneously (see Fig. 3.8), and this results in decreased resolution and a lower signal-to-noise ratio. Charging in supported catalysts is usually limited to only a few eV in standard XPS equipment, but becomes considerably more severe in monochromatic XPS. Here, the source is at a large distance from the sample, and consequently the electrons from the source do not reach the sample. However, by using a flood gun, which sprays low-energy electrons onto the sample, together with sample mounting techniques in which powders are pressed in indium foil, these charging problems may be alleviated to some extent [20].

Sensitive materials, such as certain metal salts or organometallic compounds used as catalyst precursors, may decompose during XPS analysis, particularly in equipment with standard X-ray sources. Heat and electrons generated by the source are usually responsible for damage to samples. In these cases, the monochromatic XPS offers a solution. For example, reliable spectra of the organoplatinum complexes in Figures 3.4 and 3.5 could only be obtained with a monochromatic source. Under the standard source the Pt(IV) complex indicated in Figures 3.4 and 3.5 decomposed into the Pt(II) precursor and Cl$_2$ gas.

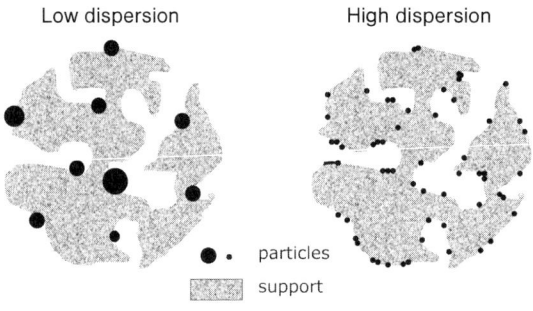

Fig. 3.9 The XPS intensity ratio of the signals from particles and the support, I_P/I_S, reflects the dispersion of the particles over the support.

3.2.6
Dispersion of Supported Particles from XPS

Because XPS is a surface-sensitive technique, it recognizes how well particles are dispersed over a support. Figure 3.9 shows, in schematic form, two catalysts with the same amount of material in supported particles, but with different dispersions. When the particles are small, almost all atoms are at the surface, while the support is covered to a large extent. In this case, XPS measures a high-intensity I_P from the particles, but a relatively low-intensity I_S for the support. Consequently, the ratio I_P/I_S is high. For poorly dispersed particles, on the other hand, the ratio I_P/I_S is low. Thus, the XPS intensity ratio I_P/I_S reflects the dispersion of a catalyst.

One can go a step further and use the I_P/I_S ratio for a quantitative estimate of the dispersion. Through the years, several methods have been proposed to predict XPS intensity ratios for supported catalysts. Angevine et al. [32] modeled their catalyst with crystallites on top of a semi-infinite support, as shown diagrammatically in Figure 3.10a. However, as the inelastic mean free path of for example SiO_2 is 3.7 nm, photoelectrons coming from particles inside pores as deep as 10 nm below the surface still contribute to the XPS signal, and the assumption of a semi infinite support is probably too simple. Indeed, the model predicts I_P/I_S ratios that may be a factor of 3 too high [33].

Kerkhof and Moulijn [33] suggested that a supported catalyst may be modeled as a stack of sheets of support material, with cubic crystals representing the supported particles. They used this stratified layer model (as illustrated in Fig. 3.10b) to calculate the intensity ratio I_P/I_S for electron trajectories perpendicular to the support sheets, assuming exponential attenuation of the electrons in the particles and the support.

Kuipers and co-workers [34] developed this model further into the randomly oriented layer model. These authors argued that powdered catalysts contain a

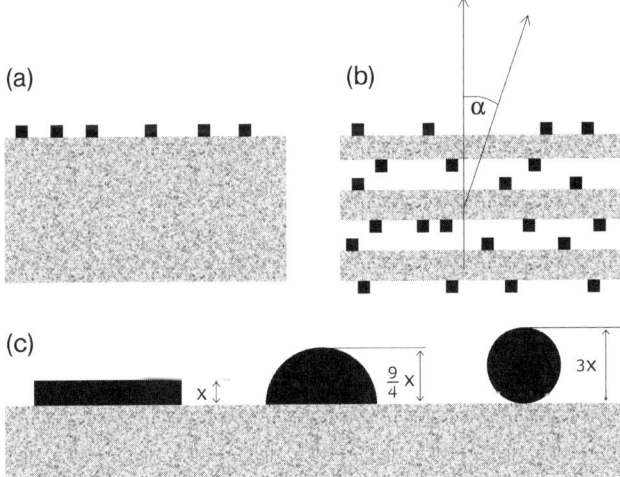

Fig. 3.10 Models used in the quantitative analysis of XPS spectra of supported catalysts. (a) Particles on a semi-infinite support. (b) Stratified layer model with cubic particles on sheets of support material, as used by Kerkhof and Moulijn [33]. (c) Particles with characteristic dimensions which have the same dispersion, giving nearly the same particle/support intensity ratio in XPS in the randomly oriented layer model according to Kuipers et al. [34].

multitude of surface elements with different orientation angles α of their surface normal with the electron analyzer (Fig. 3.10b). Thus, for samples that are randomly oriented, one needs to integrate the intensity ratio I_P/I_S over all angles α from 0 to $\frac{1}{2}\pi$ (and also over all azimuthal angles), accounting for the fact that the effective layer thickness for an electron traveling under an angle α with the surface normal becomes a factor of $1/\cos \alpha$ larger. Although angle averaging makes the model intuitively a lot more realistic, it must be noted that we are still dealing with cubic particles.

In this respect, Kuipers made an important point (as illustrated in Fig. 3.10c), namely that layers of thickness x which cover the support to a fraction θ, have the same dispersion as hemispheres of radius $2\frac{1}{4}x$, or spheres with a diameter $3x$. Even more interesting is the fact that these three particle shapes with the same surface-to-volume ratio give virtually the same I_P/I_S intensity ratio in XPS when they are randomly oriented in a supported catalyst! The authors tentatively generalized the mathematically proven result to the following statement that we quote literally: "For truly random samples the XPS signal of a supported phase which is present as equally sized but arbitrarily shaped convex particles is determined by the surface/volume ratio." Thus, in Kuipers' model the XPS intensity ratio I_P/I_S is a direct measure of the dispersion, independent of the particle shape. As the mathematics of the model is beyond the scope of this book, the interested reader is referred to the original publication [34]; here, we discuss an application of the method to ZrO_2/SiO_2 catalysts.

Zirconium oxide is of interest as a catalyst, as a support for other catalysts, and also as a diffusion barrier. In the latter application, a thin layer of ZrO_2 prevents the dissolution of rhodium in alumina supports under severe oxidative conditions [35]. Here, it is essential to apply ZrO_2 on the support at the highest possible dispersion, and to this end Meijers et al. [36] compared two methods for making ZrO_2/SiO_2 catalysts. The first method utilized the usual incipient wetness impregnation from an aqueous solution of zirconium nitrate, whereas in the second, less common procedure zirconium ethoxide $[Zr(OC_2H_5)_4]$ was anchored to the silica by a reaction between an ethoxy ligand and an OH group of the support. ZrO_2 is formed by heating (calcining) the samples in air. XPS was used to study the dispersions of these catalysts.

Figure 3.11 illustrates the XPS spectra of four ZrO_2/SiO_2 catalysts, in an energy range comprising the Zr 3d and the Si 2s signals. Because the samples are electrical insulators, the peaks are shifted and broadened but remain of course well-suited to determining the intensities of Zr and Si. Comparison of the spectra from the two differently prepared 16 wt% ZrO_2/SiO_2 catalysts reveals immediately that the preparation route via zirconium ethoxide gives the highest Zr 3d/Si 2s intensity ratio, and thus the highest dispersion (Fig. 3.11).

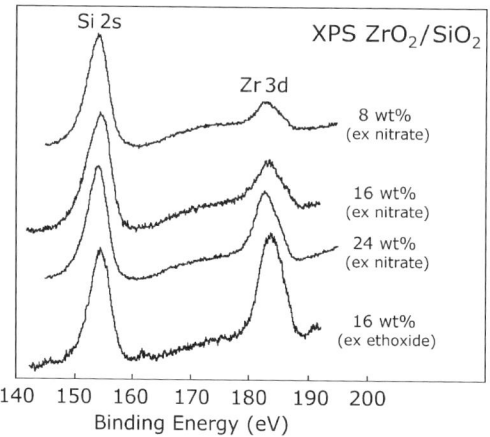

Fig. 3.11 XPS spectra in the range from 150 to 200 eV, showing the Zr 3d and Si 2s peaks of the ZrO_2/SiO_2 catalysts after calcination at 700 °C. All XPS spectra have been corrected for electrical charging by positioning the Si 2s peak at 154 eV. The spectra labeled "nitrate" correspond to the catalysts prepared by incipient wetness impregnation with an aqueous solution of zirconium nitrate; the spectrum labeled "ethoxide" to that prepared by contacting the support with a solution of zirconium ethoxide and acetic acid in ethanol. The latter preparation leads to a better ZrO_2 dispersion over the SiO_2 than the standard incipient wetness preparation does, as is evidenced by the high Zr 3d intensity of the bottom spectrum. (Adapted from [36]).

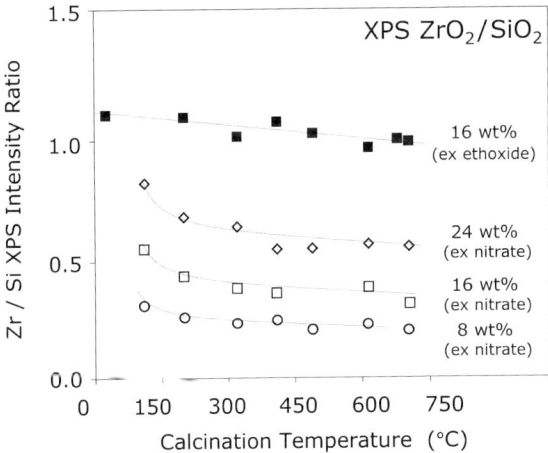

Fig. 3.12 Zr 3d/Si 2s intensity ratios calculated from the XPS spectra of ZrO_2/SiO_2 catalysts as a function of calcination temperature. (Adapted from [36]).

Figure 3.12 shows the Zr/Si intensity ratio as a function of calcination temperature. The three conventionally impregnated catalysts show a clear decrease in the I_{Zr}/I_{Si} ratio at relatively low temperatures, indicating a loss of dispersion, whereas the I_{Zr}/I_{Si} ratio of the catalyst from zirconium ethoxide decreased only slightly over a temperature range up to 700 °C.

In order to translate the XPS intensity ratios of Zr and Si into dispersions with Kuiper's model, the following input parameters are required:

- The concentrations of ZrO_2 and SiO_2.
- The densities of ZrO_2 and SiO_2.
- The cross-sections for photoemission from Zr 3d and Si 2s core levels.
- The mean free paths of Zr 3d and Si 2s electrons, each at its respective kinetic energy in ZrO_2 and SiO_2, analyzer transmissions at the prevalent kinetic energies of the electrons.
- The specific area of the support.
- The I_{Zr}/I_{Si} XPS intensity ratios.

The result of the calculations is that the calcined catalysts obtained from nitrate have dispersions of between 5 and 15% only, whereas the ZrO_2 catalyst prepared from ethoxide has a favorable dispersion of 75 ± 15% after calcination at 700 °C. The equivalent layer thickness for this system is 0.42 nm, and the support coverage about 27%. These results are summarized in Table 3.3.

It is interesting to visualize what these dispersion values from Table 3.3 actually mean. Suppose that the ZrO_2 is present in hemispherical particles, then we know from Figure 3.10 that such a particle with a radius of $2\frac{1}{4}x$ has the same dispersion as a layer of thickness x. As x is known for all catalysts in Figure 3.11, we

Table 3.3 Dispersion and support coverage in ZrO_2/SiO_2 catalysts calcined at 700 °C [36].

Loading [wt%]	Zr/Si (XPS)	Dispersion: D [%]	θ [a] [m² g⁻¹]	Effective layer thickness: d [nm]	r [b] [nm]
Catalyst from zirconium nitrate					
8	0.22 ± 0.01	12 ± 1	5	2.7	6.0
16	0.37 ± 0.04	8 ± 1	7	4.2	9.5
24	0.55 ± 0.02	7 ± 0.5	9	4.8	10.8
Catalyst from zirconium nitrate					
16	1.0 ± 0.05	75 ± 15	67	0.42	0.95

a) Area of silica covered by ZrO_2; total support area is 250 m² g⁻¹.
b) Effective radius of a half-spherical particle.

can now calculate how many half-spheres there are on 10 × 10 nm² of support area; the result is displayed in Figure 3.13. In terms of silica-support modification, the nitrate-derived catalysts can be deemed failures, but the ethoxide-derived ZrO_2 catalyst appears quite satisfactory.

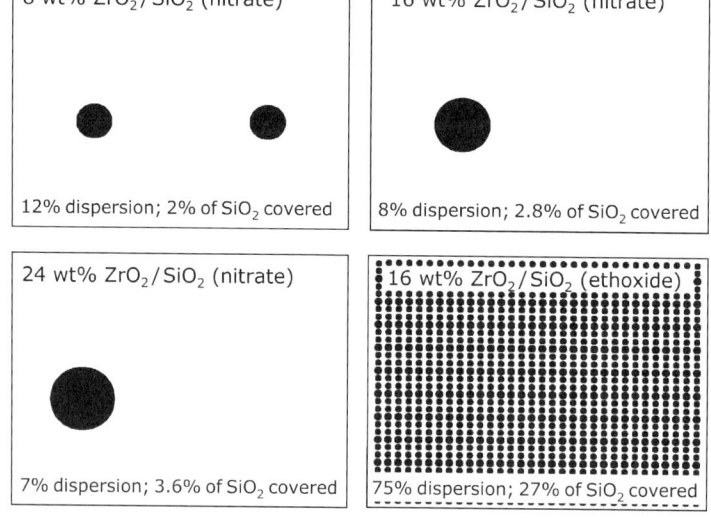

Fig. 3.13 Schematic representation of the dispersion of ZrO_2 particles and the extent to which the silica support is covered in three ZrO_2/SiO_2 catalysts prepared by impregnation with an aqueous zirconium nitrate solution, and one prepared via an exchange reaction of the support with zirconium ethoxide. The rectangles represent 100 nm² of silica support area; the circles represent a half-spherical particle of ZrO_2 seen from above. For corresponding numbers, see Table 3.3. (Adapted from [36]).

These investigations illustrate the usefulness of XPS when studying the dispersion of catalysts, and changes therein. The ZrO_2/SiO_2 system is typically a case where other techniques for determining particle size or dispersion would encounter difficulties. Selective chemisorption is effective only if a molecule can be found that adsorbs exclusively on ZrO_2 and not on the SiO_2. Transmission electron microscopy requires sufficient contrast between ZrO_2 and SiO_2, and it is doubtful whether the contrast is clear enough for the detection of highly dispersed zirconia in the ethoxide-derived catalyst. Line broadening or profile analysis in X-ray diffraction is limited to crystalline particles that are not too small. Neither is magnetic analysis applicable in this situation. Thus, XPS offers a particularly attractive alternative for determining the dispersion of those catalysts that are not accessible to investigation by the usual techniques for particle size determination.

3.2.7
Angle-Dependent XPS

Owing to the limited escape depth of photoelectrons, the surface sensitivity of XPS can be enhanced by placing the analyzer under an angle with the surface normal (the so-called take-off angle of the photoelectrons). This is illustrated in Figure 3.14 for a silicon crystal with a thin layer of SiO_2 on top [37]. The angle dependence of the Si^{4+}/Si intensity ratio can be used to determine the thickness of the oxide, and in favorable cases even the distribution of oxygen through the outer layers. The XPS intensity ratio for an overlayer on top of a substrate is readily derived from Eq. (3-2) and is equal to:

$$\frac{I_o}{I_s} = \frac{\sigma_o n_o \lambda_o(E_o)}{\sigma_s n_s \lambda_s(E_s)} \frac{1 - \exp(-d/[\lambda_o(E_o)\cos\theta])}{\exp(-d/[\lambda_o(E_s)\cos\theta])} \quad (3\text{-}9)$$

where:
I_o, I_s are the XPS intensities of overlayer and substrate;
n is the atomic densities (in mol cm^{-3});
σ is the XPS cross-section, usually taken from the tables of Scofield [23];
$\lambda(E)$ is the inelastic mean free path at the prevalent kinetic energy E through the overlayer (λ_o) or the substrate (λ_s);
θ is the take-off angle with respect to the surface normal;
d is the thickness of the overlayer.

Note that $\lambda_o(E_s)$ is the inelastic mean free path of electrons formed in the substrate traveling through the overlayer. In the case that the overlayer is a film of SiO_2 on a silicon crystal (as in Fig. 3.14), Eq. (3-9) reduces to

$$\frac{I_{SiO_2}}{I_{Si}} = \frac{n_{SiO_2} \lambda_{SiO_2}}{n_{Si} \lambda_{Si}} \frac{1 - \exp(-d/\lambda_{SiO_2} \cos\theta)}{\exp(-d/\lambda_{SiO_2} \cos\theta)} \quad (3\text{-}10)$$

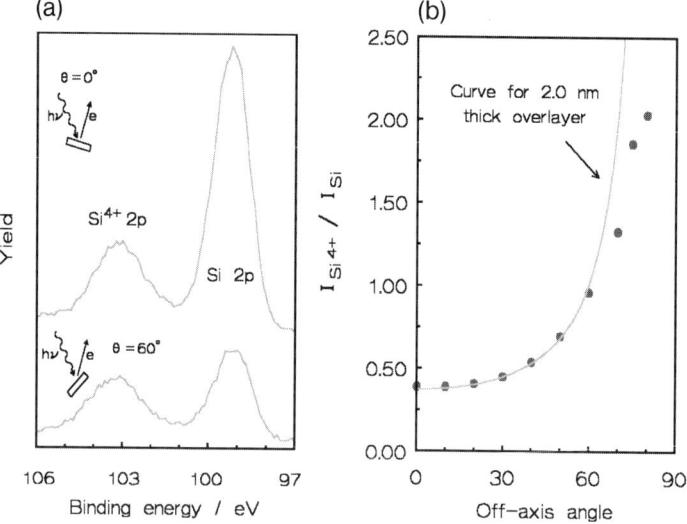

Fig. 3.14 (a) XPS spectra at take-off angles of 0° and 60°, as measured from the surface normal from a silicon crystal with a thin layer of SiO$_2$ on top. The relative intensity of the oxide signal increases significantly at higher take-off angles, illustrating that the surface sensitivity of XPS improves. (b) Plot of Si^{4+}/Si 2p peak areas as a function of take-off angle. The solid line is a fit based on Eq. (3-10), and corresponds to an oxide thickness of 2.0 nm. (From [37]).

Figure 3.14b shows the Si^{4+}/Si intensity ratio as a function of take-off angle, along with a fit based on Eq. (3-10). The agreement up to take-off angles of about 60° is excellent, but at higher angles the elastic scattering phenomena reduce the intensity of substrate electrons traveling at shallow angles from deeper regions [38]. The fit in Figure 3.14b corresponds to a homogeneously thick oxide layer of 2 nm [37].

It is possible to use Eqs. (3-9) and (3-10) to evaluate the thickness from only one XPS spectrum at a known take-off angle, θ. However, as Fadley already pointed out in his 1976 review [39], several assumptions must be made to derive Eq. (3-9). The most important of these is that the overlayer has a homogeneous composition, a uniform thickness, and a flat morphology. In particular, the latter assumption appears critical.

A surface that is rough on a scale much larger than the inelastic mean free path causes deviations in the angular dependence of the I_o/I_s ratio, for two reasons. The first reason is that some regions of the surface may not be effective in producing detectable photoelectrons because they are shaded from the detector by adjacent roughness contours. The second reason is that the true electron exit angle at a given surface point will in general be different from the experimental value as defined by the macroscopic surface plane. Gunter et al. [37] calculated the influence of types of surface roughness, as shown graphically in Figure 3.15.

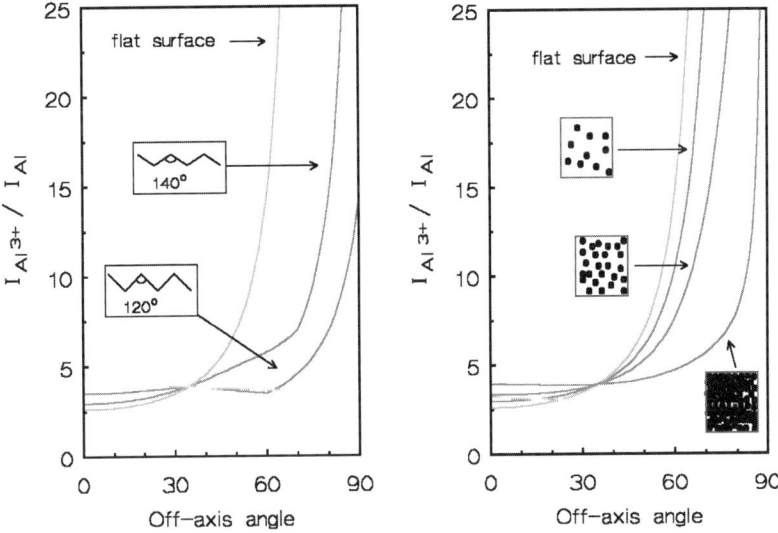

Fig. 3.15 Simulation of the effect of two types of sample roughness on the overlayer-to-substrate XPS intensity ratio. (From [37]).

The tendencies seen are quite general: for small off-axis angles (i.e., nearly perpendicular to the macroscopic surface plane), shading is relatively unimportant, and the main deviating effect is due to variations of the local take-off angle over the surface, with the result that the I_o/I_s ratio is larger than for a flat layer with the same thickness. In contrast, for higher take-off angles θ, shading becomes important. The unshaded parts of the surface will have their surface normal pointed more to the detector than the shaded parts, and consequently we would expect a lower overlayer/substrate intensity ratio. It should be noted that this implies that the least deviation from the ideal behavior described by Eq. (3-10) is to be expected for intermediate take-off angles, and not for electrons escaping at perpendicular angles! The data in Figure 3.15 suggest that the optimum angle is somewhere between 30 and 40° in a two-dimensional simulation, which has indeed been predicted [40]. If, however, the calculation is performed properly in three dimensions, the optimum angle for thickness determination appears to be 45° [41].

It is concluded that XPS is an excellent tool for determining layer thickness in the range of a few nanometers, provided that one measures the full take-off angle dependence, in order to test the applicability of the uniform layer model implicitly assumed in Eqs. (3-9) and (3-10).

Angle-dependent XPS is of course not applicable to technical catalysts, but it is extremely useful in studying model systems. For example, Baschenko et al. [42] used the technique to study oxygen adsorbates on silver. These authors observed two oxygen signals with different binding energies which, as shown in Figure 3.16, give rise to different angle dependencies of the XPS intensity. The O species with 530.5 eV binding energy is characteristic for a surface species, whereas the

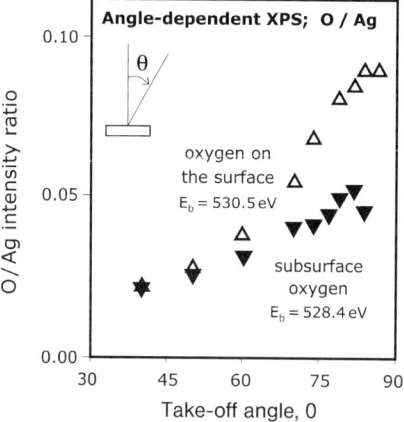

Fig. 3.16 O 1s/Ag 3d$_{5/2}$ XPS intensity ratio as a function of take-off angle for two oxygen species on polycrystalline silver. The data corresponding to an O 1s binding energy of 528.4 eV are attributed to subsurface oxygen in Ag, the other with a binding energy of 530.5 eV to oxygen atoms adsorbed on the Ag surface. (Data from [42]).

other with a binding energy of 528.4 eV shows a less-pronounced angle dependence characteristic for oxygen dissolved below the surface. Both species are present in silver under conditions prevailing in an important catalytic process, namely the reaction between ethylene and oxygen to ethylene epoxide, C_2H_4O [43]. The existence of subsurface oxygen in silver, which is believed to enhance the chemisorption of ethylene, has long been invoked on the basis of indirect mass spectroscopic data. The strength of angle-dependent XPS measurements (e.g., in Fig. 3.16) is that these provide direct spectroscopic evidence for the presence of oxygen below the surface.

In principle, this interpretation can be taken further in order to calculate which oxygen concentration profile best fits the measurements of the O/Ag ratio. Indeed, Baschenko et al. [42] performed just such a calculation and concluded that the subsurface oxygen resides mainly in the third and fourth atomic layers below the surface. These studies illustrate very well what angle-dependent XPS can achieve on catalytically relevant adsorbate systems.

Finally, angle-dependent studies as described make use of the polar angle; that is, the angle between the detector and the surface normal. When working with single crystals, it is also possible to measure XPS intensities as a function of both the polar and the azimuthal angle. Due to the forward focusing effect, the photoelectron emission is at a maximum in the directions of neighbor atoms, particularly in the case of a dense array of atoms. In this way, XPS can be used to study surface structures.

Photoelectron diffraction and forward focusing have been applied successfully when studying adsorbed species. For example, controversy whether methoxy groups ($-O-CH_3$) adsorb with the O–C bond perpendicular or tilted with respect to the surface of Cu(111) was readily solved with XPS by using the forward-focusing effect: angle-dependent XPS measurements of methoxy groups on Cu(111) showed a clear enhancement of the O 1s intensity in the direction of the surface normal, indicating that the O–C axis is perpendicular to the surface [44]. Many other interesting examples have been reviewed by Egelhoff [45].

3.2.8
In-Situ and Real Time XPS Studies

Photoelectrons do not travel very far in gases, although if the path through the gas is no longer than a few millimeters at 1 bar pressure, the attenuation of the electron intensity can be tolerated. However, a reaction cell cannot have windows, as the inelastic mean free path of electrons in solids is on the order of nanometers. Hence, the challenge is to construct a reaction cell with a small hole through which the electrons leave, while differential pumping stages around the cell maintain the vacuum in the remainder of the XPS chamber. Almost 30 years ago, in 1979, Joyner et al. [46] described such a spectrometer and used it to study species present on the surface of silver under 0.5 Torr (0.67 mbar) of oxygen [47]. In Russia, a similar instrument was recently improved upon by Bukhtiyarov and co-workers [48]. Another means of dealing with the severe attenuation of the electron intensity by a gas is to use the highly intense X-rays of a synchrotron as the source for *in-situ* XPS experiments; some examples are described in [49–51].

Another application that is enabled by the bright X-rays at a synchrotron is the measurement of XPS spectra in real time, to follow the dynamics of surface reactions, and in a recent review Baraldi et al. [52] described an impressive series of examples. Here, we discuss a study on the adsorption dynamics of CO on Rh(111), a system that also serves as an example in the sections on temperature-programmed desorption (see Chapter 2) and infrared spectroscopy (see Chapter 8).

Figure 3.17 shows real-time XPS O 1s spectra of a Rh(111) surface while it is exposed to CO gas at three different temperatures [53]. It is clear that CO first adsorbs in a state corresponding to an O 1s binding energy of 531.5 eV, while later a second state characterized by an O 1s peak at 530.1 eV becomes populated. Based on the findings of elegant studies conducted by Antonsson et al. [54], Beutler et al. [55], and Jaworowski et al. [56], we know that the O 1s at the higher binding energy corresponds to linearly adsorbed CO, and the one at lower energy to CO in the threefold hollow site (structure models for these arrangements are shown in Fig. 8.15; assignments are listed in Table 3.4).

Remarkably, the binding energies of both C 1s and O 1s are sensitive to the way in which CO binds to the surface, oxygen even more than carbon, although the coordination of the oxygen atom is the same in all bonding configurations. The explanation is that the shifts are caused by a final state effect: the binding energy shifts reflect differences in the core-ionized state [54]. Changes in the initial state of the CO in different adsorption states are small, which is in agreement with the fact that the adsorption energies of CO in different configurations are similar.

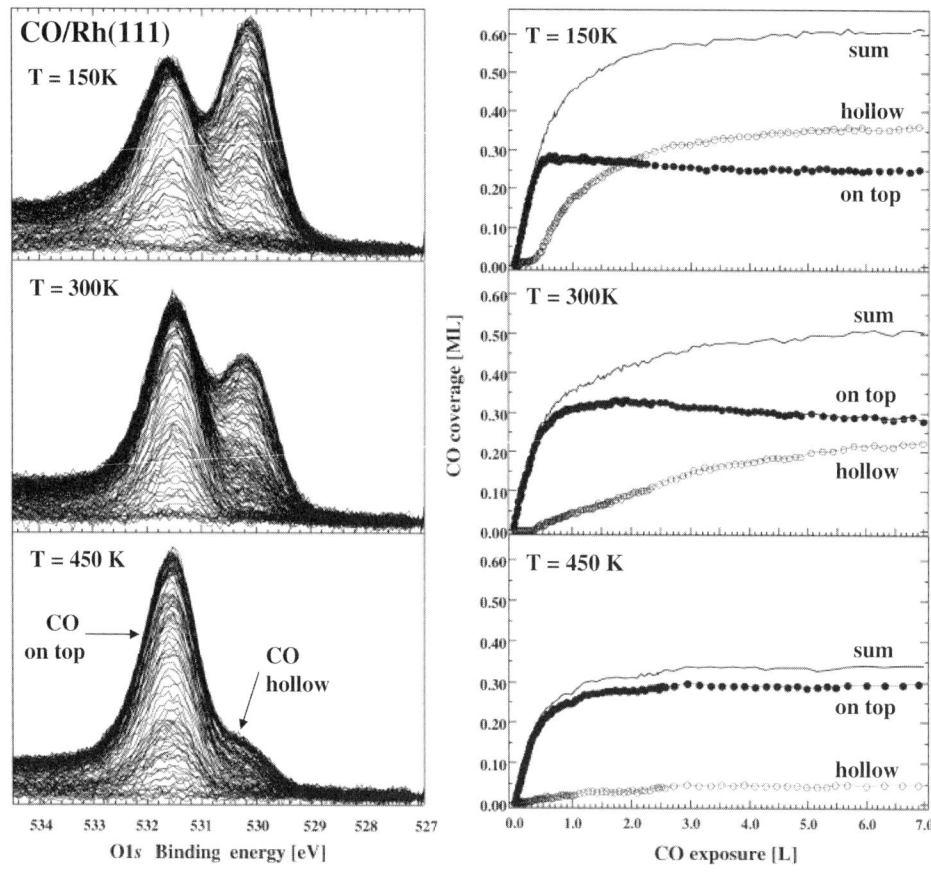

Fig. 3.17 Real-time O 1s XPS spectra during the exposure of a Rh(111) surface to CO at three different temperatures, along with uptake curves showing the XPS intensity of the two O 1s components in the spectra calibrated to a surface coverage in monolayers. (Adapted from [53]).

Table 3.4 C 1s and O 1s binding energies for CO adsorbed on metals.

C 1s [eV]	O 1s [eV]	System	Reference
Top			
285.9	532.2	CO/Ni(100)	[54]
286.00	532.1	CO/Rh(111)	[56]
Bridge			
285.5	531.3	CO/Ni(100)	[54]
Hollow			
285.35	530.5	CO/Rh(111)	[56]

Also shown in Figure 3.17 are the uptake curves of Rh(111) for CO in top and hollow positions, as derived from the intensities of the corresponding XPS peaks. At all temperatures CO begins to occupy the top sites, while the hollow sites become populated at higher coverages, due to repulsion between CO molecules when they come close to each other. In the temperature-programmed desorption (TPD) shown in Figure 2.9, this situation corresponds to the development of a shoulder at lower temperatures, due to the CO molecules in excess of 0.5 monolayers (ML), that are less strongly adsorbed. At room temperature and higher, the on-top CO is the dominant species. The low-energy electron diffraction patterns, as well as the vibrational spectra in Figure 8.15, show that the threefold site becomes populated when the (well-annealed) surface is filled for more than 50% of a ML [57]. Hence, the XPS results discussed here are in qualitative agreement with TPD, low-energy electron diffraction (LEED), and reflection absorption infrared spectroscopy (RAIRS) as reported by Linke et al. [57]. However, a CO-saturated, well-annealed Rh(111) surface is expected to have a surface coverage of 0.75 ML, with one-third of the CO in linear sites and the remainder in hollow sites [57]. In order to achieve this state of order, the molecules must be able to diffuse over the surface [58], which is probably not sufficiently possible at the low temperature of 150 K employed in the experiments shown in Figure 3.17.

Uptake curves provide information on the sticking coefficient of the molecule: if the sticking probability is high, the surface fills up rapidly. The sticking coefficient is proportional to the slope of the uptake curve. Once the CO pressure above the surface is known precisely (which is not always trivial!), the rate of collision between gas and surface follows from the kinetic theory of gases [59]. The sticking coefficient is then the fraction of molecules that adsorb upon collision, and can be calculated from the slope of the uptake curve.

Baraldi and co-workers [52] have described a wealth of dynamic XPS studies on surface reactions, including adsorption, dissociation, desorption, and even catalytic reactions, such as the epoxidation of alkenes [60], and the reduction of NO by H_2 and CO [61].

In conclusion, XPS is among the most frequently used techniques in catalysis. The advantages of XPS are that it readily provides the composition of the surface region, and that it distinguishes between the chemical states of one element. Today, XPS is becoming an increasingly important tool for studying the dispersion of supported catalysts, especially in systems where the usual methods for determining particle sizes are not applicable. Finally, angle-dependent XPS measurements on single crystals contain structural information, such as layer thickness and adsorption geometry.

3.3
Ultraviolet Photoelectron Spectroscopy (UPS)

UPS differs from XPS in that UV light is used instead of X-rays. The most frequently used sources are helium discharge lamps, which generate He I light at 21.2 eV and He II light at 40.8 eV. At these low exciting energies, photoemission

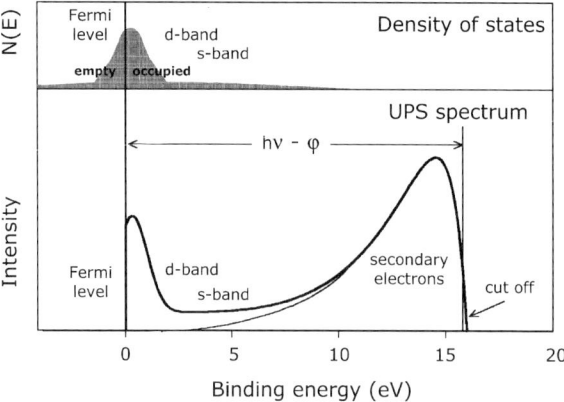

Fig. 3.18 Schematic UPS spectrum of a d-metal and the corresponding density of states.

is limited to valence electrons. Hence, UPS is particularly suited to probe bonding in metals, in molecules and in adsorbed species [5, 18].

Figure 3.18 shows a schematic UPS spectrum of a d-metal, along with the density of electron states in the metal (see also the Appendix). A d-metal has a high density of states at the Fermi level, which appears clearly as the low-binding energy onset of the spectrum and provides a convenient zero for the binding energy scale. One sees furthermore an image of the d- and the s-band. Most of the intensity at higher binding energies, however, is due to secondary electrons. These are inelastically scattered photoelectrons from the valence band, which have suffered energy loss while traveling through the solid. The exact shape of the secondary electron region is very sensitive to factors such as the structure and composition of the surface, the orientation of the sample with respect to the analyzer, and the presence of electric or magnetic fields in the spectrometer. The cut-off point corresponds to loss electrons, which had just enough kinetic energy to overcome the work function of the substrate.

Spectra as in Figure 3.18 can also be obtained as a function of polar and azimuthal angle, and with polarized UV light, enabling one to probe band structures in all directions [18]. In this chapter we limit ourselves to angle integrated measurements of the electron density of states.

UPS provides a quick measure of the macroscopic work function, as follows from the following consideration. The slowest loss electrons – that is, those contributing at the high binding energy cut-off in the spectrum – have zero kinetic energy, $E_k = 0$. Electrons from the Fermi level, on the other hand, possess the highest kinetic energy, $E_k = h\nu - \varphi$. Hence, the width, W, of a UPS spectrum equals $(h\nu - \varphi)$, and the work function becomes:

$$\varphi = h\nu - W \tag{3-11}$$

where:
- φ is the work function of the sample (not of the spectrometer this time!);
- hν is the energy of the UV photon (usually 21.2 or 40.8 eV);
- W is the width of the spectrum, as indicated in Figure 3.18.

UPS and XPS both image the density of states, although not in entirely the same way. In XPS, the photoelectrons originating from the valence band leave the sample with kinetic energies over 1 keV. In UPS, the exciting energy is on the order of 21 eV, and the kinetic energy of the electrons is low, say between 5 and 16 eV. This means that the final state of the photoelectron is within the unoccupied part of the density of states of the metal. As a result, the UPS spectrum represents a convolution of the densities of occupied and unoccupied states, which is sometimes called the "Joint Density of States".

In the formalism of quantum mechanics the probability that an electron in the initial state is transferred into the final state, is given by Fermi's Golden Rule:

$$W_{i \rightarrow f} = |\langle i|H|f \rangle|^2 \delta(h\nu - E_f + E_i) \quad (3\text{-}12)$$

where:
- $W_{i \rightarrow f}$ is the transition probability from the initial to the final state;
- $\langle i|$ is the wave function of the electron in the initial state;
- H is the Hamiltonian which describes the photon wave field;
- $f \rangle$ is the wave function of the electron in the final state;
- h is the energy of the photons;
- E_f is the energy of the final state;
- E_i is the energy of the initial state.

If hν increases to higher values (which is easily achieved in a synchrotron), $f \rangle$ shifts above the density of states into the range corresponding to entirely free electrons in vacuum, which can be considered as a continuum. Thus, if hν, and consequently E_f increase, the UPS spectrum approaches the true density of states of the metal. Of course, XPS also images the real density of state, but at lower resolution, due to the broader line width of the X-ray source.

The unoccupied part of the density of states can also be measured, by a technique called "inverse UPS" (sometimes also referred to as BIS; Bremsstrahlung Isochromat Spectroscopie). Here, a beam of low-energy electrons falls on the surface, where they pass into the unoccupied states and fall back to the Fermi level, under emission of a quantum hν. Measurement of this radiation as a function of the incident electron energy gives the density of unoccupied states. The details of this technique falls beyond the scope of this book, but the interested reader is referred elsewhere [5, 62].

UPS studies of supported catalysts are rare. However, Heber and Grünert [63, 64] recently explored the feasibility of characterizing polycrystalline oxides by He-II UPS. One attractive point of their investigations was that they employed the difference in mean free path of photoelectrons in UPS, V 2p XPS and valence band XPS (below 1 nm, around 1.5 nm, and above 2 nm, respectively) to obtain depth-profiles of the different states of vanadium ions in reduced V_2O_5 particles [64]. However, the vast majority of UPS studies concerns single crystals, for probing the band structure and investigating the molecular orbitals of chemisorbed gases. Examples for each of these applications are discussed as follows.

The first example deals with small islands of silver on a ruthenium substrate. This sample may be considered as perhaps a somewhat far-fetched model of a supported catalyst or a bimetallic surface. As metal layers are almost never in perfect registry with the substrate, they possess a certain amount of strain. Goodman and co-workers [65] used these strained metal overlayers as model systems for bimetallic catalysts. Here, we examine first the electronic properties of the Ag/Ru(001) system as studied by UPS.

Figure 3.19 shows a series of UPS spectra obtained from a Ru(001) substrate on which small amounts of Ag or Au have been evaporated. The spectrum of the clean Ru(001) substrate, being a d-metal, exhibits a high density of states at the Fermi level. The Ru 4d band is seen to extend to about 5 eV below the Fermi level. Silver and gold, on the other hand, are s-metals, with a low density of states at the Fermi level, as illustrated by the spectrum of the thick layers (top spectra of Fig. 3.19). The d-band falls between 4 and 8 eV below the Fermi level. The other spectra show what happens if the Ag or Au coverages increase submonolayers to a thick film. Photoemission from the Ru substrate is rapidly attenuated, while the d-band of Ag develops from narrow states in the submonolayer regime, via progressively broader bands at coverages between one to four ML, until the energy

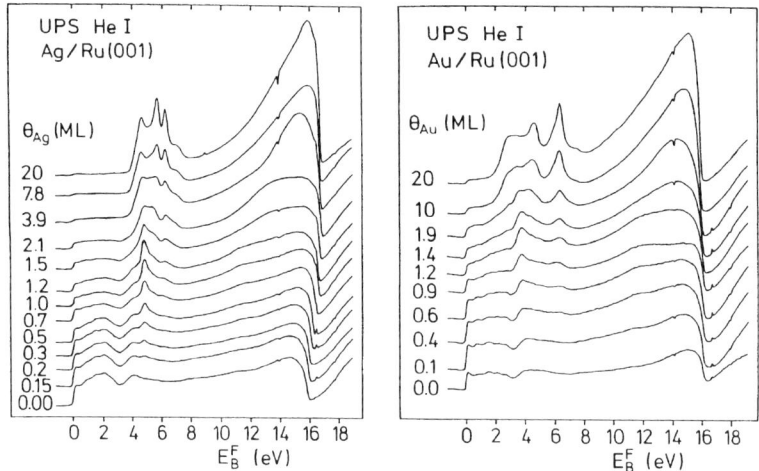

Fig. 3.19 Ultraviolet photoelectron spectroscopy (UPS) spectra of silver and gold layers on a Ru(001) substrate show the evolution of the d-band as a function of the silver dimensions. Note also the changes in work function reflected in the width of the spectra of Ag. (Figure courtesy of K. Wandelt, Bonn [66]).

distribution characteristic of silver and gold are reached for a film thickness between four and 10 ML [66].

Of direct interest for photoemission of supported catalysts is that similar increases in the width of d-bands have been observed by Mason in UPS spectra of small metal particles deposited on amorphous carbon and silica substrates [67]. Theoretical calculations by Baetzold et al. [68] indicate that the bulk density of states is reached if Ag particles contain about 150 atoms, which corresponds to a hemispherical particle of 2 nm in diameter. Concomitant with the appearance of narrowed d-bands in small particles is the occurrence of an increase in core level binding energies of up to 1 eV. The effect is mainly an initial and only partly a final state effect [67], although many authors have invoked final state–core hole screening effects as the only reason for the increased binding energy.

Finally, Figure 3.19 illustrates the effect of the work function on the width of UPS spectra. Silver has a lower work function (4.72 eV) than ruthenium (5.52 eV), and consequently the width of the spectra increases from 15.7 eV for Ru(001) to 16.7 eV for a thick Ag film, in agreement with Eq. (3-11). The work function of gold, however, is almost equal to that of Rh(001).

Another important application of UPS is the study of adsorbates. Figure 3.20 shows schematically what is observed in the UPS spectrum of an adsorbed gas: occupied molecular orbitals of the adsorbate with ionization potentials between 0 and $h\nu - \varphi$ become visible. If one compares their binding energies with those in an UPS spectrum of a physisorbed multilayer of the same gas, one recognizes readily which of the molecular orbitals are involved in the chemisorption bond.

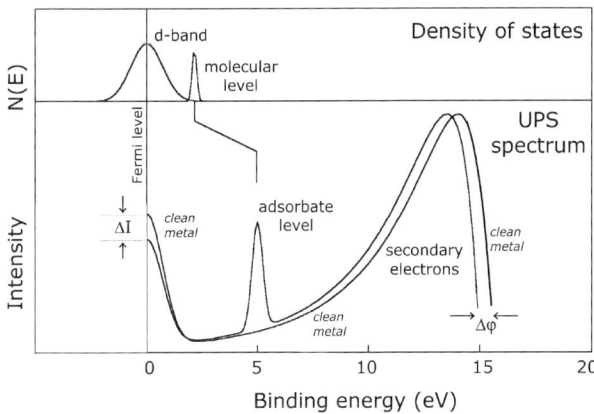

Fig. 3.20 UPS spectra of adsorbed species reveal the binding energies of electrons in the orbitals of the adsorbate. The densities of states at the top are those of the d-band of the metal and an adsorbate level when they have no interaction. Upon adsorption, the adsorbate level has broadened and shifted to lower energy (see also the Appendix). Note the attenuation of the d-band signal and the work function increase caused by adsorption.

For example, the adsorbate level in Figure 3.20 has shifted a few eV with respect to its position indicated in the density of states picture (taken as the position in a physisorbed gas), indicating that the level is involved in the chemisorption bond.

The width of the spectrum changes if the work function changes upon adsorption. In Figure 3.20, the work function has become larger after chemisorption, indicating that the adsorbed species enlarges the dipole layer which constitutes the surface contribution to the work function (see Appendix). Examples of such adsorbates are chlorine and oxygen atoms, which become negatively charged, or CO, which is a dipole with the negative charge pointing away from the surface. The adsorbed gas attenuates the photoemission of the substrate. Figure 3.20 illustrates this by the decreased intensity of the d-band contribution. This effect should *not* be taken as evidence that electrons from the d-band have flown into unoccupied orbitals of the adsorbate [69].

Figure 3.21 shows the UPS spectrum of CO adsorbed onto a clean and a potassium-promoted surface of an iron single crystal [70]. The molecular orbitals of CO are clearly visible. The 5σ orbital has shifted to lower energy in comparison to the spectrum of gas phase CO, indicating that it is involved in the formation of a chemisorption bond. We would also like to see bonding effects in the $2\pi^*$ level of CO, near the Fermi level. Unfortunately, this region of the spectrum is heavily distorted due to the attenuation of the iron signal by the adsorbed layer of CO. Hence, the small contribution from the $2\pi^*$ orbital is not distinguished.

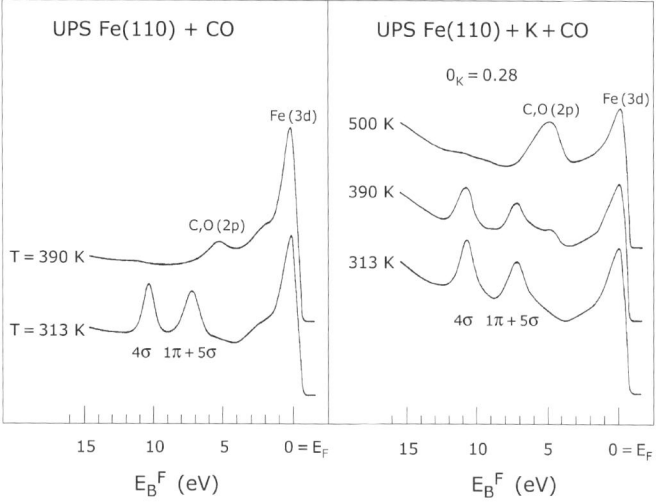

Fig. 3.21 UPS spectra of CO chemisorbed on iron show that the 5σ orbital has shifted down to higher binding energy as a result of chemisorption. CO largely desorbs from clean iron upon heating to 390 K. Potassium enhances the bond between CO and the metal and promotes the dissociation of CO at higher temperatures. (Adapted from [70]).

The UPS spectra of Figure 3.21 indicate that heating of CO on clean Fe(110) to 390 K leads mainly to desorption, and only a fraction of the CO dissociates. Substantially less CO desorbs from the potassium-promoted surface, however, and after heating to 500 K, all CO on the surface has dissociated. Thus, potassium enhances CO bonding to the surface and promotes its dissociation. Such promoter effects are discussed in more detail in Chapter 9 and in the Appendix.

3.3.1
Photoemission of Adsorbed Xenon

The photoemission of adsorbed xenon (abbreviated as PAX) is a site-selective titration technique, in which the UPS spectrum of physisorbed Xe reveals the nature of the Xe adsorption site [71]. As we are dealing with a weakly physisorbing atom, these experiments must be conducted at cryogenic temperatures on the order of 50–60 K. First, we explain the theory behind the PAX method, and then illustrate the technique with an example.

Figure 3.22a shows the UPS spectrum of xenon adsorbed on ruthenium, containing the 5p doublet of Xe superimposed on the density of states of the ruthenium d-band [72]. The $5p_{1/2}$ component of the doublet is sharper than the $5p_{3/2}$ component, due to a multiplet-like splitting in the latter. Spectra as in Figure 3.22a immediately reveal the Xe $5p_{1/2}$ binding energy with respect to the Fermi level. However, the true binding energy is of course measured with respect to the vacuum level. The basic assumption of the PAX method is that the true Xe

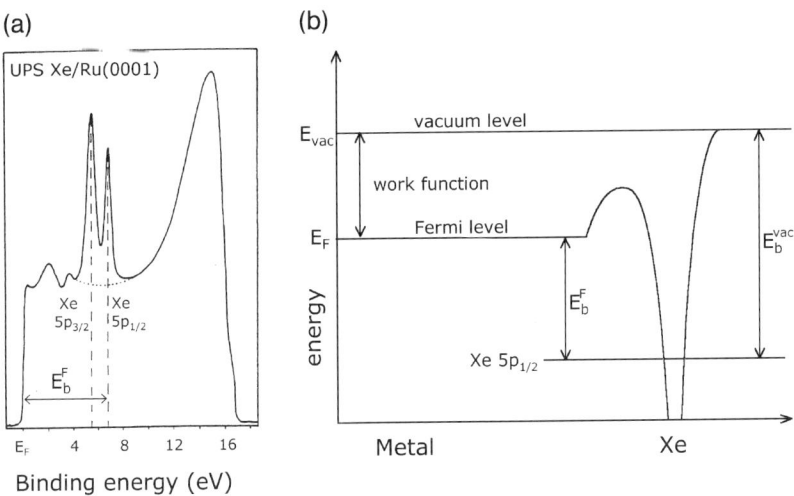

Fig. 3.22 UPS spectrum of Xe physisorbed on Ru(001) showing the superposition of the Xe 5p levels with the d-band of Ruthenium. The position of the Xe $5p_{1/2}$ peak with respect to the Fermi level of Ru is a measure of the work function of the adsorption site, as the potential diagram indicates. (From [72]).

binding energy with respect to the vacuum level is the same for all substrates. This assumption appeared valid for a large number of surfaces. As explained in Figure 3.22b, the difference between the two binding energies is simply the work function of the substrate, however, measured locally at the adsorption site of the Xe atom:

$$\varphi_{\text{local}} = E_{\text{Xe}}^{\text{vac}} - E_{\text{Xe}}^{\text{F}} \qquad (3\text{-}13)$$

where:

φ_{local} is the local work function;
$E_{\text{Xe}}^{\text{vac}}$ is the binding energy of Xe $5p_{1/2}$ electrons with respect to the vacuum level (12.3 ± 0.15 eV for a large number of substrates);
E_{Xe}^{F} is the binding energy of Xe $5p_{1/2}$ electrons with respect to the Fermi level.

It may not immediately be obvious that Eq. (3-13) is indeed correct. One expects that the Xe atom becomes slightly polarized upon adsorption, which affects the binding energy. As the extent of polarization changes from one substrate to another, the value of $E_{\text{Xe}}^{\text{vac}}$ is also expected to vary. Apparently, such effects are small and fall within the experimentally determined range for $E_{\text{Xe}}^{\text{vac}}$ of 12.3 ± 0.15 eV. Care should be taken, however, with the application of Eq. (3-13) to Xe adsorption sites adjacent to adsorbed atoms or steps, where the polarization of the Xe may be quite different from that on a flat surface.

We can check whether Eq. (3-13) applies to the Xe/Ru(001) spectrum: according to Figure 3.22a, the binding energy of the Xe $5p_{1/2}$ level is 6.8 eV with respect to the Fermi level of Ru(001), which results in a local work function of $(12.3 - 6.8) = 5.5$ eV. This value is in good agreement with the macroscopic work function for Ru(001), of 5.52 eV [71]. Thus, by measuring the binding energy of adsorbed Xe, we can determine the local work function of the adsorption site with a resolution of a few angstroms. This makes PAX a very useful technique for the titration of different sites on a heterogeneous surface, as the following example shows.

A bimetallic Ag/Ru(001) surface was prepared by evaporating a small amount of silver onto the clean Ru(001) substrate, and subsequent annealing. This procedure yields mono atomically thick Ag islands of a slightly distorted (111) structure on top of the Ru surface [73] (UPS spectra of such surfaces were discussed earlier in this section, whereas the thermal desorption of silver atoms from these surfaces was described in Chapter 2; see Figs. 2.9 to 2.11). Photoemission spectra of adsorbed Xe on this sample in Figure 3.23 show one Xe 5p doublet at low Xe coverages, with the $5p_{1/2}$ peak at 6.7 eV, characteristic of Xe on Ru(001). At higher Xe coverages, a second doublet becomes apparent, with the Xe $5p_{1/2}$ peak at 7.6 eV, associated with Xe on Ag. The sequential population by Xe of Ru before Ag sites reflects that the adsorption energy of Xe is larger on Ru than on Ag.

The top spectrum of Figure 3.23 corresponds to a complete monolayer of Xe on the sample. It can be used for a quantitative titration of sites: the area under the

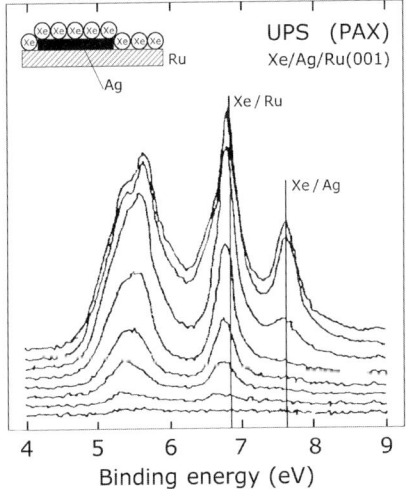

Fig. 3.23 Photoemission of adsorbed xenon (PAX) spectra at 60 K of 0.27 monolayer of Ag deposited on Ru(001) followed by annealing, showing that Xe first populates Ru, and then Ag. The top spectrum corresponds to a complete monolayer of Xe on the Ag/Ru(001) sample and can be used for quantitation. (From [73]).

two peaks indicates that about 27% of the Xe in the monolayer is adsorbed on silver, indicating that 27% of the ruthenium surface is covered by silver islands. As a peak at 7.25 eV, which would be characteristic of Ag–Ru boundary sites, is not observed, the results suggest that Ag is present in fairly large islands.

Its capability to titrate sites on heterogeneous surfaces makes PAX (in principle) a particularly attractive technique for investigating the surfaces of catalysts. Unfortunately, the technique has its limitations, because the Xe $5p_{1/2}$ peak has a finite linewidth of about 0.4 eV. Moreover, if a surface possesses more than three to four different adsorption sites, the spectra may become too complicated, rendering analysis by curve fitting unwarranted. In fortunate cases, however, where all adsorption sites are populated sequentially, one may be able to identify many sites. Further applications in this respect may be found in the literature [71–74].

Goodman and co-workers reported an elegant variation of the PAX technique based on the de-excitation of metastable helium, which makes it even more surface sensitive [75]. In metastable atom electron spectroscopy (MAES; also referred to as metastable impact electron spectroscopy; MIES), the signal is due to the Auger decay which occurs when an excited atom collides with a surface or an adsorbed molecule [76]. De-excitation proceeds by an electron from the outer surface into the core hole of the impinging core-ionized atom, from which the excited electron leaves as an Auger electron. The latter thus carries information on the electron density of states, or on the orbital energies of the adsorbed atom (see Fig. 3.24 for a scheme of the transitions involved). Goodman showed that this method yields high-quality spectra of adsorbed xenon monolayers which hardly contain contributions from the underlying substrate [75]. Hence, MAX –

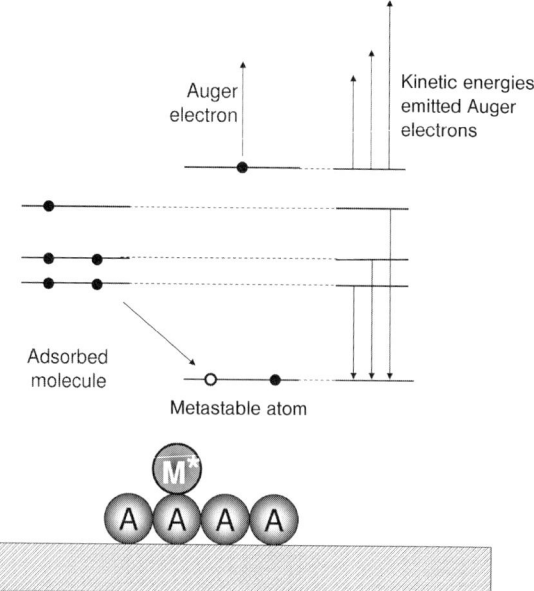

Fig. 3.24 Electron energy scheme explaining the principle behind metastable atom electron spectroscopy. An excited atom M* collides gently with an adsorbed molecule A; the metastable atom is de-excited by electron transfer from the adsorbate to the core hole of M*; the excess energy is carried away by an Auger electron. The spectrum of these Auger electrons represents the orbital energies of the adsorbed molecules, or the surface density of states of the substrate when it is empty.

which represents MAES of adsorbed xenon – may be an interesting alternative in cases where PAX spectra are affected by complicated background features from the substrate.

In conclusion, UPS provides an image of the density of states of the valence band. Compared to XPS, UPS is much more surface-sensitive, and it has higher energy resolution than XPS owing to the narrow UV lines of helium. UPS readily reveals the work function of the surface under study, and is particularly suited to studying the occupied molecular orbitals of adsorbed species. UPS of adsorbed noble gases (PAX) is useful for the titration of sites on a heterogeneous surface, and also for the determination of local work functions. UPS is a typical surface science technique, which is best applied to single crystals.

3.4
Auger Electron Spectroscopy

We have already met the Auger process in Figures 3.2 and 3.24 as a way in which core-ionized atoms relax to lower energy. However, Auger electrons are not simply a byproduct of XPS. Rather, these highly surface-sensitive, primarily element-specific electrons form the basis of a spectroscopy which is highly appreciated in the fields of materials and surface science, namely Auger electron spectroscopy (AES) [5, 17, 19–21].

AES is carried out by exciting the sample with a beam of primary electrons with kinetic energy between 1 and 10 keV. These electrons create core holes in the atoms of the sample, whereupon the excited atom relaxes by filling the core hole with an electron from a higher shell. The energy is either liberated as an X-ray photon (X-ray fluorescence) or in an Auger transition by the emission of a second electron, the Auger electron. As indicated in Figure 3.2, the kinetic energy of the Auger electron is entirely determined by the electron levels involved in the Auger process and not by the energy of the primary electrons.

The notation of Auger transitions uses the X-ray level nomenclature of Table 3.1. For example, KL_1L_2 stands for a transition in which the initial core hole in the K-shell is filled from the L_1-shell, while the Auger electron is emitted from the L_2-shell. Valence levels are indicated by "V", as in the carbon KVV transition.

3.4.1
Energy of Auger Peaks

In electron-excited spectra, Auger electrons are seen as small peaks on an intense background of secondary electrons originating from the primary beam (Fig. 3.25). In order to enhance the visibility of the small Auger peaks, spectra are often presented in the derivative mode. Although the actual kinetic energy of the Auger electron corresponds to the center of the integral peak (E_o in Fig. 3.25), it is customary to report Auger energies as the negative excursion of the derivative.

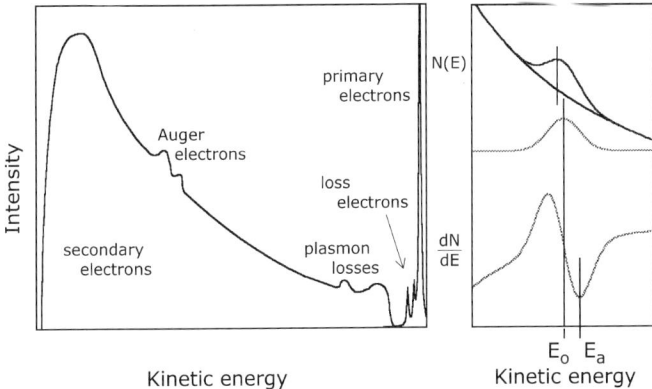

Fig. 3.25 Energy spectrum of electrons coming from a surface irradiated with a beam of primary electrons. Electrons have lost energy to vibrations and electronic transitions (loss electrons), to collective excitations of the electron sea (plasmon losses), and to all kinds of inelastic processes (secondary electrons). The element-specific Auger electrons appear as small peaks on an intense background, and are better visible in a derivative spectrum.

The energy of an Auger transition, for example KLM, is to a first approximation given by:

$$E_{KLM} \approx E_K - E_L - E_M - \delta E - \varphi \qquad (3\text{-}14)$$

where:

E_{KLM} is the kinetic energy of the Auger electron (i.e., E_o in Fig. 3.25);
E_K is the binding energy of an electron in the K-shell, etc.;
φ is the work function;
δE is the energy shift caused by relaxation effects.

The δE term accounts for the relaxation effects involved in the decay process, which leads to a final state consisting of a heavily excited, doubly ionized atom.

Auger energies are as element-specific as binding energies in XPS. However, chemical information such as oxidation states is much more difficult to abstract.

Three effects can occur:

- Similarly to XPS, a change in oxidation state shifts all core levels and as the kinetic energy of the Auger electron is, according to Eq. (3-14), a difference between *three* core levels, one might expect that the Auger energy measures the same chemical shift as does XPS. Unfortunately, relaxation processes are much more complicated, and consequently Auger shifts correlate less successfully with oxidation states as binding energies in XPS do.

- Chemical bonding information is also obtained when the Auger transition involves valence levels, as with the KVV Auger transitions of carbon and oxygen. As Figure 3.26 shows, the fine structure of the C KVV signal can be used as a fingerprint of the state of the carbon, but is furthermore difficult to interpret.

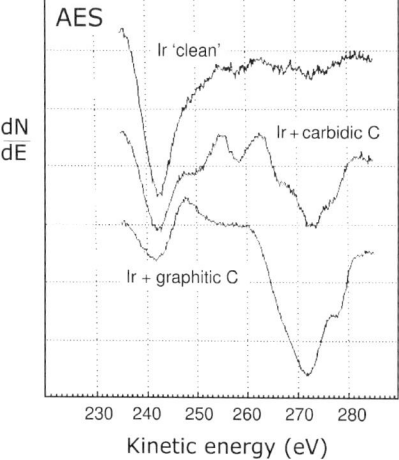

Fig. 3.26 The spectrum of the carbon KVV Auger transition contains chemical information and can be used as a fingerprint of the state of the carbon.

- Changes in energy loss mechanisms operating when the Auger electron is emitted due to chemical bonding, affect the low-energy side of the Auger peaks.

In general, XPS is the preferred technique for studying oxidation states and ligand effects. However, there is a good reason why one should be interested in deriving chemical information from AES. Whereas the spatial resolution of XPS is at best a few micrometers, Auger spectra can be obtained from spots with diameters as small as a few nanometers. It would be extremely interesting to have oxidation state information on the same scale as well!

3.4.2
Intensity of Auger Peaks

The Auger process is preceded by the creation of a core hole, formed by irradiating the atom with X-rays, ions or, as done in Auger spectroscopy, with electrons of kinetic energy E_p (0.5–5 keV). The ionization probability Q_i to form a core hole in a level of energy E_i depends on the energy of the core level as $1/E_i^2$, as well as on the ratio E_p/E_i. The latter dependence (see Fig. 3.27a) is easy to understand: in order to remove an electron with a binding energy E_i one needs a primary electron of higher energy, hence $Q = 0$ when $E_p \ll E_i$. If the primary energy is too high, the electron simply goes too fast to have efficient interaction with a bound electron in a core level. As a result, the ionization probability maximizes for $E_p/E_i \approx 2–3$.

A core-ionized atom has two possibilities to lower its energy, namely Auger decay and X-ray fluorescence (this is described in more detail in Chapter 7). The Auger yields for processes following core hole creation in the K and L shells are shown in Figure 3.27b. Clearly, Auger processes are the dominant decay mode in light elements.

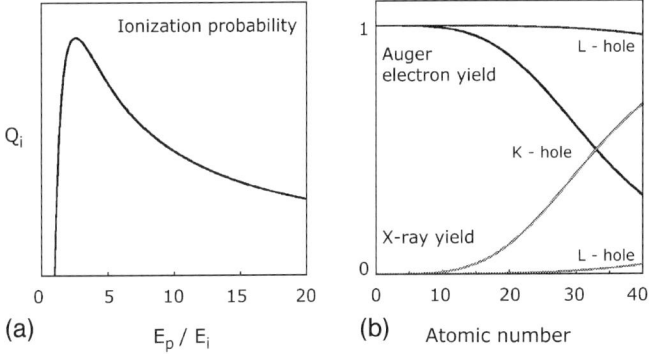

Fig. 3.27 (a) The probability Q_i to create a core hole in a level with binding energy E_i with a primary electron of energy E_p maximizes for $E_p/E_i \approx 2–3$. (b) Auger decay is the preferred mode of de-excitation in light elements, while X-ray fluorescence becomes more important for heavier elements.

The $1/E_i^2$ dependence of the ionization probability forms the reason that Auger transitions starting from a core hole in the K or 1s level are only used in light elements up to phosphorus ($Z \leq 15$), while transitions involving an L-hole are used for elements with $12 \leq Z \leq 40$. Note that these are also the ranges where the Auger yields are close to 1. Another fortunate coincidence is that the kinetic energies of most transitions of interest are not higher than 1 keV, which is favorable for the surface sensitivity of the technique.

A quantitative analysis of Auger spectra is easily possible, but less straightforward than in XPS. The main reason for this is that the effective intensity of primary electrons in the sample may be higher than that of the primary beam, due to backscattering effects. For example, the Auger yield measured from a thin film of a few atomic layers of molybdenum deposited on a tungsten substrate is about 20% higher than that of bulk molybdenum, because the heavier tungsten is an efficient backscatterer. We account for this by multiplying the primary electron intensity i_p by a factor of $(1 + R)$, where R is the backscattering factor. In general, R is higher for heavier than for lighter elements, but it may also be matrix-dependent. The general expression for the Auger electron current i_A becomes:

$$i_A = i_p(1+R)QY \int_0^\infty n(z) e^{-z/\lambda \cos \theta} \, dz \qquad (3\text{-}15)$$

where:
- i_A is the Auger electron current;
- i_p is the primary electron current;
- R is the backscattering factor;
- Q is the ionization probability (see Fig. 3.27a);
- Y is the probability of Auger decay (see Fig. 3.27b);
- $n(z)$ is the concentration of the element at depth z;
- λ is the inelastic mean free path of the Auger electron;
- θ is the take-off angle of the Auger electron measured from the surface normal.

For a homogeneous sample with the detector perpendicular to the sample surface, Eq. (3-15) reduces to

$$i_A = i_p(1+R)QYn\lambda \qquad (3\text{-}16)$$

In order to obtain the measured intensity, one needs to correct Eqs. (3-15) and (3-16) for the transmission and the acceptance angle of the detector.

3.4.3
Application of AES in Catalytic Surface Science

AES provides element-specific information on the surface region. For many catalytically relevant elements (C, Cl, S, Pt, Ir, Rh), the main Auger electrons have

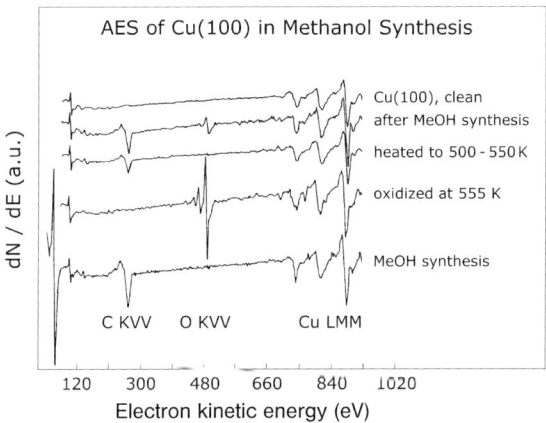

Fig. 3.28 Auger electron spectra (AES) of a Cu(100) with Cu LMM peaks between 700 and 950 eV, after methanol synthesis, showing signals due to carbon (275 eV) and oxygen (505 eV), measured at room temperature, after heating to reaction temperature (500–550 K), oxidation at 555 K and subsequent methanol synthesis. (Adapted from [77]).

energies in the range of 100 to 300 eV, where the mean free path of the electrons is at its minimum. Thus, for many elements, AES is considerably more surface-sensitive than XPS.

A report by Szanyi and Goodman [77] on the synthesis of methanol over a copper single crystal provides a good example of how AES is often used in surface science studies of catalytic reactions. These authors investigated the formation of methanol from a mixture of CO_2, CO and H_2 on Cu(100) at temperatures between 500 and 550 K and pressures of approximately 1 bar, in a reactor coupled to a UHV system. Figure 3.28 shows the Auger spectrum of the clean copper surface; the characteristic Cu LMM peaks appear between 700 and 900 eV. After methanol synthesis, two additional peaks are present – the C KVV peak with the negative excursion at 275–280 eV and the O KVV peaks at about 510 eV (the energies given as E_a in Fig. 3.25). As AES is mainly element-sensitive, it is not clear whether carbon and oxygen are present in molecular form (CO, CO_2, CH_3O, CHO, etc.) or as atomic species. Vibrational spectroscopy would be the best choice to obtain this type of information. After heating to reaction temperature in UHV, all oxygen has disappeared but some carbon is left, implying that there is dissociated carbon on the surface.

Treatment of the Cu surface in 2 Torr of oxygen at 555 K gives a strong O KVV signal, whereas the Cu LMM spectrum differs slightly from that of reduced copper, indicating the limited sensitivity of Auger spectroscopy to chemical effects. The oxidized copper surface is twice as active as reduced copper in the methanol synthesis. Interestingly, this surface contains significantly more carbon than the initially reduced Cu surface after methanol synthesis and subsequent heating (compare the spectra in Fig. 3.28).

A question, which has occupied many catalytic scientists, is whether the active site in methanol synthesis consists exclusively of reduced copper atoms, or contains copper ions [78, 79]. The results of Szanyi and Goodman suggest that ions are involved, as the preoxidized surface is more active than the initially reduced one. However, the activity of these single crystal surfaces expressed in turnover frequencies (i.e., the activity per Cu atom at the surface) is a few orders of magnitude lower than those of the commercial $Cu/ZnO/Al_2O_3$ catalyst, indicating that support-induced effects play a role. Here, the stabilization of ionic copper sites is a likely possibility. Returning to Auger spectroscopy, Figure 3.28 illustrates how many surface scientists have used the technique in a qualitative way to monitor the surface composition.

3.4.4
Scanning Auger Spectroscopy

As electron beams are easily collimated and deflected electrostatically, AES can be used to image the composition of surfaces (this process is termed scanning Auger microscopy; SAM). The best-obtained resolution is now around 25 nm, which is satisfactory for many applications in materials science. For investigating supported catalysts, however, one needs to see details on the order of 1 nm or better. Hence, scanning Auger studies have been useful only incidentally in catalyst characterization. Performing scanning Auger spectroscopy in an electron microscope (see Chapter 7) offers the opportunity to obtain higher lateral resolution. By using a scanning transmission electron microscopy (STEM) instrument, Liu et al. [80] were able to distinguish 2-nm Ag particles from an alumina support by using the Ag MNN Auger signal. Such applications, however, are by no means a matter of routine.

One disadvantage of AES – and of scanning AES in particular – is that the intense electron beam easily causes damage to sensitive materials such as polymers, insulators, and adsorbate layers. The charging of electrically insulating samples may also present problems in this respect.

3.4.5
Depth-Sensitive Information from AES

The strong point of AES is that it provides a quick measurement of the surface composition of conducting samples. Owing to the short data collection times (seconds per scan per peak), AES can be combined with sputtering to measure concentrations as a function of depth. Factors that may complicate the interpretation of depth profiles are preferential sputtering (when one element is eroded at a higher rate than the others), sputter-induced segregation or mixing of atoms in the surface region, and the fact that the probing depth of AES differs from element to element. Figure 3.29 shows the AES depth profile measured from a flat ZrO_2/SiO_2 model catalyst, with a support consisting of a thin SiO_2 layer on top of

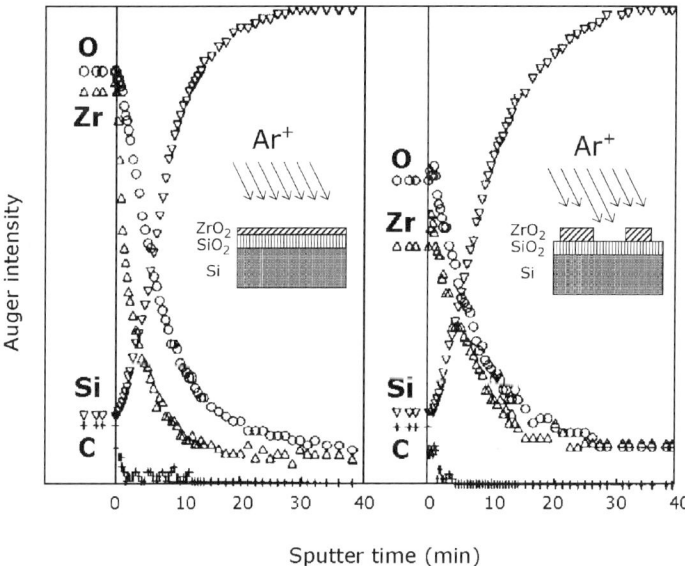

Fig. 3.29 Auger sputter depth profile of a layered $ZrO_2/SiO_2/Si$ model catalyst. While the sample is continuously bombarded with argon ions which remove the outer layers of the sample, the Auger signals of Zr, O, Si and C are measured as a function of time. The depth profile is a plot of Auger peak intensities against sputter time. The profile indicates that the outer layer of the model catalyst contains carbon. Next, Zr and O are sputtered away, but note that oxygen is also present in deeper layers where Zr is absent. The left pattern is characteristic for a layered structure, and confirms that the zirconium is present in a well-dispersed layer over the silicon oxide. The right pattern is consistent with the presence of zirconium oxide in larger particles. (From [81]).

a Si single crystal. The layered nature of one of the samples is clearly revealed [81].

In conclusion, AES is mainly used to study the elemental surface composition of conducting samples. It can be used to create chemical maps of heterogeneous surfaces, and also to study the vertical distribution of elements as a function of depth. The technique is of great importance in surface science and materials science, though less important in the characterization of supported catalysts.

References

1 H. Hertz, *Ann. Phys. (Leipzig)* **31** (1887) 983.
2 J.J. Thomson, *Philos. Mag.* **48** (1899) 547.
3 A. Einstein, *Ann. Phys. (Leipzig)* **17** (1905) 132.
4 G. Margaritondo, *Physics Today*, April 1988, p. 66.
5 G. Ertl and J. Küppers, *Low Energy Electrons and Surface Chemistry.* VCH, Weinheim, 1985.
6 K. Siegbahn, C. Nordling, A. Fahlman, H. Nordberg, K. Hamrin, J. Hedman, G. Johansson, T. Bergmark, S.E. Karlsson, J. Lindgren, and B. Lindberg,

Electron Spectroscopy for Chemical Analysis; Atomic, Molecular and Solid State Structure Studies by means of Electron Spectroscopy. Almquist and Wiksells, Stockholm, 1967.
7 D.W. Turner, C. Baker, A.D. Baker, and C.R. Brundle, *Molecular Photoelectron Spectroscopy.* Wiley, London, 1970.
8 P.K. Ghosh, *Introduction to Photoelectron Spectroscopy.* Wiley, New York, 1983.
9 W.E. Spicer, *Phys. Rev.* **112** (1958) 114; *Phys. Rev.* **125** (1962) 1297.
10 D.E. Eastman and J.K. Cashion, *Phys. Rev. Lett.* **27** (1971) 1520.
11 P. Auger, *J. Phys. Radium* **6** (1925) 205.
12 J.J. Lander, *Phys. Rev.* **91** (1953) 1382.
13 L.A. Harris, *J. Appl. Phys.* **39** (1968) 1419.
14 P.W. Palmberg, G.K. Bohn, and J.C. Tracy, *Appl. Phys. Lett.* **15** (1969) 254.
15 C.J. Powell, D.M. Hercules, and A.W. Czanderna, in: *Ion Spectroscopies for Surface Analysis*, A.W. Czanderna and D.M. Hercules (Eds.), Plenum, New York, 1991, p. 417.
16 G.A. Somorjai, *Chemistry in Two Dimensions, Surfaces.* Cornell University Press, Ithaca, 1981.
17 S. Hüfner, *Photoelectron Spectroscopy – Principles and Applications.* Springer, Berlin, 1996.
18 B. Feuerbacher, B. Fitton, and R.F. Willis (Eds.), *Photoemission and the Electronic Properties of Surfaces.* Wiley, New York, 1978.
19 G.C. Smith, *Surface Analysis by Electron Spectroscopy: Measurement and Interpretation.* Plenum, New York, 1994.
20 D. Briggs and M.P. Seah (Eds.), *Practical Surface Analysis: Auger and X-Ray Photoelectron Spectroscopy.* Wiley, New York, 1996.
21 J.F. Watts and J. Wolstenholme, *An Introduction to Surface Analysis by XPS and AES.* Wiley, Chichester, 2003.
22 J.F. Moulder, W.M.F. Stickle, P.R.E. Sobol, and K.H.D. Bomben, *Handbook of X-ray Photoelectron Spectroscopy, a Reference Book of Standard Spectra for Identification and Interpretation of XPS Data.* Perkin Elmer, 1995.
23 J.H. Scofield, *J. Electron. Spectrosc. Rel. Phenom.* **8** (1976) 129.
24 M.P. Seah and W.A. Dench, *Surf. Interface Anal.* **1** (1979) 2.
25 S. Tanuma, C.J. Powell, and D.R. Penn, *Surf. Interface Anal.* **11** (1988) 577.
26 A. Jablonski and C.J. Powell, *Surface Sci. Rep.* **46** (2002) 33.
27 J.C. Muijsers, J.W. Niemantsverdriet, I.C.M. Wehman-Ooyevaar, D.M. Grove, and G. van Koten, *Inorg. Chem.* **31** (1992) 2655.
28 R.D. van de Grampel, W. Ming, A. Gildenpfennig, W.J.H. van Gennip, J. Laven, J.W. Niemantsverdriet, H.H. Brongersma, G. de With, and R. van der Linde, *Langmuir* **20** (2004) 6344.
29 D.A. Shirley, *Chem. Phys. Lett.* **16** (1972) 220.
30 K. Siegbahn, C. Nordling, G. Johansson, J. Hedman, P.F. Heden, K. Hamrin, U. Gelius, T. Bergmark, L.O. Werme, R. Manne, and Y. Baer, *ESCA Applied to Free Molecules.* North Holland, Amsterdam, 1969.
31 D.A. Shirley, R.L. Martin, F.R. McFeely, S. Kowalczyk, and L. Ley, *Faraday Soc. Disc.* **60** (1975) 7.
32 P.J. Angevine, J.C. Vartulli, and W.N. Delgass, *Proceedings 6th International Congress of Catalysis 1976*, (1977) 2.
33 F.P.J.M. Kerkhof and J.A. Moulijn, *J. Phys. Chem.* **83** (1979) 1612.
34 H.P.C.E. Kuipers, H.C.E. van Leuven, and W.M. Visser, *Surf. Interface Anal.* **8** (1986) 235.
35 R.W. Joyner, *Zirconium in Catalysis.* Magnesium Elektron, Twickenham, UK, 1992.
36 A.C.Q.M. Meijers, A.M. de Jong, L.M.P. van Gruijthuijsen, and J.W. Niemantsverdriet, *Appl. Catal.* **70** (1991) 53.
37 P.L.J. Gunter, A.M. de Jong, J.W. Niemantsverdriet, and H.J.H. Rheiter, *Surf. Interface Anal.* **19** (1992) 161.
38 A. Jablonski and S. Tougaard, *J. Vac. Sci. Technol. A* **8** (1990) 106.
39 C.S. Fadley, *Progr. Solid State Chem.* **11** (1976) 265.
40 P.L.J. Gunter and J.W. Niemantsverdriet, *Appl. Surface Sci.* **89** (1995) 69.
41 P.L.J. Gunter, O.L.J. Gijzeman, and J.W. Niemantsverdriet, *Appl. Surface Sci.* **115** (1997) 342.
42 O.A. Baschenko, V.I. Bukhtiyarov, and A.I. Boronin, *Surface Sci.* **271** (1992) 493.
43 R.A. van Santen and H.P.C.E. Kuipers, *Adv. Catal.* **45** (1988) 265.

44 A.V. de Carvalho, M.C. Asensio, and D.P. Woodruff, *Surface Sci.* **273** (1992) 381.
45 W.F. Egelhoff, *Crit. Rev. Solid State Mater. Sci.* **16** (1990) 213.
46 R.W. Joyner, M.W. Roberts, and K. Yates, *Surface Sci.* **87** (1979) 501.
47 R.W. Joyner and M.W. Roberts, *Chem. Phys. Lett.* **60** (1979) 459.
48 V.I. Bukhtiyarov, V.V. Kaichev, and I.P. Prosvirin, *Topics Catal.* **32** (2005) 3.
49 D.F. Ogletree, H. Bluhm, G. Lebedev, C.S. Fadley, Z. Hussain, and M. Salmeron, *Rev. Sci. Instr.* **73** (2002) 3872.
50 M. Hävecker, R.W. Mayer, A. Knop-Gericke, H. Bluhm, E. Kleimenov, A. Liskowski, D. Su, R. Follath, F.G. Requejo, D.F. Ogletree, M. Salmeron, J.A. Lopez-Sanchez, J.K. Bartley, G.J. Hutchings, and R. Schlögl, *J. Phys. Chem. B* **107** (2003) 4587.
51 V.I. Bukhtiyarov, A.I. Nizovskii, H. Blum, M. Hävecker, E. Kleimenov, A. Knop-Gericke, and R. Schlögl, *J. Catal.* **238** (2006) 260.
52 A. Baraldi, G. Comelli, S. Lizzit, M. Kiskinova, and G. Paolucci, *Surface Sci. Rep.* **49** (2003) 169.
53 L. Rumiz, Thesis, University of Trieste 1997 (referenced in [52]).
54 H. Antonsson, A. Nilsson, N. Martensson, I. Panas, and P.E.M. Siegbahn, *J. Electron. Spectrosc. Rel. Phenom.* **54/55** (1990) 601.
55 A. Beutler, E. Lundgren, R. Nyholm, J.N. Andersen, B.J. Setlik, and D. Heskett, *Surface Sci.* **396** (1998) 117.
56 A.J. Jaworowski, A. Beutler, F. Strisland, R. Nyholm, B. Setlik, D. Heskett, and J.N. Andersen, *Surface Sci.* **431** (1999) 33.
57 R. Linke, D. Curulla, M.J.P. Hopstaken, and J.W. Niemantsverdriet, *J. Chem. Phys.* **115** (2001) 8209.
58 R.M. van Hardeveld, M.J.P. Hopstaken, J.J. Lukkien, P.A.J. Hilbers, A.P.J. Jansen, R.A. van Santen, and J.W. Niemantsverdriet, *Chem. Phys. Lett.* **302** (1999) 98.
59 P.W. Atkins, *Physical Chemistry*, Fifth Edition. Oxford University Press, Oxford, 1994.
60 J.J. Cowell, A.K. Santra, R. Lindsay, R.M. Lambert, A. Baraldi, and A. Goldoni, *Surface Sci.* **437** (1999) 1.

61 E.D. Rienks, J.W. Bakker, A. Baraldi, S.A.C. Carabineiro, S. Lizzit, C.J. Weststrate, and B.E. Nieuwenhuys, *J. Chem. Phys.* **119** (2003) 6245.
62 V. Dose, *Progr. Surface Sci.* **13** (1983) 225.
63 M. Heber and W. Grünert, *Topics Catal.* **15** (2001) 3.
64 M. Heber and W. Grünert, *J. Phys. Chem. B* **104** (2000) 5288.
65 J.A. Rodriguez and D.W. Goodman, *J. Phys. Chem.* **95** (1991) 4196.
66 K. Markert, P. Dolle, J.W. Niemantsverdriet, and K. Wandelt, unpublished results.
67 M.G. Mason, *Phys. Rev. B* **27** (1983) 748; M.G. Mason, in: *Cluster Models for Surface and Bulk Phenomena*, G. Pacchioni, P.S. Bagus and F. Parmigiani (Eds.), Nato ASI Series B, Vol. 283, Plenum, New York, 1992, p. 115.
68 R.C. Baetzold, M.G. Mason, and J.F. Hamilton, *J. Chem. Phys.* **72** (1980) 366.
69 W. Weiss and E. Umbach, *Surface Sci. Lett.* **249** (1991) L333.
70 G. Brodén, G. Gafner, and H.P. Bonzel, *Surface Sci.* **84** (1979) 295.
71 K. Wandelt, *J. Vac. Sci. Technol. A* **2** (1984) 802.
72 K. Wandelt, J.W. Niemantsverdriet, P. Dolle, and K. Markert, *Surface Sci.* **213** (1989) 612.
73 K. Wandelt, K. Markert, P. Dolle, A. Jablonski, and J.W. Niemantsverdriet, *Surface Sci.* **189/190** (1987) 114.
74 J. Feydt, A. Elbe, G. Meister, and A. Goldmann, *Surface Sci.* **445** (2000) 115.
75 Y.D. Kim, J. Stultz, T. Wei, and D.W. Goodman, *J. Phys. Chem. B* **107** (2003) 592.
76 Y. Harada, S. Mazuka, and H. Ozaki, *Chem. Rev.* **97** (1997) 1897.
77 J. Szanyi and D.W. Goodman, *Catal. Lett.* **10** (1991) 383.
78 G.C. Chinchen, M.S. Spencer, K.W. Waugh, and D.A. Whan, *J. Chem. Soc., Faraday Trans. I* **83** (1987) 2193.
79 L.E.Y. Nonneman and V. Ponec, *Catal. Lett.* **7** (1990) 213.
80 J. Liu, G.G. Hembree, G.E. Spinnler, and J.A. Venables, *Surface Sci.* **262** (1992) L111.
81 L.M. Eshelman, A.M. de Jong, and J.W. Niemantsverdriet, *Catal. Lett.* **10** (1991) 201.

4
The Ion Spectroscopies

Keywords

Secondary ion mass spectrometry (SIMS)
Secondary neutral mass spectrometry (SNMS)
Ion-scattering spectroscopy (ISS)
Rutherford backscattering (RBS)
Low-energy ion scattering (LEIS)

4.1
Introduction

Techniques based on the interaction of ions with solids, such as secondary ion mass spectrometry (SIMS) and low-energy ion scattering (LEIS) have undoubtedly been accepted in catalyst characterization, but are by no means as widely applied as for example X-ray photoelectron spectroscopy (XPS) or X-ray diffraction (XRD). Nevertheless, SIMS, with its unsurpassed sensitivity for many elements, may yield unique information on whether or not elements on a surface are in contact with each other. LEIS is a surface technique with true outer layer sensitivity, and is highly useful for determining to what extent a support is covered by the catalytic material. Rutherford backscattering (RBS) is less suitable for studying catalysts, but is indispensable for determining concentrations in model systems, where the catalytically active material is present in monolayer (ML)-like quantities on the surface of a flat model support.

SIMS, LEIS, and RBS have in common that they all are *ion in–ion out* techniques. However, the physical phenomena underlying these techniques are entirely different: SIMS is based on the sputtering action of the incident ion beam, LEIS on the scattering of these ions by atoms at or just below the surface, and RBS on the scattering of high-energy ions by the nuclei of atoms throughout the outer micrometers of the solid. SIMS and LEIS are conducted with typical surface science laboratory equipment (ion guns, mass spectrometers and energy analyzers), but for RBS one needs an accelerator to produce a beam of high-energy ions. As the inelastic mean free path of ions with energies in the keV range is small, the ion spectroscopies are necessarily carried out in vacuum.

Spectroscopy in Catalysis: An Introduction, Third Edition
J. W. Niemantsverdriet
Copyright © 2007 WILEY-VCH Verlag GmbH & Co. KGaA, Weinheim
ISBN: 978-3-527-31651-9

In this chapter we describe briefly the physical phenomena, such as sputtering, scattering, neutralization, and reionization that are involved in ion spectroscopy. For a detailed description of the interactions of ions with solids, the reader is referred to the textbooks by Feldman and Mayer [1], Benninghoven, Rüdenauer and Werner [2], and Czanderna and Hercules [3].

4.2
Secondary Ion Mass Spectrometry (SIMS)

SIMS is by far the most sensitive surface technique, but also the most difficult one to quantify. SIMS is very popular in materials research for making concentration depth profiles and chemical maps of the surface. The principle of SIMS is conceptually simple: a primary ion beam (Ar^+, 0.5–5 keV) is used to sputter atoms, ions and molecular fragments from the surface, and these are consequently analyzed with a mass spectrometer. It is as if one were to scratch some material from the surface and place it in a mass spectrometer to see what elements are present. The theory behind SIMS is far from simple, however. In particular, the formation of ions upon sputtering in or near the surface is hardly understood. The interested reader will find a wealth of information on SIMS in the textbooks by Benninghoven et al. [2] and Vickerman et al. [4, 5], while many applications have been described by Briggs et al. [6].

When a surface is exposed to a beam of ions, energy is deposited in the surface region of the sample by a *collision cascade*. Some of the energy will return to the surface and stimulate the ejection of atoms, ions, and multi atomic clusters (Fig. 4.1). In SIMS, secondary ions (positive or negative) are detected directly with a

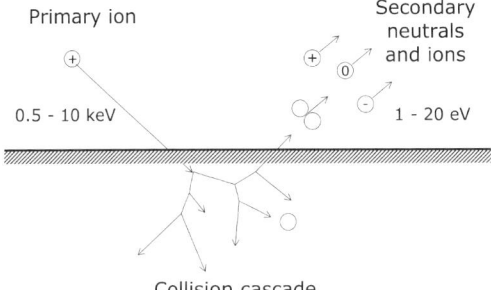

Fig. 4.1 The principle of secondary ion mass spectrometry (SIMS). Primary ions with an energy between 0.5 and 10 keV cause a collisional cascade below the surface of the sample. Some of the branches end at the surface and stimulate the emission of neutrals and ions. In SIMS, the secondary ions are detected directly with a mass spectrometer, whereas in SNMS the secondary neutrals are ionized before they enter the mass spectrometer.

mass spectrometer. In the more recently developed secondary neutral mass spectrometry (SNMS), the secondary neutrals are ionized, for example by electron impact ionization, before their mass is analyzed. Most SIMS instruments use a quadrupole mass spectrometer, which has the advantage that it is relatively inexpensive. However, for higher sensitivity and higher mass resolution, magnetic sector or time-of-flight (TOF) mass spectrometers are used. With such instruments it is for example possible to distinguish between the masses of Al^+ (26.98) and $C_2H_3^+$ (27.024), which is not possible with a quadrupole-based instrument [5].

SIMS is, strictly speaking, a destructive technique, but not necessarily a damaging one. In the dynamic mode, which is used for making concentration depth profiles, several tens of monolayers are removed per minute. In static SIMS, however, the rate of removal corresponds to one monolayer per several hours, implying that the surface structure does not change during the measurement (between seconds and minutes). In this case one can be sure that the molecular ion fragments are truly indicative of the chemical structure on the surface. Static SIMS is a very gentle, non-damaging technique, which causes less damage to a surface than does for example Auger spectroscopy or even standard XPS. In support of this optimistic statement, it must be mentioned that Benninghoven and co-workers have been able to observe the intact emission of very large biomolecules such as vitamin B_{12} (1356 amu) adsorbed onto silver [2].

The advantages of SIMS are its high sensitivity (detection limit of parts per millions for certain elements), its ability to detect hydrogen, and the emission of molecular fragments which often bear tractable relationships with the parent structure on the surface. The disadvantages are that secondary ion formation is a poorly understood phenomenon, and that quantitation is often difficult. A major drawback is the *matrix effect*, whereby secondary ion yields of one element can vary tremendously with its chemical environment. This matrix effect and the elemental sensitivity variation of five orders of magnitude across the periodic table make quantitative interpretation of SIMS spectra of technical catalysts extremely difficult.

Figure 4.2 shows the SIMS spectrum of a promoted iron–antimony oxide catalyst used in selective oxidation reactions. Note the simultaneous occurrence of single ions (Si^+, Fe^+, Cu^+, etc.) and molecular ions (SiO^+, $SiOH^+$, FeO^+, SbO^+, $SbOSi^+$). Also clearly visible are the isotope patterns of copper (two isotopes at 63 and 65 amu), molybdenum (seven isotopes between 92 and 100 amu), and antimony (121 and 123 amu). Isotopic ratios play an important role in the identification of peaks, because all peak intensities must agree with natural abundances. Figure 4.2 also illustrates the differences in SIMS yields of the different elements: although iron and antimony are present in comparable quantities in the catalyst, the iron intensity in the spectrum is about 25 times as high as that of antimony!

For a good understanding of SIMS spectra it is important to have at least a qualitative understanding of phenomena such as sputtering, ionization and neutralization, ion-induced electron and light emission, and the energy distribution of sputtered particles.

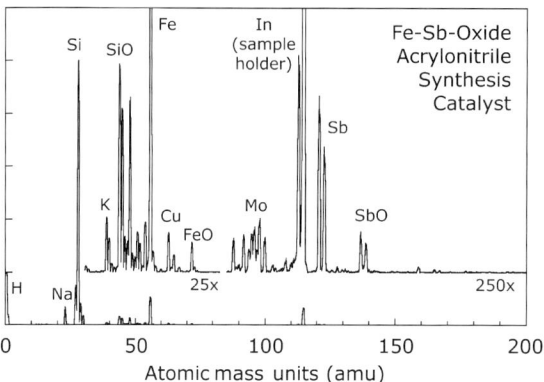

Fig. 4.2 SIMS spectrum of a promoted Fe–Sb oxide catalyst, used for the selective oxidation of propylene and ammonia to acrylonitrile. The spectrum was taken with a 5 keV beam of Ar$^+$ ions under dynamic conditions. (Figure courtesy of H.J. Borg, Eindhoven).

4.2.1
Theory of SIMS

The signal intensity of an elemental positive or negative secondary ion is given by:

$$I_s^\pm = I_p Y R^\pm c^{\text{surf}} T \quad (4\text{-}1)$$

where:
- $I_s^\pm$ is the intensity of positive or negative secondary ions (expressed as a rate in counts per second);
- I_p is the flux of primary ions;
- Y is the sputter yield of the element, equal to the number of atoms ejected per incident ion;
- $R^\pm$ is the probability that the particle leaves the surface as a positive or negative ion;
- c^{surf} is the fractional concentration of the element in the surface layer (a number between 0 and 1);
- T is the transmission of the mass spectrometer, typically 10^{-3} for a quadrupole and 10^{-1} for a TOF instrument.

The essential quantities that determine the yield of secondary ions in a SIMS spectrum are thus the sputter yield, Y, and the ionization probability, $R^\pm$.

Sputtering is a reasonably well understood phenomenon [7]. Sputter yields depend on the properties of the sample as well as on those of the incident ions. Sputter yields of the elements vary roughly between 1 and 10 (see Fig. 4.3), with a few exceptions to the low side, such as bismuth with a sputter yield around 0.1

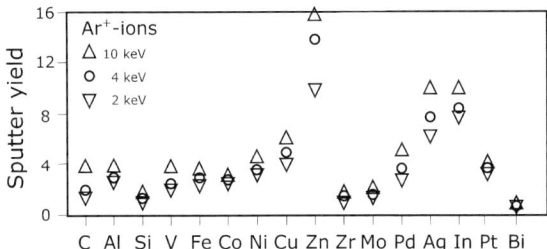

Fig. 4.3 Sputter yields Y of selected elements under bombardments with Ar$^+$ ions of 2, 4 and 10 keV energy. (Data from [4]).

under SIMS conditions, and to the high side, such as zinc, which has a sputter yield of around 15 under 5 keV argon bombardment.

For a given element, the sputter yield depends on the surface morphology: surfaces that are rough on the scale comparable to the dimensions of the sputter cascade give higher yields of secondary particles than flat surfaces [4]. For single crystals, sputter yields have been observed to vary with the plane exposed [8].

The sputter yield depends furthermore on the mass of the primary ion, its energy, and the angle of incidence (see Fig. 4.4). The trends are qualitatively easy to understand:

- Mass: A heavy ion such as xenon deposits more energy into a surface and sputters more efficiently than a lighter ion such as argon.

- Energy: Increasing the energy of the primary ions initially increases the sputter yield. At high energy, however, ions penetrate more deeply into the solid and dissipate their energy further away from the surface. The result is that fewer

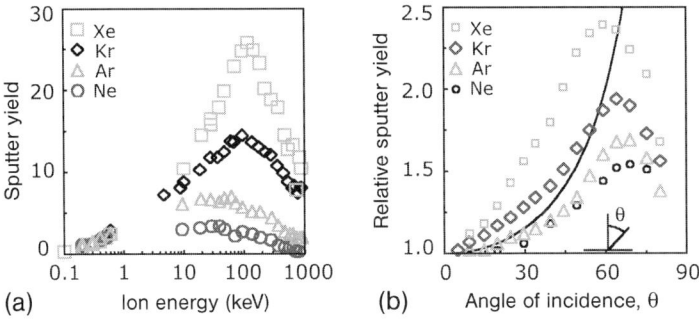

Fig. 4.4 (a) Sputter yields of copper under bombardment with Ne, Ar, Kr and Xe ions as a function of energy (data from [2]). (b) Relative sputter yields of polycrystalline copper as a function of incident angle measured from the surface normal. The primary ion energy is 1.05 keV. The dashed line represents $1/\cos\theta$. (Adapted from [9]).

collision cascades reach the surface and the sputter yield decreases at high energy. Thus, the sputter yield goes through a maximum.

- Angle: The penetration depth of primary ions with certain energy into the sample will be at maximum when the angle of incidence is perpendicular to the surface (if we exclude channeling phenomena). Hence, the sputter yield will increase when the angle of incidence, θ, measured with respect to the surface normal, increases. We expect that the sputter yield varies with $1/\cos\theta$, and this is indeed approximately the case. At glancing angles, however, the ions may scatter back from the surface into the vacuum. As a result, the sputter yield maximizes at angles around 70° (Fig. 4.4b).

Sputtering can be performed not only with ions but also with atoms. For metal samples, the sputter yields are the same. In fact, an incident ion neutralizes rapidly when it enters the metal and hence, there is no difference in the sputtering effect of argon atoms and argon ions. For insulators or semi-conductors, sputter yields are up to a factor of 2.5 higher for ions than for atom beams [4]. Thus, for investigating oxide-supported catalysts or zeolites, measuring SIMS in the fast atom bombardment mode (FABSIMS) gives lower sputter yields than in the standard ion beam SIMS. In spite of the less-efficient sputtering, the use of an atom source offers an advantage when the sample is an insulator, as it minimizes charging phenomena (this point is discussed later in the chapter).

4.2.2
Electron and Photon Emission under Ion Bombardment

Inherently connected to the interaction of ions with a solid is the emission of electrons from the sample. Typical yields are 0.1 to 0.2 electron per incident argon ion of 2 keV, and 0.2 to 0.5 electron at 5 keV [2]. For the heavier krypton, the values are about equal. Two factors contribute to electron emission:

- *Potential emission*: this is due to an Auger process, in which the ion just above the surface is neutralized by a valence electron from the sample; the energy gain is taken up by a second electron which is emitted from the sample surface. This process is discussed in more detail in connection with neutralization in LEIS.

- *Kinetic emission*: at higher kinetic energy above a certain threshold energy, the impact of an ion can cause the emission of an electron from an inner shell. The core-ionized atom may subsequently decay by the Auger process, which leads to the emission of another electron.

In addition to emitting electrons, a solid bombarded with ions in the keV range emits electromagnetic radiation from the near infrared to the near ultraviolet, with a photon yield of typically 10^{-4} per incident ion for a metal, and 10^{-2} to 10^{-1} for insulators. If the primary beam is intense (as in the dynamic SIMS range), and the sample is an insulator, one observes a bright glow at the point where the beam hits the sample. With conductors, the effect is not or only hardly observable.

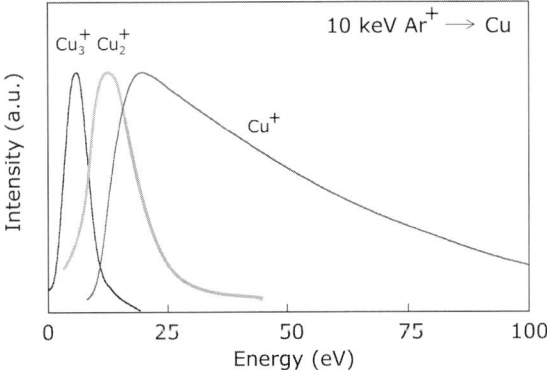

Fig. 4.5 Energy distribution of secondary Cu^+, Cu_2^+ and Cu_3^+ ions during bombardment of copper with 10 keV Ar^+ ions; curves have been normalized to the same height. (Data from [11]).

4.2.3
Energy Distribution of Secondary Ions

It is important to know the energy distribution of secondary ions because it has consequences for their detection, especially in the case of insulators. As Figure 4.5 shows, the energy distribution of elemental secondary ions has usually a peak between 15 and 30 eV, and falls off rapidly at higher energy, but then exhibits a low-level tail to a few hundred eV. The maximum in the energy distribution is hardly affected by the energy of the primary ions. The high-energy tail, however, increases in intensity when the primary energy goes up. Figure 4.5 has a straightforward interpretation. Most of the primary ions deposit their energy of a few keV well below the surface. The energy of the final collisions in the cascades that cause the emission of secondary particles is on the order of 1 to 30 eV. A primary ion beam of higher energy penetrates deeper into the solid and, as a result, the cascades start both deeper and at higher energy. Hence, both factors virtually cancel each other. The high-energy tails of the secondary ion distribution, however, are the result of shorter collision cascades closer to the point of impact of the primary ion. The intensity of this part of the distribution depends on the energy of the primary ion [10, 11].

The energy distributions of molecular ions peak at significantly lower energies and do not tail to high energies, partly because the transfer of high energies to a cluster will lead to bond breaking. In addition, some of the transferred energy goes into vibrational and rotational modes of the clusters and does not contribute to translational energy. The data in Figure 4.5 show that secondary Cu_2^+ and Cu_3^+ ions indeed leave a copper target at lower energy than single Cu^+ ions do. As quadrupole mass spectrometers operate with relatively narrow energy windows of typically 10 eV, it is evident that the setting of the window is essential for the detection of molecular ions.

The angular dependence of the secondary ion intensity is expected to follow a simple cosine law, in particular for randomly oriented polycrystalline surfaces. The explanation for this is that, upon impact, the collision cascade takes care of an isotropic distribution of the energy through the sample. Hence, the intensity of collision cascades that arrive at the surface under an angle θ with the surface normal varies as $\cos \theta$. For single crystals, however, anisotropic emission may occur due to channeling of the beam, while the focusing of collisions along close-packed directions in the crystal may lead to anisotropy in the secondary ion emission.

4.2.4
The Ionization Probability

The formation of secondary ions is the most difficult feature in SIMS. Whereas sputtering is relatively well understood, the process of sputter ionization is not, and a theory that describes the process of secondary ion formation satisfactorily does not yet exist. However, a number of trends can be rationalized.

For the formation of positive ions from initially neutral elements, a clear relationship exists between the ionization probability, R^+, and the ionization potential, I. Elements such as Na, K, Mg, and Ca, all of which have a low ionization potential, give high yields in positive SIMS, whereas elements with a high ionization potential (e.g., N, Pt, Au) give low yields. A similar relationship exists between the probability for the formation of negative ions and the electron affinity, ε_A: electronegative elements such as O, F, and Cl give high intensities in negative SIMS. The noble metals Pt and Au also appear usually much more intense in negative than in positive SIMS.

The ionization probabilities $R^{\pm}$ vary over some five decades across the elements in the Periodic Table. In addition, they vary with the chemical environment of the element. This effect – which usually is referred to as the "matrix effect" – makes quantitation of SIMS spectra extremely difficult. As shown in Table 4.1, positive secondary ion yields from metal oxides are typically two orders of magnitude higher than those of the corresponding metals. A similar increase in yields from metals is observed following the adsorption of gases such as oxygen or carbon monoxide.

As noted previously, the ionization probability – which accompanies sputtering – is at best qualitatively understood. Several attempts have been undertaken to develop models for secondary ion formation, and in this respect the interested reader may consult the literature for reviews [2, 4]. Here, we will briefly describe one model that accounts quantitatively for a number of observations on metals – the perturbation model of Nørskov and Lundquist [12]. This model assumes that the formation of a secondary ion occurs just above the surface, immediately after emission. Then:

- The probability for the formation of positive secondary ions increases when the ionization potential becomes smaller.

Table 4.1 Ionization potential (I), work function (φ), sputter yield (Y), and secondary ion yield of positive ions (R⁺ Y) of selected elements and their oxides.

Element	I [eV]	φ [eV]	Y	R^+Y	R^+Y_{oxide}
Mg	7.64	3.66	4.5	0.01	0.9
Al	5.98	4.28	2.8	0.007	0.7
Si	8.15	4.85	1.1	0.0084	0.58
Ti	6.28	4.33	1.2	0.0013	0.4
V	6.74	4.3	2.0	0.001	0.3
Cr	6.76	4.5	3.0	0.0012	1.2
Mn	7.43	4.1	4.0	0.0006	0.3
Fe	7.87	4.5	2.4	0.0015	0.35
Ni	7.63	5.15	3.2	0.0006	0.045
Cu	7.72	4.05	4.0	0.0003	0.007
Mo	7.13	4.6	1.3	0.00065	0.4
Ta	7.70	4.25	1.25	0.00007	0.02
W	7.98	4.55	1.25	0.00009	0.035
Pd	8.33	5.12	2.8	–	–
Ag	7.57	4.26	6.4	–	–
Pt	8.96	5.65	3.4	–	–
Au	9.22	5.1	5.5	–	–

- When the secondary ion is close to the surface, there is a chance that it will be neutralized by an electron from the surface of the metal. This process becomes more likely if the work function of the metal is lower; in other words, a high work function prevents neutralization and is favorable for positive ion emission.

Both ion formation and neutralization just above the surface are more likely if the velocity, v, of the departing particle is small, or in other words, when its residence time in the interaction zone just above the surface is long.

These features are recognized in the following expression for the ionization probability:

$$R^+ \propto e^{\text{const}((\varphi-I)/v)} \tag{4-2}$$

where:
R^+ is the ionization probability;
φ is the work function of the sample;
I is the ionization potential of the sputtered particle;
v is the velocity of the sputtered particle.

Equation (4-2) accounts qualitatively for the observed variations of secondary ion yields with ionization potential. It also describes correctly that the yields of positive secondary ions from metals rise when molecules such as CO or oxygen, which increase the work function, cover the surface. Although the model elegantly predicts a number of trends correctly, it is not sufficiently detailed to serve as a basis for the quantitative analysis of technical samples.

4.2.5
Emission of Molecular Clusters

Molecular cluster ions are very useful because they reveal which elements are in contact in the sample. Of course, this presupposes that such clusters are emitted intact and are not the result of recombination processes above the surface. Oechsner [13] collected evidence that direct cluster emission processes predominate in the case when relatively strong bonds exist between neighbor atoms. Direct emission becomes even more likely if the two atoms differ significantly in mass, and when the heavier atom receives momentum from the sputter cascade in the solid. Thus, there is little doubt that clusters of the type ZrO^+, $FeCl_3^-$, MoS^+, CH_3^+, $PdCO^+$, or Rh_2NO^+ (which we will encounter in the applications later) stem from direct emission processes and reflect bonds present in the sample [2, 4]. Some evidence exists, however, that atomic recombination may play a role in the SIMS of metals, and in alloys where the two constituents have comparable mass [13].

4.2.6
Conditions for Static SIMS

The time necessary to remove one monolayer during a SIMS experiment depends not only on the sputter yield, but also on the type of sample under study. We will make an estimate for two extremes. First, the surface of a metal contains about 10^{15} atoms per cm^2. If we use an ion beam with a current density of 1 nA cm^{-2}, we then need some 150 000 seconds (about 40 hours) to remove one monolayer if the sputter yield is 1, and 4 hours if the sputter rate is 10. However, if we are working with adsorbed macromolecules and polymers, we need significantly lower ion doses to remove a monolayer. It is believed [4] that one impact of a primary ion affects an area of about 10 nm^2, which is equivalent to a circle of about 3.5 nm diameter. Hence, if the sample consists for example of a monolayer film of polymer material, a dose of 10^{13} ions cm^{-2} could, in principle, be sufficient to remove or alter all material on the surface. With a current density of 1 nA this takes about 1500 seconds or 25 minutes only. For adsorbates such as CO adsorbed onto a metal surface, we estimate that the monolayer lifetime is at least a factor of 10 higher than that for polymer samples. Thus, for static SIMS, one needs primary ion current densities on the order of 1 nA cm^{-2} or less, and one should be able to collect all spectra of one sample within a few minutes.

4.2.7
Charging of Insulating Samples

If the sample under study is an insulator (e.g., an oxide-supported catalyst), the arrival of positively charged primary ions and the emission of electrons by either potential or kinetic emission leads to positive charging of the sample. This has two negative effects:

- As the sample has a positive potential, V, the energy of all positive secondary ions will be increased by an amount eV, whereas the energy of negative secondary ions will be the same amount lower; as a consequence, these ions may shift largely outside the detection window.
- A high potential on the sample may deflect the primary ions.

One can compensate for charging by using a so-called "flood gun", which sprays low-energy electrons onto the sample. Charging can also be minimized by using a beam of atoms instead of ions as primary particles. In this case, the kinetic emission of electrons is the only source of charging, if we ignore the low yields of secondary ions.

4.2.8
Applications on Catalysts

SIMS investigations in catalysis can be allocated approximately into three categories [14, 15]:

- The characterization of technical catalysts.
- Studies on model systems which both allow for optimum SIMS analysis and realistically simulate aspects of technical catalysts, chosen to address specific questions.
- Surface science studies dealing with the adsorption and reaction of gases on single crystal surfaces.

Here, we discuss examples from all three categories.

Matrix effects and inhomogeneous sample charging seriously hinder the quantitative analysis of SIMS on technical catalysts. Although full quantitation is almost impossible in this area, the interpretation of SIMS data on a more qualitative base offers nevertheless unique possibilities. Molecular cluster ions may be particularly informative about compounds present in a catalyst.

Coverdale et al. [16], while investigating the structure of alumina-supported rhenium oxide catalysts active for olefin metathesis reactions, observed a range of negative cluster ions from ReO^- through $Re_3O_8^-$ in their SIMS spectra. The relative intensities of these ions correlated well with those obtained from a crystalline Re_2O_7 reference, indicating the presence also of this oxide in the catalysts. If rhenium were present as discrete ReO_4^- units, as had been suggested in the literature, secondary ion clusters with more than one Re atom would not have been expected [16].

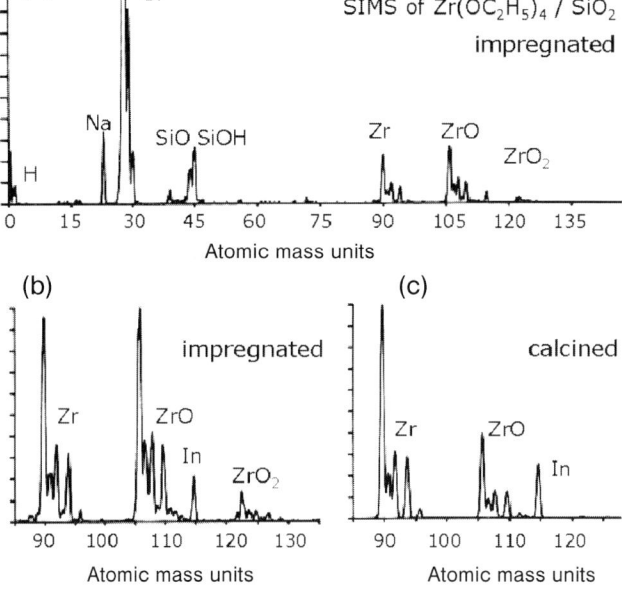

Fig. 4.6 (a) Positive SIMS spectrum of a 9 wt% ZrO_2/SiO_2 catalyst prepared from zirconium ethoxide; (b) after impregnation and drying; and (c) after calcination in air at 400 °C. (From [17]).

Another study on the preparation of supported oxides illustrates how SIMS can be used to follow the decomposition of catalyst precursors during calcination. Here, we discuss the formation of zirconium dioxide from zirconium ethoxide on a silica support [17]. ZrO_2 is catalytically active for a number of reactions such as isosynthesis, methanol synthesis, and catalytic cracking, but is also of considerable interest as a barrier against the diffusion of catalytically active metals such as rhodium or cobalt into alumina supports at elevated temperatures.

The $Zr(OC_2H_5)_4$ precursor was applied on the silica by a condensation reaction between the ethoxide ligands of zirconium and OH groups on the support. Figure 4.6a shows the SIMS spectrum of a freshly prepared Zr catalyst from ethoxide. It contains peaks of H^+, C^+, O^+, Na^+, Si^+, K^+, Ca^+, SiO^+, $SiOH^+$, Zr^+, ZrO^+, In^+, and ZrO_2^+. The relatively high intensities of Na, K and Ca are a well-known artifact of SIMS and have no significance; the low ionization potential of these elements forms the reason that they dominate most SIMS spectra of technical samples, even if they are present at impurity levels. The In peak arises from the indium foil in which the catalyst powder was pressed. The most useful information is in the relative intensities of the Zr^+, ZrO^+, $ZrOH^+$ and ZrO_2^+ ions. This is illustrated in Figures 4.6b and c, which show the isotope patterns of these ions of a freshly dried and a calcined catalyst, respectively. Note that the SIMS spectrum of the fresh catalyst contains small (but significant) contributions

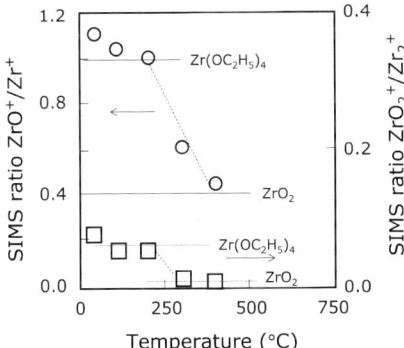

Fig. 4.7 SIMS ZrO^+/Zr^+ and ZrO_2^+/Zr^+ intensity ratios as a function of calcination temperature for the same catalyst as in Figure 4.6, along with ratios measured from zirconium ethoxide and oxide reference compounds. (Data from [17]).

from $ZrOH^+$ ions; see for example the peaks at 107 amu ($^{90}ZrOH^+$ and $^{91}ZrO^+$) and 111 amu ($^{94}ZrOH^+$). $ZrOH^+$ is most probably a fragment ion from zirconium ethoxide. In the spectrum of the catalyst that was oxidized at 400 °C, the isotopic pattern in the ZrO range resembles that of Zr, indicating that no more ZrOH species are present.

Interpretation of the spectra in Figure 4.6 is best done by comparison with those of zirconium ethoxide and zirconium oxide reference compounds. Figure 4.7 contains the ZrO^+/Zr^+ and ZrO_2^+/Zr^+ ratios from the SIMS spectra of the reference compounds, and of the catalysts as a function of the calcination temperature. The figure clearly shows that catalysts calcined at temperatures up to 200 °C have ZrO^+/Zr^+ and ZrO_2^+/Zr^+ ratios about equal to those measured from a zirconium ethoxide reference compound. However, samples calcined above 300 °C have intensity ratios close to that of ZrO_2.

Thus, the SIMS intensity ratios sensitively reflect the transition from zirconium ethoxide to zirconium oxide, and indicate that this reaction takes place at temperatures between 300 and 400 °C. Infrared spectroscopy measured in transmission confirms the disappearance of ethoxide groups at the same temperatures at which the SIMS ZrO^+/Zr^+ and ZrO_2^+/Zr^+ intensity ratios change to those characteristic of ZrO_2. Whilst the infrared spectra might also be due to free ethoxide ligands on the support, the inherent advantage of SIMS is that it confirms – albeit indirectly – that the ethoxide ligands are connected to zirconium [17].

One strong point of SIMS is its ability to detect elements that are present in trace amounts, and as such the technique is highly suited to the detection of poisons on a catalyst caused by contaminants in the reactor feed. Chlorine, for example, poisons the iron catalyst used in ammonia synthesis because it suppresses the dissociation of nitrogen molecules. Plog et al. [18] used SIMS to show that chlorine impurities may coordinate to potassium promoters, as evidenced by a KCl_2^- signal, or to iron, visible by an $FeCl_3^-$ peak. The SIMS intensity ratio

FeCl$_3^-$/Cl$^-$ correlated remarkably well with the decrease in catalytic activity, whereas no correlation existed between the KCl$_2^-$/Cl$^-$ ratio, indicating that potassium may act as an attractant for chlorine where it is relatively harmless.

Whereas chlorine is a poison for the ammonia synthesis over iron, it serves as a promoter in the epoxidation of ethylene over silver catalysts, where it increases the selectivity to ethylene oxide at the cost of the undesired total combustion to CO$_2$. In this case, an interesting correlation was observed between the AgCl$_2^-$/Cl$^-$ ratio from SIMS, which reflects the extent to which silver is chlorinated, and the selectivity towards ethylene oxide [18]. In both examples, the molecular clusters reveal which elements are in contact in the surface region of the catalyst.

A useful property of all mass spectroscopic techniques is that they can distinguish between isotopes, thereby providing interesting opportunities to study reaction mechanisms, also using SIMS. Iron-antimony oxides (see Fig. 4.2) are used for the partial oxidation of hydrocarbons, for example of propylene and ammonia to acrylonitrile. These reactions are believed to proceed through the so-called Mars–van Krevelen mechanism, in which lattice oxygen reacts with the hydrocarbon, leaving a vacancy behind at the surface. The latter is filled by oxygen from the gas phase [19]. Aso et al. [20] addressed the question of which of the components in the catalyst is responsible for oxygen activation – iron or antimony. These authors exposed an active catalyst to ^{18}O$_2$ and followed the SIMS intensities of oxygen-containing clusters as a function of time. The fact that the ^{18}O content of the SbO$_x^-$ clusters was initially higher than in the FeO$_x^-$ clusters strongly suggests that antimony activates the oxygen, which subsequently replenishes vacancies associated with iron, and perhaps also antimony.

Several other successful applications of SIMS in catalyst characterization have been described in the literature [14, 15]. Of particular note are the extensive studies of Vickerman et al. [21] on automotive exhaust catalysts, a SIMS analysis of catalyst coatings in microreactors by Gnaser et al. [22], imaging studies which reveal the lateral distributions of elements [23, 24], and an excellent study on the reactive species present on a supported catalyst during reaction. These will be discussed in a little more detail in the following paragraphs.

Kruse and co-workers [25] reported one of the few studies in which SIMS revealed information on the mechanism of a catalytic reaction on a real catalyst. In being interested in the sites where synthesis gas (CO + H$_2$) reacts to methanol on lanthanum-promoted Pd/SiO$_2$, these authors compared SIMS spectra of a catalyst before and after the reaction, which was carried out in a high-pressure cell attached to a SIMS/XPS spectrometer. The SIMS spectra indicated the formation of species such as formate (HCOO) and methoxy (OCH$_3$), which appeared to be mainly present on the support, as evidenced by the presence of a SiOCH$_3$ fragment. Some methoxy was found in association with the lanthanum promoter, but combinations with palladium were not observed, and in fact, not expected, because experiments with Pd(111) single crystals had revealed that a stabilizing background pressure of methanol would be required to observe methoxy species on the metal, even at room temperature [26]. However, even in the presence of gas-phase methanol, no signals due to Pd-OCH$_3$ or PdOOCH could be measured, which led the authors to conclude that methanol formation occurred on the promoted support, while the role of the noble metal was to provide atomic hydrogen. These investigations were worthy of special mention because mechanistic studies with SIMS are usually carried out with model systems and single crystals.

4.2.9
Model Catalysts

Many technical catalysts consist of particles on an oxide support. Of course, these electrically insulating systems, in which most of the active particles are hidden inside pores, are not particularly suited to SIMS analysis. However, interesting new possibilities arise if one replaces the porous, insulating oxide by a planar model support, consisting of a thin oxide layer of a few nanometers thickness on a conducting substrate [27]. Borg et al. [28] utilized this approach to study the decomposition of $RhCl_3$-derived complexes on alumina, modeled by a 4-nm thin layer of Al_2O_3 on an aluminum foil. A clearly observable $RhCl^-$ signal in the SIMS spectrum of a freshly impregnated catalyst reflected the presence of chlorine-containing rhodium complexes on the support. The cluster disappeared from the spectra after reduction in H_2 at 200 °C, while the elemental Cl^- signal still had substantial intensity, indicating that the rhodium chlorine cluster had disintegrated but that chlorine was still present on the support. These investigations are described further in the case study on rhodium catalysts in Chapter 9.

In using a similar approach, Thüne et al. [29] applied static SIMS to show that Cr/SiO_2 model catalysts which are active for ethylene polymerization contain only monochromates. Secondary ions with more than one Cr ion in the cluster, such as $Cr_2O_4^-$ and $Cr_2O_5^-$, disappeared from the spectra after the catalysts had been calcined; only $CrSiO_x^-$ ions remained. Aubriet et al. [30] studied the anchoring of chromium acetyl acetonate, $Cr(acac)_3$ to a planar $SiO_2/Si(100)$ model support with static SIMS. Chromium polymerization catalysts are discussed further in Chapter 9.

Another SIMS study on model systems concerns molybdenum sulfide catalysts. The removal of sulfur from heavy oil fractions is carried out over molybdenum catalysts promoted with cobalt or nickel, in a process called hydrodesulfurization (HDS) [19]. Catalysts are prepared in the oxidic state, but must be sulfided in a mixture of H_2S and H_2 in order to be active. SIMS sensitively reveals the conversion of MoO_3 into MoS_2, in model systems of MoO_3 supported on a thin layer of SiO_2 [31].

The top spectrum of Figure 4.8 shows the positive SIMS pattern of MoO_3, consisting of elemental Mo^+ (92–100 amu) and molecular MoO^+ (108–116 amu) and MoO_2^+ clusters (124–132 amu). The other spectra in Figure 4.8 show that sulfidation at relatively low temperatures (25 °C) already results in significant uptake of sulfur by the catalyst. This is evidenced by the increase in the $(Mo + 32)^+$ signal, associated with MoO^+ and MoS^+ clusters, and the decrease in the $(Mo + 16)^+$ signal, which is uniquely associated with MoO^+. Sulfidation at 150 °C increases the sulfur and decreases the oxygen content further, while the

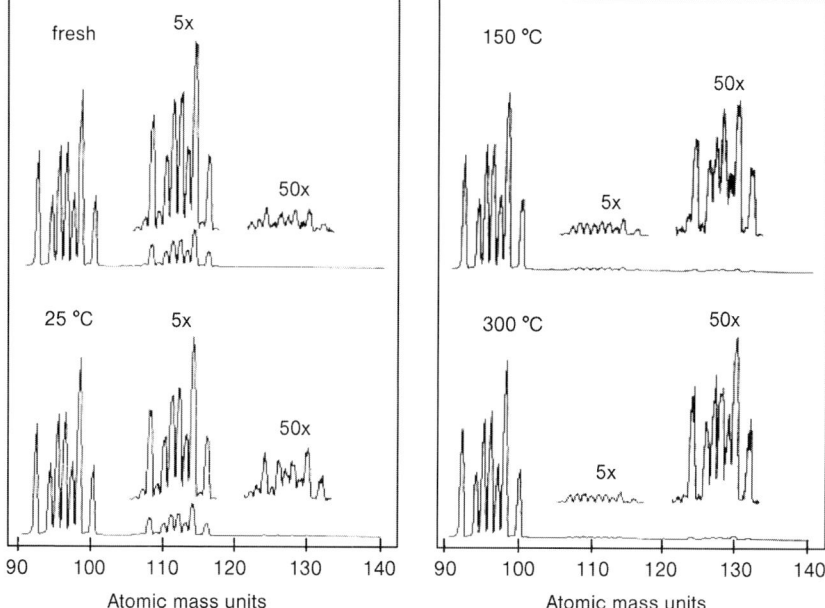

Fig. 4.8 Positive static SIMS spectra of a $MoO_3/SiO_2/Si(100)$ model catalyst, after impregnation (top) and after sulfidation in 10% H_2S in H_2 at the indicated temperatures, showing the transition from MoO_3 to MoS_2. The patterns are those of Mo^+ (92–100 amu), MoO^+ (108–116 amu), and $MoS^+ + MoO_2^+$ (124–132 amu). (From [31]).

pattern measured after sulfidation at 300 °C resembles that of a MoS_2 reference compound, indicating that sulfidation is complete. The data in Figure 4.8 illustrate that, although the analysis is somewhat hindered by the mass interference of O_2- and S-containing clusters, the sulfidation can be followed by inspecting relative intensity ratios. Combination of these results with XPS and RBS data revealed important information on the intermediate oxysulfides that occur in the initial stages of sulfidation [31].

Studies on metal and alloy foils are yet another step away from real catalysts, but can nevertheless provide highly relevant information. Noteworthy here are the investigations conducted by Ott et al. [32] on the behavior of FeRu alloys in CO hydrogenation. SIMS was used to determine the outer layer composition of the alloys, and the results could be successfully correlated with catalytic tests. The characterization of carbonaceous deposits on Pt, Ir, and PtIr alloys with SIMS fall also in this category [33]. Here, the negative $C_2H_n^-$ ($n = 0$–2) provided valuable information on the hydrogen content of carbon on the surface, whereas the intensity ratio of $C_4H_n^-$ to $C_2H_n^-$ clusters reflected the extent of polymerization of the carbon into larger units. The latter play a role in the deactivation of catalysts in hydrocarbon conversions such as catalytic reforming.

4.2.10
Single Crystal Studies

In applications of SIMS on single crystals, matrix effects are largely circumvented. Excellent quantitation has been achieved by calibrating SIMS yields of adsorbates by means of other techniques such as electron energy loss spectroscopy or thermal desorption spectroscopy [4, 6]. Although the absolute SIMS intensities depend sensitively on adsorbate coverage through the work function [Eq. (4-2)], these matrix effects cancel out if one takes intensity ratios. The ratio Ni_2H^+/Ni^+, for example, varies linearly with the coverage of hydrogen atoms on a nickel surface [34]. A similar correlation has been established for CO adsorption on several metals. Vickerman and co-workers [4, 35] showed that the relative intensities of Pd_3CO^+/Pd_3^+, Pd_2CO^+/Pd_2^+ and $PdCO^+/Pd^+$ may be used to derive the fractions of CO adsorbed in threefold, twofold, and single positions on palladium.

An example of monitoring adsorbed species on surfaces with SIMS is shown in Figure 4.9, which shows positive SIMS spectra for the interaction of NO with the Rh(111) surface [36]. The lower curve shows the adsorption of molecular NO (peak at 236 amu) on Rh(111) at 120 K. The middle curve shows the situation after heating the sample to 400 K. The presence of the peaks at 220 amu (Rh_2N^+) and 222 amu (Rh_2O^+) and the absence of the Rh_2NO^+ (236 amu) indicate that NO has dissociated. Heating the sample at 400 K in H_2 causes the removal of atomic oxygen and thus the disappearance of the Rh_2O^+ at 222 amu, as the upper curve shows. As the data collection is rapid, dissociation of NO can be monitored

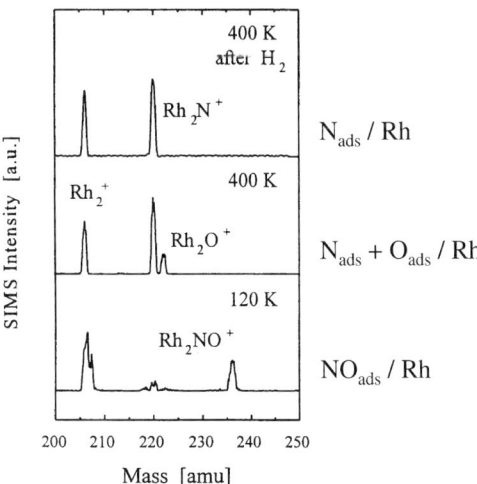

Fig. 4.9 SIMS spectra of a Rh(111) surface after adsorption of 0.12 monolayer of NO at 100 K, where NO adsorbs molecularly, and after heating to 400 K, illustrating dissociation of NO into N- and O-atoms. Note that the yields of O and N differ in SIMS, due to different ionization energies of the respective clusters. (Adapted from [36]).

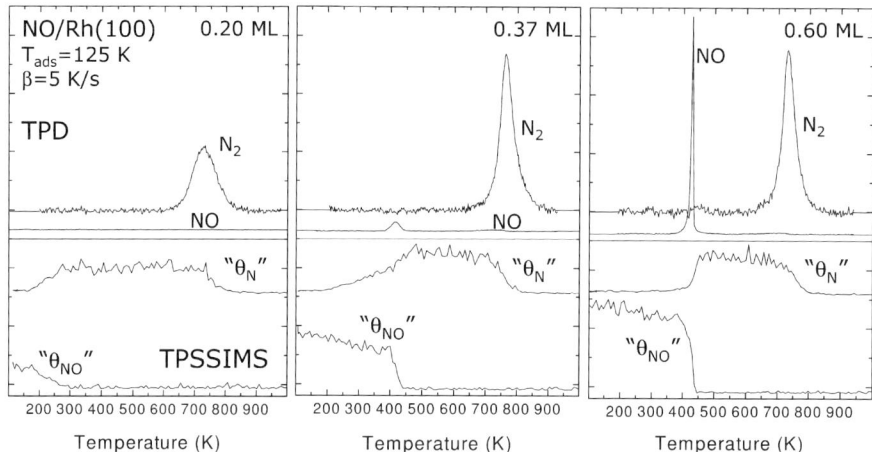

Fig. 4.10 Dissociation of NO on the Rh(100) surface monitored in real time by temperature-programmed SIMS and desorption (TPSIMS and TPD), showing the desorption of N_2 and NO, and $Rh_2(NO)^+/Rh_2^+$ and Rh_2N^+/Rh_2^+ TPSSIMS ion intensity ratios, representing the surface coverages of NO and N-atoms respectively, at low, middle and high initial NO coverage. (Adapted from [38]).

in real time, by monitoring the intensities of the Rh_2NO^+ peak for adsorbed NO, and Rh_2N^+ and Rh_2O^+ for the dissociation products [37, 38].

The dissociation of NO is a crucial elementary step in the mechanism of NO reduction by CO in automotive exhaust catalysis. The overall reaction here is:

$$CO + NO \rightarrow \tfrac{1}{2} N_2 + CO_2$$

As NO dissociation produces two atoms from one molecule, the reaction can only proceed when the surface contains empty sites adjacent to the adsorbed NO molecule. In addition, the reactivity of the molecule is affected by lateral interactions with neighboring species on the surface. Figure 4.10 clearly illustrates all of these phenomena [38]. The experiment starts at low temperature (175 K) with a certain amount (expressed in fraction of a monolayer, ML) of NO on the Rh(100) surface. During temperature programming, the SIMS intensities of characteristic ions of adsorbed species are followed, along with the desorption of molecules into the gas phase, as in temperature-programmed desorption (TPD) or temperature-programmed reaction spectroscopy (TPRS) (see Chapter 2).

At a low NO coverage of 0.2 ML, dissociation is observed between 170 and 280 K, whereas the N-atoms combine and desorb as N_2 at between 600 and 800 K. Oxygen atoms remain on the surface and are removed afterwards by flashing to higher temperatures. If the NO coverage is higher, however, the NO dissociation is seen to retard to higher temperatures, indicating that lateral interac-

Table 4.2 Activation energy (E_a) for several elementary reaction steps involved in the CO + NO reaction on Rh(111) [37] and Rh(100) [38].

Elementary reaction step	E_a on Rh(111) [kJ mol^{-1}]	E_a on Rh(100) [kJ mol^{-1}]
$NO_{ads} + {}^* \rightarrow N_{ads} + O_{ads}$	65	37
$NO_{ads} \rightarrow NO_{gas} + {}^*$	113	106
$N_{ads} + N_{ads} \rightarrow N_{2,gas} + 2^*$	118	215

tions with NO, N and O on the surface slow down the dissociation reaction. At the onset of the reaction, NO dissociates in the presence of other NO molecules, but as the reaction progresses the environment of a dissociating molecule changes to atomic species. Hence, the measurement indicates that repulsion between NO and N and O atoms has a much stronger effect on the dissociation rate of NO than interactions between NO molecules has (also if one accounts for the difference in total coverage of almost a factor of two near completion of the reaction). At high NO coverages the dissociation reaction is blocked (see Fig. 4.10). All chemistry is suppressed until some NO desorbs around 400 K, after which dissociation follows instantaneously. The latter forms a clear illustration of the fact that ensembles of more than one adsorption site are necessary for dissociating molecules.

Analyzing the rates of dissociation and desorption (as described in Chapter 2) yields kinetic parameters for the elementary steps (see Table 4.2). The attractive feature of static SIMS is that it provides a way to measure the kinetics of reactions on the surface, which would be very difficult to do with any other technique. Of course, such experiments require that SIMS operates under truly static conditions [38].

Another example of this type of investigation is the SIMS study of the H–D isotope exchange reaction in ethylidyne on platinum, as reported by Creighton et al. [39]. Ethylidyne ($\equiv CCH_3$), adsorbed on three metal atoms, is a common species that is formed when ethylene adsorbs onto Group VIII metals such as rhodium, palladium, iridium, and platinum. The CH_3 group gives a clear signal at 15 amu in positive SIMS. When ethylidyne is exposed to D_2, the CH_3 group exchanges its H-atoms for deuterium. During the course of the reaction, $\equiv C-CH_2D$ and $\equiv C-CHD_2$ appear as intermediates towards the fully deuterated $\equiv C-CD_3$. All species are distinguishable by their mass, and the experiment forms a beautiful illustration of the kinetics of consecutive surface reactions through two intermediates [40].

As a final example, we show how SIMS can be used to identify the rate-determining step in a sequence of elementary reactions [41]. Imagine the following situation: we have a Rh(111) surface partially covered by N-atoms which we heat up to 400 K under a low, constant pressure of H_2 with the aim of forming NH_3. We expect the following reactions:

$$H_2 + 2^* \rightarrow 2H_{ads}$$

$$N_{ads} + H_{ads} \rightarrow NH_{ads} + {}^*$$

$$NH_{ads} + H_{ads} \rightarrow NH_{2,ads} + {}^*$$

$$NH_{2,ads} + H_{ads} \rightarrow NH_{3,ads} + {}^*$$

$$NH_{3,ads} \rightarrow NH_3 + {}^*$$

where * indicates a vacant site on the surface.

The corresponding SIMS spectrum in Figure 4.11 shows that the surface contains N-atoms and NH_2 species, whereas adsorbed NH and NH_3 species are not observed. A separate study on the adsorption of NH_3 on Rh(111) revealed that $Rh_2(NH_3)^+$ peaks are clearly observable when NH_3 is adsorbed [42]. Repeating the experiment with D_2 instead of H_2 enhances the mass separation between peaks and confirms the presence of N-atoms and ND_2 species. The interpretation is that the rate-determining step in the hydrogenation of N-atoms is the reaction from NH_2 to NH_3. The presence of N-atoms may indicate that these atoms are present in islands and that hydrogenation takes places at the boundaries of these islands [41]. Again, the unique aspect of this type of investigation is that SIMS can be used to monitor all surface species during the reaction on the surface.

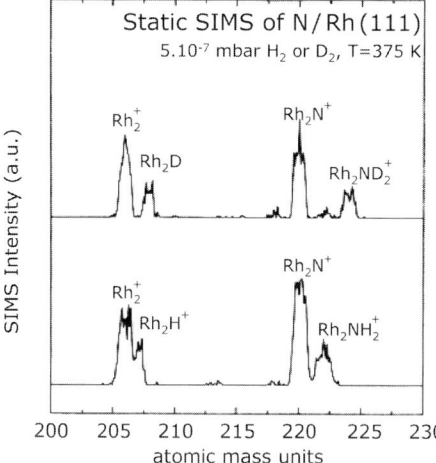

Fig. 4.11 SIMS spectra of a Rh(111) surface covered by 0.1 monolayer of N-atoms, during hydrogenation in H_2 and D_2 showing the presence of H (Rh_2H^+ at 207 amu), N (Rh_2N^+ at 220 amu), and NH_2 ($Rh_2NH_2^+$ at 222 amu) as reaction intermediates. The experiment with D_2 confirms these assignments. The results indicate that the hydrogenation of NH_2 is the rate-determining step in the hydrogenation of N-atoms to NH_3. (From [41]).

4.2.11
Concluding Remarks

In summary, although SIMS is among the less-frequently used techniques in catalysis, its applications in catalysis fall roughly into three categories. In the characterization of technical catalysts, the extreme sensitivity of SIMS enables the detection of promoters or unwanted contaminations, which might act as a poison. In addition, careful comparison of characteristic SIMS patterns with those of reference compounds can be helpful in compound identification. In general, not only charging but also matrix effects mean that the success of SIMS on technical catalysts may be unpredictable, although a number of useful applications do exist. Model systems such as particles on a flat conducting support, polycrystalline foils and single crystals, however, offer excellent opportunities for using SIMS to its full potential.

4.3
Secondary Neutral Mass Spectrometry (SNMS)

In SNMS, sputtered neutrals are post-ionized before they enter the mass spectrometer. In contrast to SIMS, SNMS does not suffer from the matrix effects associated with the ionization probability of sputtered particles, as here the sensitivity for a certain element is mainly determined by its sputter yield. As sputtering is relatively well understood, excellent quantitation of SNMS has been demonstrated. Moreover, SNMS as a technique was developed much later than SIMS, and has not yet been fully exploited for catalysts.

The data listed in Table 4.1 show that sputtered particles from clean metal surfaces are predominantly neutral. As we have seen in Section 4.2, sputter yields fall generally in the range of 1 to 10, whereas the yields of secondary ions – the product of Y and R – are almost all below 0.01 for metallic samples; hence, typically 99% of the sputtered particles from metals are neutral. The major advantage of SNMS is that the signal intensity of secondary neutrals is [compare Eq. (4-1)]:

$$I_s^o = I_p Y(1 - R^+ - R^-)c^{surf}\eta$$
$$\approx I_p Y c^{surf}\eta \qquad (4\text{-}3)$$

where:
- I_s^o is the intensity of secondary neutrals (expressed as a rate in counts per second);
- I_p is the flux of primary ions;
- Y is the sputter yield, the number of atoms ejected per incident ion;
- $R^{\pm}$ is the probability of forming a positive or negative ion;
- c^{surf} is the fractional concentration of the element in the surface layer;
- η is an instrumental factor, equal to the transmission of the mass spectrometer times the ionization efficiency.

As extensive tables of reliable sputter yields are available, SNMS is much more suitable for quantitative work than SIMS. Interestingly, ionic solids also provide significant yields of secondary neutrals, stressing that neutralization processes at the surface are also important for non-metallic samples.

In order to determine their mass, secondary neutrals need to be ionized before entering the mass spectrometer. In the original form of SNMS, developed by Oechsner [43], post ionization was carried out in a plasma. Although this provides efficient ionization (1–10% of the neutrals), it has the disadvantage that molecular species are entirely fragmented. Detection limits are favorable, in the order of parts per millions. Post ionization by electron impact in the entry region of the quadrupole mass spectrometer, as described by Lipinsky et al. [44], leaves molecular clusters largely intact, but the process is much less efficient in producing ions (efficiency of 10^{-4} to 10^{-3}). Consequently, SNMS is much less sensitive than SIMS, and measurement under static conditions is not possible. Typical detection limits of SNMS by electron impact post ionization are in the order of 0.1 to 1%. Only a very few examples of SNMS in catalysis are available [45].

An alternative for the low detection efficiencies of the emitted particles is to ionize them with a UV laser beam, either in a resonant or non-resonant manner [46]. In this way, the ionization efficiency is increased about 1000-fold and the attractive perspective of performing SNMS under static conditions at sensitivities comparable to those of SIMS comes into the reckoning. As yet, however, we are not aware of any applications in the fields of catalyst characterization or in catalytic surface chemistry.

4.4
Ion Scattering: The Collision Process

In ion scattering a beam of ions with energy E_i scatters elastically from atoms in a solid. As the wavelength associated with ions is typically a few orders of magnitude smaller than the distance between atoms in a lattice, a quantum mechanical description is not necessary. In fact, the scattering process is similar to that between billiard balls (although with balls of different mass), and is adequately described by the laws of classical mechanics. The geometry of the collision is shown diagrammatically in Figure 4.12. The energy of the outgoing ion is determined by the laws of energy and momentum conservation, and reveals the mass of the target atom from which it scattered [1, 3]:

$$K_M = \frac{E_f}{E_i} = \left(\frac{(M^2 - M_{ion}^2 \sin^2 \theta)^{1/2} + M_{ion} \cos \theta}{M + M_{ion}} \right)^2 \qquad (4\text{-}4)$$

where:
K_M is the kinematic factor, equal to E_f/E_i;
E_i is the energy of the incoming ion;
E_f is the energy of the scattered ion;

M_{ion} is the atomic mass of the incoming ion;
M is the atomic mass of the scattering atom in the sample;
θ is the scattering angle (as defined in Fig. 4.12).

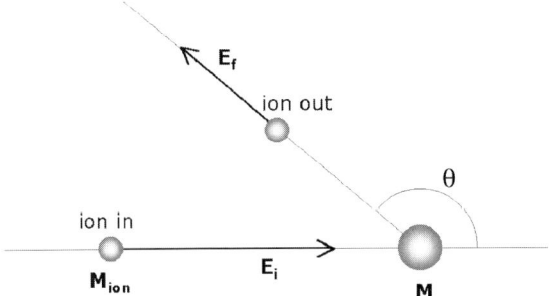

Fig. 4.12 Geometry of an ion-scattering experiment. An incident ion with mass M_{in} and energy E_i collides with a sample atom of mass M_s and loses energy. Its final kinetic energy E_f, after scattering over an angle θ, is a measure for the mass of the target atom.

Two special cases of Eq. (4-4) occur for $\theta = 90°$ and $\theta = 180°$:

$$\frac{E_f}{E_i} = \frac{M - M_{ion}}{M + M_{ion}}, \theta = 90°; \quad \frac{E_f}{E_i} = \left(\frac{M - M_{ion}}{M + M_{ion}}\right)^2, \theta = 180° \quad (4\text{-}5)$$

Thus, for a given choice of incident ions, the kinematic factor depends only on the scattering angle θ and on the mass of the target atom, M (Fig. 4.13), and for a given scattering geometry the energy spectrum represents a mass spectrum. In agreement with the expression for the energy, Figure 4.13 shows that the primary ions lose less energy to heavy atoms than to light atoms. Optimum discrimina-

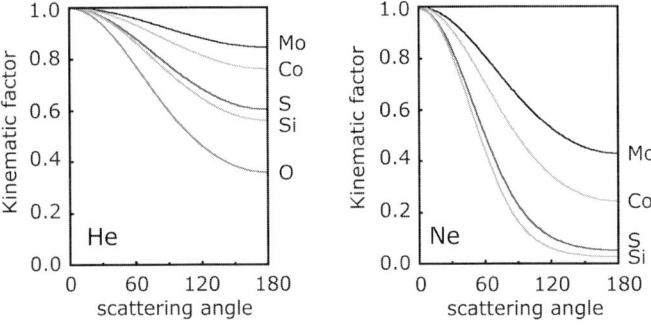

Fig. 4.13 Kinematic factors E_f/E_i for He and Ne scattering as a function of scattering angle for the elements present in a hydrotreating catalyst, calculated with Eq. (4-4).

tion between masses is obtained at angles in the range of 150 to 180°, and for incident ions that are not too light, as the different kinematic factors for He and Ne scattering in Figure 4.13 illustrate. The energy resolution depends of course on the type of detector that is used, but for helium scattering one can take as a rule of thumb that elements with atomic mass up to about 40 can in practice always be distinguished from elements which differ by as little as one unit in atomic mass.

The intensity of the scattered ions is determined by the scattering cross-section, and for this we need to distinguish between high- and low-energy ion scattering. Although low-energy ion scattering is more relevant for catalyst characterization, we start with a discussion of high-energy ion scattering, because the theory involved in the latter is somewhat easier to explain.

4.5
Rutherford Backscattering Spectrometry (RBS)

In RBS, a mono energetic beam of typically 2 to 4 MeV He^+ ions is directed onto a sample, where the ions collide elastically with atoms in a layer of a few microns thickness. Sputtering as in SIMS and LEIS plays no role, as can be inferred from Figure 4.4, which shows that the sputter yield of 1 MeV ions is close to zero. Ions scattering off surface atoms have a final energy E_f [as given by Eq. (4-4)], but ions which collide more deeply in the sample return with kinetic energies lower than E_f, due to inelastic energy losses either on their entry, or after collision. As a result, the RBS spectrum of an elemental bulk sample resembles that shown in Figure 4.14b; the edge energy equals $K_M \cdot E_i$, where K_M is the kinematic factor of the element and E_i the energy of the incident ion.

A different situation arises if the direction of the incoming beam is parallel to the lattice planes inside a single crystal. In the ideal case, scattering occurs only at the surface and the remaining ions channel deeply into the crystal. The spectrum shown in Figure 4.14a consists of a single peak at energy $E_f = K_M \cdot E_i$. The intensity at lower energies is a measure for the number of defects in the crystal. Channeling is an important tool for structure investigations in single crystals, and the interested reader is referred to the excellent textbook of Feldman and Mayer [1].

Figure 4.14c presents the case of interest for planar model catalysts. If the sample consists of small particles of a heavy element on top of a lighter substrate, as in a supported catalyst, the RBS spectrum shows a single peak for the heavy element well separated from the continuum spectrum of the substrate. The area of the peak, as well as the height of the continuum, is a measure for the number of atoms hit by the incident beam.

The intensity of a peak in RBS is determined by the cross-section σ for scattering. At MeV energies, the helium ion penetrates deeply into the atom and approaches the nucleus of the target atom to within 10^{-4} nm – that is, well within

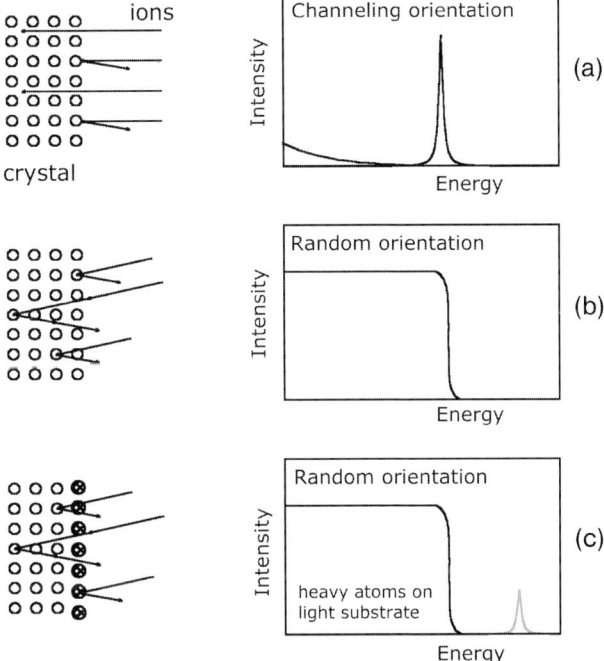

Fig. 4.14 Schematic representation of Rutherford backscattering (RBS). (a) The incident ions are directed such that they either scatter back from surface atoms or channel deeply into the crystal. (b) The ions scatter back from target atoms throughout the outer micrometers and suffer inelastic losses, causing the energy of the backscattered ions to tail to zero. (c) Scattering from the heavy outer layer gives a sharp peak separated from the spectrum of the substrate as in (b).

the radius of the K-electron shell. This means that the scattering event depends only on the Coulomb repulsion between the two nuclei, whereas screening by the electrons (which is important in LEIS) plays no role. Thus, the scattering cross-section is a function of charge on the nucleus of the target atom and of the energy of the incoming helium ion. In addition, σ depends on the scattering geometry:

$$\sigma = \frac{Z_{He}^2 Z^2 e^4}{4 E_i^2 \sin^4 \frac{\theta}{2}} \tag{4-6}$$

where:
σ is the scattering cross-section;
Z is the atomic number, equal to the number of protons of an element;

- e is the unit charge (Ze equals the charge of the nucleus);
- E_i is the energy of the incoming He ions;
- θ is the scattering angle (see Fig. 4.12).

One recognizes in Eq. (4-6) the Coulomb interaction energy between a helium nucleus with charge $Z_{He} \cdot e$ and a target nucleus with charge $Z \cdot e$. The expression shows that optimum signal intensity is obtained with scattering angles close to 180° and low energies. The Z^2-dependence of the cross-section makes RBS much more sensitive for heavy than for light elements.

Deviations from Rutherford behavior occur at higher energies E_i, when the helium ion comes so close to the target nucleus that nuclear reactions begin to take place. A well-known example is that of 3.04 MeV He ions on oxygen, where the scattering cross-section is an order of magnitude higher than the Rutherford value. This reaction can be used to increase the sensitivity for the detection of oxygen.

Rutherford backscattering is conducted using helium ions with an energy of a few MeV, and consequently a Van der Graaf accelerator or a cyclotron is needed to produce the ions. The detector used for measuring the high energies of the back-scattered ions is a simple solid-state device, consisting of a silicon disc covered by a gold film. It is based on the collection of electron-hole pairs created by the incident ion, the number of pairs being a function of particle energy. The energy resolution is typically between 10 and 20 keV, which means that 2 MeV helium easily distinguishes between all isotopes up to atomic mass 40, but cannot for example discriminate between Ir and Pt.

We illustrate the use of RBS with a study on the sulfidation of molybdenum hydrodesulfurization catalysts supported on a thin layer of SiO_2 on silicon [31]. As explained in connection with the SIMS experiments on this model system (see Fig. 4.8), the catalyst is sulfided by treating the oxidic MoO_3/SiO_2 precursor in a mixture of H_2S and H_2. RBS is used to determine the concentrations of Mo and S.

The lower spectrum in Figure 4.15 is that of the fresh $MoO_3/SiO_2/Si(100)$ model catalyst (a part is shown enlarged). The peak at 3.4 MeV has a kinematic factor of about 0.85. As the scattering angle is 170°, the reader can use Figure 4.13 to verify that this peak belongs to Mo. In the same way, one may check that the continuum below $E_f = 2.3$ MeV is due to ^{28}Si. Note that small peaks due to the Si isotopes at 29 and 30 amu are also just visible. The structure around 2 MeV is caused by non-Rutherford scattering behavior. The Mo intensity corresponds to a loading of $1.2 \pm 0.1 \cdot 10^{15}$ atoms cm^{-2}. It would be difficult to obtain such an accurate number from any other technique in this concentration range.

Two spectra of sulfided catalysts in Figure 4.15 show that sulfur can also be detected, albeit at lower sensitivity, due to the Z^2-dependence of the cross-section in Eq. (4-6). In Figure 4.16, the atomic S/Mo ratio is plotted against the sulfidation temperature; this indicates that sulfur uptake by MoO_3 is already significant at room temperature, and increases to the expected S/Mo ratio of 2 above 100 °C. The combination of these results with SIMS and XPS data has led to a detailed

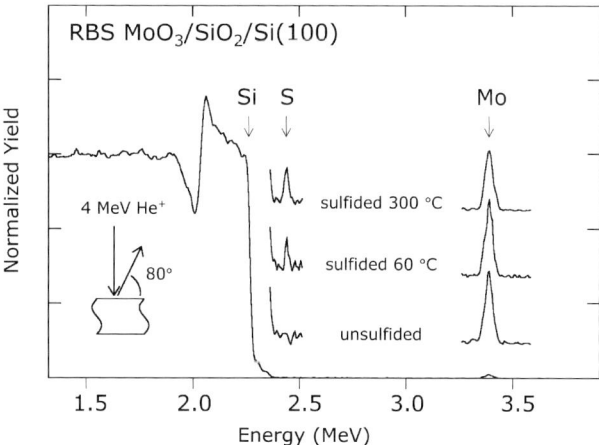

Fig. 4.15 RBS spectra of a MoO$_3$ model catalyst supported on a flat SiO$_2$/Si(100) model support, before, and after sulfidation in a mixture of H$_2$S and H$_2$ at room temperature, and at 300 °C. (Adapted from [31]).

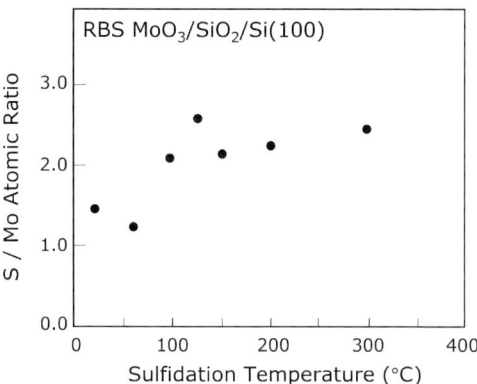

Fig. 4.16 The atomic S/Mo ratio in the MoO$_3$ model catalyst of Figure 4.14, as determined with RBS, plotted versus sulfidation temperature. (From [31]).

mechanism for the sulfidation of silica-supported molybdenum catalysts [31] (see also Chapter 9).

Other successful applications of RBS on flat supported model catalysts include systems such as Rh/Al$_2$O$_3$/Al [47, 48] and ZrO$_2$ [49], PtCo [50] and Cr on SiO$_2$/Si(100) [51]. The reason why RBS is so effective with these systems is that they consist of heavy elements on top of a lighter support, with the fortunate consequence that peaks due to the elements of interest appear on a background of almost zero.

4.6
Low-Energy Ion Scattering (LEIS)

Low-energy ion scattering, also referred to as ion scattering spectroscopy (ISS), is the surface-sensitive counterpart of the Rutherford backscattering technique [1, 3, 52]. In LEIS, one uses a beam of noble gas ions with an energy between 0.1 and 10 keV. The kinematics of the scattering process is described by Eq. (4-4), as illustrated in Figures 4.12 and 4.13. In contrast to RBS, which probes a layer of several micrometers, LEIS is exclusively sensitive for the outer surface layers. This may at first sight be surprising, because keV ions penetrate a solid to a considerable extent. The reason for the favorable surface sensitivity of the technique is twofold:

- The incident ion is neutralized inside the solid. For example, 1 keV He^+ ions have a neutralization probability of about 99% in passing through one layer of substrate atoms. Hence, the majority of ions that reach the detector must have scattered off the outermost layer (or were reionized upon leaving the solid, as will be seen later).

- The detection of low-energy particles requires an electrostatic analyzer. As a consequence, only ions can be detected and backscattered neutrals are not measured, unlike in RBS.

Thus, in order to understand ion scattering with low-energy ions, we need to understand not only the scattering process but also the probabilities of neutralization and reionization.

Here, we discuss briefly the factors that determine the intensity of the scattered ions. During collision, a low-energy ion does not penetrate the target atom as deeply as in RBS. As a consequence, the ion feels the attenuated repulsion by the positive nucleus of the target atom, because the electrons screen it. In fact, in a head-on collision with Cu, a He^+ ion would need to have about 100 keV energy to penetrate within the inner electron shell (the K or 1s shell). An approximately correct potential for the interaction is the following modified Coulomb potential [1]:

$$V(r) = \frac{Z_{ion}eZe}{r} \cdot \frac{0.885 a_o}{2r(\sqrt{Z_{ion}} + \sqrt{Z})^{2/3}}$$

$$= V_{Coulomb}(r) \cdot \text{screening function} \quad (4\text{-}7)$$

where a_o is the Bohr radius of the atom. The difference between high- and low-energy ion scattering is that RBS is governed by the *pure* Coulomb interaction between the nuclei, as expressed by $V_{Coulomb}(r)$, whereas low-energy ions scatter of a *screened* Coulomb potential. The screening function in Eq. (4-7) becomes less useful at small *r*-values, as it does not yield the true Coulomb potential inside the K-shell, and for this range the Moliere approximation should be used [1]. Screened potentials as in Eq. (4-7) form the basis for calculating the trajectories of scattered ions.

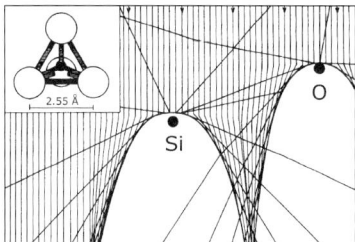

Fig. 4.17 Trajectory calculations for the scattering of a parallel beam of 1 keV He$^+$ ions by SiO$_2$; the inset shows a SiO$_4$ tetrahedron. (From [52]).

The simulated scattering process of a parallel He$^+$ beam by a SiO$_2$ surface shown in Figure 4.17 illustrates two important points [52]. The first point is that the interaction of *keV* ions with matter is a highly local event, determined by separate scattering centers. The second point is that shadow cones develop behind the scattering atoms in the target, which may block neighbors from view. If the sample is a single crystal, one may use this property to determine the position of adsorbate atoms on the surface, and several such applications are discussed by Feldman and Mayer [1].

A consequence of the screening of the nucleus by electrons is that the cross-section for low-energy ion scattering varies less steeply with the atomic number than in RBS, where the intensity depends on the atomic number squared. Another difference from RBS is that the scattering intensity is determined not only by the cross-section but also by the probability that an ion is neutralized.

4.6.1
Neutralization

When an ion closely approaches the surface, an electron from the valence band may neutralize the ion by jumping to the empty state of the ion. The energy gained by this process equals the ionization potential of the primary ion minus the binding energy of the electron in the target. The target atom can use this energy to emit another electron from the valence band. The complete process is a two-center Auger transition, which is believed to be the main route for the neutralization of an ion near a surface. It is not difficult to see that the maximum energy the Auger electron may have equals $(I - 2\varphi)$, where φ is the work function. The minimum energy of the Auger electron arises if the two electrons both come from the bottom of the band. However, this minimum cannot be observed as it is obscured by secondary electrons that suffered energy loss in inelastic collisions.

A second type of neutralization occurs through a resonant process, in which an electron from the sample tunnels to the empty state of the ion, which should then be at about the same energy. Resonance neutralization becomes likely if the electron affinity of the ion is somewhat larger than the work function of the sample, or when the ion has an unfilled core level with approximately the same energy as an occupied level in the target atom. The latter takes place when He^+ ions approach indium, lead, or bismuth atoms. The inverse process can lead to re-ionization.

The probability that neutralization takes place depends on the energy of the ion, simply because a slow ion spends a longer time in the vicinity of an atom. The maximum distances at which neutralization processes are thought to occur are on the order of 0.2 nm for Auger and 0.5 nm for resonance neutralization.

The LEIS technique owes its excellent surface sensitivity to the high neutralization probability of the rare gases. The fraction of He^+ ions that survives a single collision without being neutralized is only between 10^{-4} and 10^{-2}. This implies automatically that the probability of a He^+ ion to penetrate the surface, scatter off deeper atoms, and return as an ion is practically zero. However, a finite probability exists that the backscattered neutral He atom is ionized upon leaving the sample, and this is the reason that a LEIS spectrum still contains some information on the state of a sample below its surface.

Charge exchange processes, as discussed above, are important for a good understanding of LEIS, and also of SIMS. Unfortunately, the subject is still not yet completely understood, and this has to date impeded quantitative analysis with both techniques. Nevertheless, a quantitative interpretation of LEIS spectra is clearly possible if one uses appropriate calibration standards.

4.6.2
Applications of LEIS in Catalysis

The fact that LEIS provides quantitative information on the outer layer composition of multicomponent materials makes this technique an extremely powerful tool for the characterization of catalysts. Figure 4.18 shows the LEIS spectrum of an alumina-supported copper catalyst, taken with an incident beam of 3 keV $^4He^+$ ions. Peaks due to Cu, Al and O and a fluorine impurity are readily recognized. The high intensity between about 40 and 250 eV is due to secondary (sputtered) ions. The fact that this peak starts at about 40 eV indicates that the sample has charged positively. Of course, the energy scale must be corrected for this charge shift before kinematic factors E_f/E_i are determined.

An interesting feature of LEIS is that atoms below the surface can be distinguished from those in the outer layer by a broadening of the peaks to lower energies. This occurs because an ion scattered in the second layer has a considerable chance of losing an additional amount of energy on its way back through the first layer. (Note the high step in the background on the low-energy side of the Al peak of the support.) In addition, concentration–depth profiles of the constituents can be investigated by using the sputtering action of the ion beam.

Several authors believe that comparing LEIS spectra of one sample measured at different primary ion energies may also provide a means of determining which atoms are in the outer layer, and which are in the subsurface region [53–55]. Because these arguments are based on a rather persistent misunderstanding, the

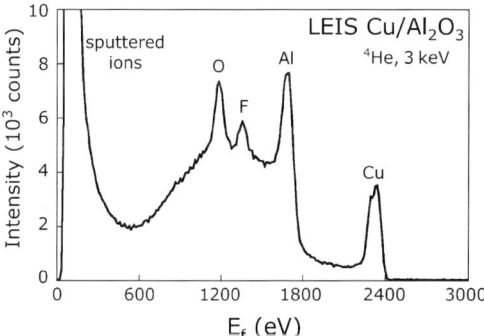

Fig. 4.18 The low-energy ion scattering (LEIS) spectrum of a Cu/Al$_2$O$_3$ catalyst illustrates that ions lose more energy in collisions with light elements than with heavy elements. Note the step in the background at the low kinetic energy side of the peaks. The high peak at low energy is due to sputtered ions. The low-energy cut-off of about 40 eV is indicative of a positively charged sample. (Figure courtesy of J.P. Jacobs and H.H. Brongersma, Eindhoven).

matter is discussed in some detail here. The type of measurement is exemplified by the spectra of SiO$_2$ at two energies (see Fig. 4.19) [56]. The spectra were scaled on the oxygen signal to stress that the Si peak is significantly less intense at a primary ion energy of 1 keV. As the surface of SiO$_2$ is terminated by O^{2-} ions, with Si^{4+} ions residing below the surface, and moreover, because the ionic radius of O^{2-} (0.14 nm) is much larger than that of Si^{4+} (0.04 nm), the interpretation of this phenomenon has long been that the low Si/O intensity ratio is due to the shielding of the inward Si by the outward O. In fact, several oxides such as Al$_2$O$_3$, TiO$_2$, and ZrO$_2$ show the same behavior, and shielding has been invoked in all cases to the explain their spectra [53–55].

Van Leerdam and Brongersma [52, 56] proved that the interpretation is not correct, however. In spite of the larger ionic radius, the scattering cross-section of oxygen is smaller than that of Si, owing to the lower nuclear charge on O. Also, the trajectory calculations in Figure 4.17 indicate that Si is not blocked from view to a large extent for scattering angles above 120°. The low intensity of Si in the spectrum of SiO$_2$ (and also of Al, Ti, and Zr in LEIS spectra of Al$_2$O$_3$, TiO$_2$, and ZrO$_2$, respectively) appears to be an intrinsic property of these elements, and is attributed to a higher neutralization probability of the incident ion at low kinetic energy.

LEIS is most often carried out with a beam of helium ions, but when a catalyst contains heavier elements, it may be advantageous to use neon ions to distinguish better between the elements (see Fig. 4.13). Of course, elements with a mass smaller than neon cannot be detected in that case. Figure 4.20 illustrates this with He- and Ne-LEIS spectra of a multicomponent oxide catalysts containing vanadium, molybdenum, niobium and tellurium. These mixtures are of inter-

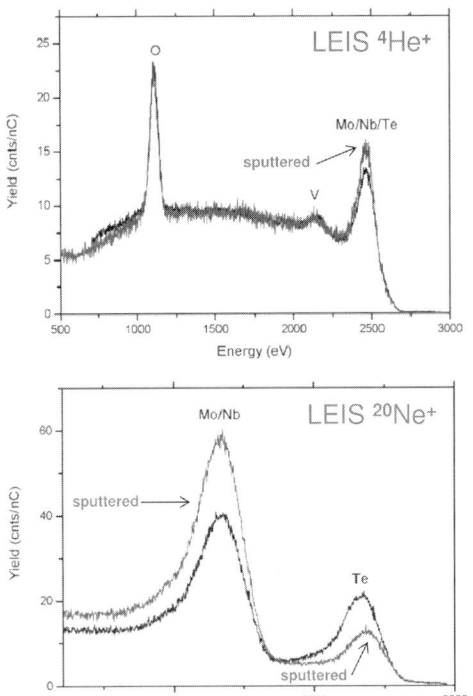

Z		M
8	O	16.00
23	V	50.94
41	Nb	92.90
42	Mo	95.94
52	Te	127.60

Fig. 4.19 LEIS spectrum of SiO_2 taken with He ions of 1 and 3 keV, illustrating the lower sensitivity of LEIS for Si at low incident ion energies. This effect has erroneously been attributed to the shielding of silicon by oxygen ions, but is in fact an intrinsic property of scattering by silicon. (From [57]).

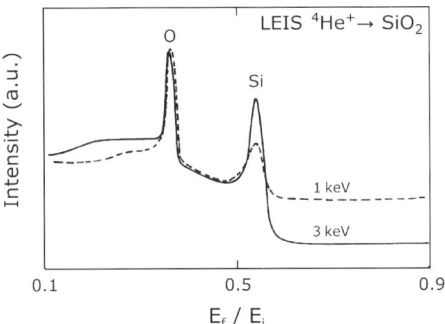

Fig. 4.20 Helium and neon LEIS spectra of a mixed metal oxide containing V, Mo, Nb, and Te, illustrating the better propensity of neon scattering to distinguish elements of higher mass. Light sputtering leads to decrease of the Te peak, indicating that this element resides in the outer layer of the catalyst. (Adapted from [56]).

est for the selective oxidation of propane to acrylic acid and acrylonitrile [57]. The He^+ spectra yield one peak for Mo, Nb and Te, but the Ne^+ spectra easily distinguish Te from Mo and Nb. The latter two are neighbors in the Periodic Table, and their signals remain unresolved. Figure 4.20 also illustrates how sputtering can be used to reveal additional information. Following removal of the equivalent of one monolayer, the Te-signal has decreased, while the Mo–Nb peak becomes stronger, indicating that the tellurium promoter resides mainly on top of the oxides.

LEIS has been applied to study the surface composition of Co–Mo and Ni–Mo hydrodesulfurization catalysts [58–60], Fe-based Fischer–Tropsch [61], automotive exhaust catalysts [62], ammonia synthesis catalysts [63], and of model systems such as Pt evaporated on TiO_2 [64] and Au and Pd on thin SiO_2 layers [65]. In this respect, the review of Horrell and Cocke [66] describes several applications.

In conclusion, LEIS is extremely surface-sensitive, and suitable for determining the outer layer compositions of catalysts and alloys, provided that proper calibration can be achieved. The shape of the background contains information over the vertical distribution of elements in the surface region.

References

1 L.C. Feldman and J.W. Mayer, *Fundamentals of Surface and Thin Film Analysis*. North Holland, New York, 1986.

2 A. Benninghoven, F.G. Rüdenauer, and H.W. Werner, *Secondary Ion Mass Spectrometry, Basic Concepts, Instrumental Aspects, Applications and Trends*. Wiley, New York, 1987.

3 A.W. Czanderna and D.M. Hercules (Eds.), *Ion Spectroscopies for Surface Analysis*. Plenum, New York, 1991

4 J.C. Vickerman, A. Brown, and N.M. Reed (Eds.), *Secondary Ion Mass Spectrometry, Principles and Applications*. Clarendon, Oxford, 1989.

5 J.C. Vickerman and D. Briggs (Eds.), *ToF-SIMS, Surface Analysis by Mass Spectrometry*. IM Publications and Surface Spectra Ltd, Chichester and Manchester, 2001.

6 D. Briggs, A. Brown, and J.C. Vickerman, *Handbook of Static Secondary Ion Mass Spectrometry*. Wiley, Chichester, 1989.

7 P. Sigmund, in: *Sputtering by Particle Bombardment I*, R. Behrisch (Ed.), *Topics in Applied Physics*. Vol. 47, Springer, Berlin, 1981 p. 9; P. Sigmund, *Phys. Rev.* **184** (1969) 383.

8 H.E. Roosendaal, in: *Sputtering by Particle Bombardment I*, R. Behrisch (Ed.), *Topics in Applied Physics*. Vol. 47, Springer, Berlin, 1981, p. 219.

9 H. Oechsner, *Z. Phys.* **261** (1973) 37.

10 K. Wittmaack, *Surface Sci.* **90** (1979) 557.

11 E. Dennis and R.J. MacDonald, *Radiat. Eff.* **13** (1972) 243.

12 J.K. Nørskov and B.I. Lundqvist, *Phys. Rev. B* **19** (1979) 5661.

13 H. Oechsner, in: *Secondary Ion Mass Spectrometry III*, A. Benninghoven, J. Giber, J. Laszlo, M. Riedel, and H.W. Werner (Eds.), Springer, Berlin, 1982, p. 106.

14 H.J. Borg and J.W. Niemantsverdriet, in: *Catalysis: A Specialist Periodical Report, Volume 11*, J.J. Spivey and A.K. Agarwal, Royal Society of Chemistry, Cambridge, 1994, p. 1.

15 M.J.P. Hopstaken, R. Linke, W.J.H. van Gennip, and J.W. Niemantsverdriet, in: *ToF-SIMS, Surface Analysis by Mass Spectrometry*. J.C. Vickerman and D. Briggs (Eds.). IM Publications and

Surface Spectra Ltd, Chichester and Manchester, 2001, p. 697.

16 A.K. Coverdale, P.F. Dearing, and A. Ellison, *J. Chem. Soc., Chem. Commun.* (1983) 567.

17 A.C.Q.M. Meijers, A.M. de Jong, L.M.P. van Gruijthuijsen, and J.W. Niemantsverdriet, *Appl. Catal.* **70** (1991) 53.

18 C. Plog, W. Gerhard, O. Inacker, E. Hamer, and M. Seidl, *Proceedings, 8th International Congress of Catalysis (Berlin 1984)*, Vol. II. Verlag Chemie, Weinheim, 1985, p. 581.

19 B.C. Gates, J.R. Katzer, and G.C.A. Schuit, *Chemistry of Catalytic Processes*. McGraw-Hill, New York, 1979.

20 I. Aso, T. Amamoto, N. Yamazoe, and T. Seiyama, *Chem. Lett. (Japan)* (1980) 1435.

21 J.C. Vickerman, A. Oakes, and H. Gamble, *Surface Interf. Anal.* **29** (2000) 349.

22 H. Gnaser, W. Bock, E. Rowlett, Y. Men, C. Ziegler, R. Zapf, and V. Hessel, *Nucl. Instr. Methods Phys. Res. B* **219–220** (2004) 880.

23 J. Grams, J. Goralski, and T. Paryjczak, *Surface Sci.* **549** (2004) L21.

24 J. Rynkowski, D. Rajski, I. Szyszka, and J.R. Grzechowiak, *Catal. Today* **90** (2004) 159.

25 C. Sellmer, R. Prins, and N. Kruse, *Catal. Lett.* **47** (1997) 83.

26 M. Rebholz and N. Kruse, *J. Chem. Phys.* **95** (1991) 7745.

27 P.L.J. Gunter, J.W. Niemantsverdriet, F.H. Ribeiro, and G.A. Somorjai, *Catal. Rev. – Sci. Eng.* **39** (1997) 77.

28 H.J. Borg, L.C.A. van den Oetelaar, and J.W. Niemantsverdriet, *Catal. Lett.* **17** (1993) 81.

29 P.C. Thüne, R. Linke, W.J.H. van Gennip, A.M. de Jong, and J.W. Niemantsverdriet, *J. Phys. Chem. B* **105** (2001) 3073.

30 F. Aubriet, J.-F. Muller, C. Poleunis, P. Bertrand, P.G. di Croce, and P. Grange, *J. Am. Soc. Mass Spectrom.* **17** (2006) 406.

31 A.M. de Jong, H.J. Borg, V.G.F.M. Soudant, V.H.J. de Beer, J.A.R. van Veen, and J.W. Niemantsverdriet, *J. Phys. Chem.* **97** (1993) 6477.

32 G.L. Ott, W.N. Delgass, N. Winograd, and W.E. Baitinger, *J. Catal.* **56** (1979) 174.

33 J.W. Niemantsverdriet and A.D. van Langeveld, *J. Vac. Sci. Technol.* **A6** (1988) 1134.

34 X.-Y. Zhu and J.M. White, *J. Phys. Chem.* **92** (1988) 3970.

35 N.M. Reed and J.C. Vickerman, in: *Fundamental Aspects of Heterogeneous Catalysis Studied by Particle Beams*, H.H. Brongersma and R.A. van Santen (Eds.), NATO ASI Series B, Physics Vol. 265, Plenum, New York, 1991, p. 357.

36 R.M. van Hardeveld, H.J. Borg, and J.W. Niemantsverdriet, *J. Mol. Catal. A Chemical* **131** (1998) 199.

37 H.J. Borg, J.F.C.-J.M. Reijerse, R.A. van Santen, and J.W. Niemantsverdriet, *J. Chem. Phys.* **101** (1994) 10052.

38 M.J.P. Hopstaken and J.W. Niemantsverdriet, *J. Phys. Chem. B* **104** (2000) 3058.

39 J.R. Creighton, K.M. Ogle, and J.M. White, *Surface Sci.* **138** (1984) L137; K.M. Ogle and J.M. White, *Surface Sci.* **165** (1986) 234.

40 I. Chorkendorff and J.W. Niemantsverdriet, *Concepts of Modern Catalysis and Kinetics*. Wiley-VCH, Weinheim, 2003.

41 R.M. van Hardeveld, R.A. van Santen, and J.W. Niemantsverdriet, *J. Phys. Chem. B* **101** (1997) 998.

42 R.M. van Hardeveld, R.A. van Santen, and J.W. Niemantsverdriet, *Surface Sci.* **369** (1996) 23.

43 H. Oechsner, in: *Thin Film and Depth Profile Analysis*, H. Oechsner (Ed.), Springer, Berlin, 1984, p. 63.

44 D. Lipinsky, R. Jede, O. Ganschow, and A. Benninghoven, *J. Vac. Sci. Technol.* **A3** (1985) 2007.

45 C. Schild, A. Wokaun, and A. Baiker, *Surface Sci.* **269/270** (1992) 520.

46 C.H. Becker, in: *Ion Spectroscopies for Surface Analysis*, A.W. Czanderna and D.M. Hercules (Eds.), Plenum, New York, 1991, p. 273.

47 H.J. Borg, L.C.A. van den Oetelaar, L.J. van IJzendoorn, and J.W. Niemantsverdriet, *J. Vac. Sci. Technol.* **A10** (1992) 2737.

48 C. Linsmeier, H. Knözinger, and E. Taglauer, *Surface Sci.* **275** (1992) 101.

49 A.M. de Jong, L.M. Eshelman, L.J. van IJzendoorn, and J.W. Niemantsverdriet, *Surface Interf. Anal.* **18** (1992) 412.

50 A. Borgna, B.G. Anderson, A.M. Saib, H. Bluhm, M. Hävecker, A. Knop-Gericke, A.E.T. Kuiper, Y. Tamminga, and J.W. Niemantsverdriet, *J. Phys. Chem. B* **108** (2004) 17905.

51 E.M.E. van Kimmenade, A.E.T. Kuiper, Y. Tamminga, P.C. Thüne, and J.W. Niemantsverdriet, *J. Catal.* **223** (2004) 134.

52 H.H. Brongersma and G.C. van Leerdam, in: *Fundamental Aspects of Heterogeneous Catalysis Studied by Particle Beams*, H.H. Brongersma and R.A. van Santen (Eds.), NATO ASI Series B, Physics Vol. 265, Plenum, New York, 1991, p. 283.

53 R.C. McCune, *J. Vac. Sci. Technol.* **16** (1979) 1569; R.C. McCune, *J. Vac. Sci. Technol.* **18** (1981) 700.

54 G.C. Nelson, *J. Vac. Sci. Technol.* **15** (1978) 702.

55 P.J. Martin and R.L.P. Netterfield, *Surface Interf. Anal.* **10** (1987) 13.

56 G.C. van Leerdam, Ph.D. Thesis, Eindhoven University of Technology, 1991.

57 V.V. Guliants, R. Bhandari, A.R. Hughett, S. Bhatt, B.D. Schuler, H.H. Brongersma, A. Knoester, A.M. Gaffney, and S. Han, *J. Phys. Chem. B* **110** (2006) 6129.

58 F. Delannay, E.N. Haeussler, and B. Delmon, *J. Catal.* **66** (1980) 469.

59 S. Kasztelan, J. Grimblot, and J.P. Bonnelle, *J. Phys. Chem.* **91** (1987) 1503.

60 L.E. Makovsky, J.M. Stencel, F.R. Brown, R. Tischer, and S.S. Pollack, *J. Catal.* **89** (1984) 334.

61 U. Lindner and H. Papp, *Appl. Surf. Sci.* **32** (1988) 75.

62 J.M.A. Harmsen, W.P.A. Jansen, J.H.B.J. Hoebink, J.C. Schouten, and H.H. Brongersma, *Catal. Lett.* **74** (2001) 133.

63 S.M. Davis, *Catal. Lett.* **1** (1988) 85.

64 G.B. Hoflund, A.L. Grocan, Jr., and D.A. Asbury, *J. Catal.* **109** (1988) 226.

65 K. Luo, T. Wei, C.-W. Yi, S. Axnanda, and D.W. Goodman, *J. Phys. Chem. B* **109** (2005) 23517.

66 B.A. Horrell and D.L. Cocke, *Catal. Rev. – Sci. Eng.* **29** (1987) 447.

5
Mössbauer Spectroscopy

Keywords

Mössbauer absorption spectroscopy (MAS)
Mössbauer emission spectroscopy (MES)

5.1
Introduction

Although Mössbauer spectroscopy is a technique that is relatively little used in catalysis, it has yielded very useful information on a number of important catalysts, such as the iron catalyst for Fischer–Tropsch and ammonia synthesis, and the cobalt–molybdenum catalyst for hydrodesulfurization reactions. The technique is limited to those elements that exhibit the Mössbauer effect. Iron, tin, iridium, ruthenium, antimony, platinum and gold are those relevant for catalysis. Through the Mössbauer effect in iron, one can also obtain information on the state of cobalt. Mössbauer spectroscopy provides valuable information on oxidation states, magnetic fields, lattice symmetry and lattice vibrations. Several books on Mössbauer spectroscopy [1–3] and reviews on the application of the technique on catalysts [4–9] are available, with Millet's [9] review being the most recent at the time of writing this revision.

Mössbauer spectroscopy is a nuclear technique, but why should such a technique be useful for the study of catalysts? The answer is simple: the nucleus, being at the heart of the atom, feels precisely what the state of the atom is. Mössbauer spectroscopy analyzes the energy levels of the nucleus with extremely high accuracy, and in this way it reveals for example what the oxidation state of the atom is, or how large the magnetic field is at the nucleus. In this way, we can determine in straightforward manner the compound to which the atom belongs.

The great advantage of Mössbauer spectroscopy for catalyst research is that it uses γ-radiation of high penetrating power, such that the technique can be applied *in situ*. An economic advantage is that the technique is relatively inexpensive, with equipment costs being about a factor of ten less than for electron microscopy or photoelectron spectroscopy.

Spectroscopy in Catalysis: An Introduction, Third Edition
J. W. Niemantsverdriet
Copyright © 2007 WILEY-VCH Verlag GmbH & Co. KGaA, Weinheim
ISBN: 978-3-527-31651-9

5 Mössbauer Spectroscopy

In this chapter, we will first describe what the Mössbauer effect is, and then explain why it can only be observed in the solid state and in a limited number of elements. Next, we discuss the so-called hyperfine interactions between the nucleus and its environment, which make the technique so informative. After some remarks on spectral interpretation, we will pass systematically through a number of examples which illustrate the type of information that Mössbauer spectroscopy yields about catalysts.

5.2
The Mössbauer Effect

Let us consider the following experiment (see Fig. 5.1). Suppose we have two identical atoms, one with its nucleus in the excited state, and the other with its nucleus in the ground state. The excited nucleus decays to the ground state by emitting a photon with energy of typically some 10 to 100 keV. This photon falls on the nucleus of the second atom, which is in the ground state. The question is,

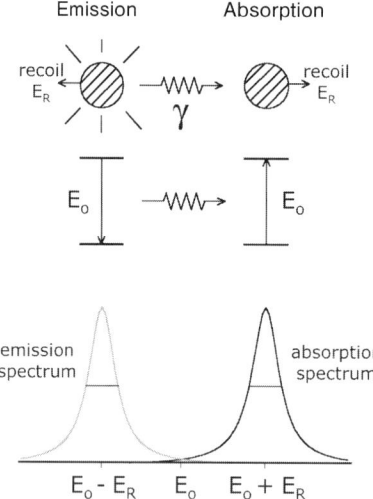

Fig. 5.1 Resonant absorption of γ-radiation by a nucleus can only take place in the solid state due to recoil effects. The excited nucleus of a free atom emits a γ-photon with an energy $E_o - E_R$, whereas the nucleus in the ground state of a free atom can only absorb a photon if it has an energy equal to $E_o + E_R$. As the linewidth of nuclear transitions is extremely narrow, the emission spectrum does not overlap with the absorption spectrum. In a solid, a considerable fraction of events occurs recoil free ($E_R = 0$), and here the emission spectrum overlaps completely with the absorption spectrum (provided that the source and absorber have the same chemical environment).

will the second nucleus absorb the photon to achieve an excited state? The answer is no, because *recoil energy* is involved.

We analyze the experiment in terms of an energy balance. Suppose the energy difference between the excited state and the ground state of the nucleus is E_o. When the excited nucleus decays and emits a photon, it will recoil – just as a gun from which a bullet is fired. This recoil energy, E_R, can easily be calculated from mechanics, and equals $E_o/2mc^2$, where m is the mass of the nucleus and c the velocity of light. Thus, the energy of the emitted photon, E_γ equals $E_o - E_R$. Similarly, if the photon hits the second nucleus, it will recoil also with an energy E_R, and the energy available to excite the nucleus is only $E_o - 2E_R$, whereas an energy equal to E_o would be needed. For illustration, if the nuclei are those of iron, $E_o = 14.4$ keV and E_R is on the order of a few meV. The natural line width of the transition, however, is much smaller, in the case of iron as small as $4.6 \cdot 10^{-9}$ eV. Hence, the experiment to absorb the emitted γ-photon fails, due to recoil effects.

In a solid lattice, the atom cannot recoil as if it were free. Here, the recoil energy is taken up by vibrations of the lattice as a whole. Lattice vibrations are quantized, just as the vibrations or rotations of a molecule are; the quantum is called a *phonon*. If the recoil energy due to emission of the γ-quantum is larger than the phonon energy, the lattice will simply take up the recoil energy in portions equal to the phonon energy and our Mössbauer experiment fails again. If – and this is the situation of interest to us – the recoil energy is smaller than the energy of the lattice vibrations, a situation arises which can only occur in quantum mechanics: a number of emission or absorption events takes place without exchange of recoil energy. The fraction of recoil-free events follows from the correspondence principle which relates quantum mechanics to every day life: the average value of the recoil energy over a large number of events must be equal to E_R. The important thing to realize is that the recoil energy is taken up in portions which are larger than E_R, and thus there must also be emission and absorption events for which the recoil energy is zero.

The conclusion of all this is the following. If we place our two atoms in a lattice and perform the experiment under conditions where recoil energy of the photon emission and absorption are significantly smaller than the energy of the lattice vibrations, a fraction of the photons emitted by the source nucleus will be absorbed by the nucleus in the absorber. This is the Mössbauer effect, named after Rudolf L. Mössbauer, who discovered it in 1957 and subsequently received the Nobel Prize in 1961 [10].

The intensity of the Mössbauer effect is determined by the recoil-free fraction, or f factor, which can be considered as a kind of efficiency. It is determined by the lattice vibrations of the solid to which the nucleus belongs, the mass of the nucleus, and the photon energy, E_o and is given by:

$$f = e^{-k_\gamma^2 \langle x^2 \rangle} \tag{5-1}$$

which, if we express $\langle x^2 \rangle$ in the Debye model, becomes:

$$f = \exp\left[-\frac{3}{2}\frac{E_R}{k\theta_D}\left(1 + 4\frac{T^2}{\theta_D^2}\int_0^{\theta_D/T}\frac{x\,dx}{e^x - 1}\right)\right] \quad (5\text{-}2)$$

where:
- f is the recoil-free fraction;
- k_γ is the wavenumber of the γ-radiation, equal to $2\pi/\lambda$;
- $\langle x^2 \rangle$ is the mean squared displacement of atoms from their average position due to lattice vibrations;
- E_R is the recoil energy of the nucleus upon emission of a γ-quantum;
- k is Boltzmann's constant;
- θ_D is the Debye temperature; $k\theta_D$ corresponds to the maximum energy of the phonons;
- T is the temperature.

Equation (5-1) expresses that the recoil-free fraction is larger for low-energy (i.e., low k_γ) transitions as in Fe and Sn than for high-energy transitions such as in Ir, Pt, and Au. Both expressions indicate that the recoil-less fraction is higher for a rigid lattice ($\langle x^2 \rangle$ small, θ_D high) than for a structure with soft vibrational modes ($\langle x^2 \rangle$ large, θ_D low). Figure 5.2 shows plots of Eq. (5-2) for the Mössbauer effect in iron; the Debye temperature of bulk α-Fe is 470 K [11], while that of Fe_2O_3 and Fe_3O_4 is on the order of 500 K [12]. Surface atoms are coordinatively unsaturated and have a greater freedom to vibrate. As a result, the surface Debye temperature of well-crystallized solids is generally between one-third and two-thirds of the bulk value [13, 14]. However, Debye temperatures as low as 50 K have been reported in passivation layers on top of small iron catalyst particles [15].

There is a second condition that must be satisfied in order to observe the Mössbauer effect, namely that one needs nuclei in the excited state as a source for the γ-photons. Figure 5.3 shows how this is achieved in the case of the most fre-

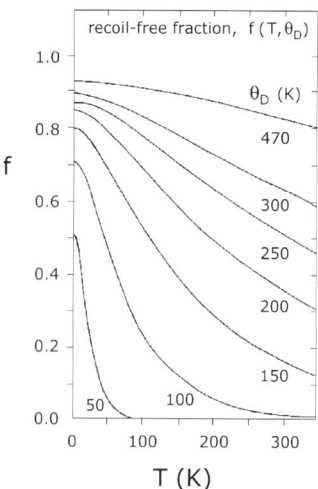

Fig. 5.2 The recoil-free fraction, f, of iron as a function of temperature for different values of the Debye temperature, θ_D. Bulk iron compounds have Debye temperatures on the order of 450–500 K; surface phases, however, have significantly lower Debye temperatures, implying that measurements may have to be carried out at lower temperatures.

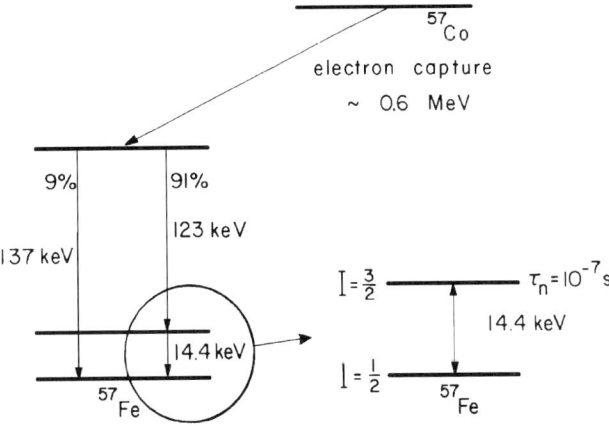

Fig. 5.3 The decay of ^{57}Co to ^{57}Fe. The encircled part is the transition commonly used for Mössbauer spectroscopy of iron-containing samples.

quently used nucleus for Mössbauer spectroscopy, the ^{57}Fe isotope. In this case, ^{57}Co is used as a source. This isotope is made in a nuclear accelerator, and decays with a half-life of 270 days to an excited level of ^{57}Fe, which in its turn decays rapidly (half-life 10^{-8} s) to the excited level required. The last step in the decay process generates the γ-quantum of approximately 14.4 keV used in Mössbauer spectroscopy. The parallel decay process that leads to the emission of a 137 keV photon can also be used, but the corresponding recoil-less fractions are much smaller, as Eq. (5-1) predicts. Thus, a necessary condition for an observable Mössbauer effect is that one has a source which decays to the excited state of the nucleus we want to study with a sufficiently long lifetime such that experiments are practical. The actual transition used for the Mössbauer effect should follow instantaneously. Only a limited number of elements satisfy this condition. For catalysis, iron, tin, antimony, ruthenium, iridium and platinum are the most important elements (see Table 5.1) [4].

Table 5.1 Mössbauer nuclei, sources, half-life times and energies.

Isotope	Source	Half-life	Energy [keV]
^{57}Fe	^{57}Co	270 days	14.4
^{119}Sn	^{119m}Sn	245 days	23.9
^{121}Sb	^{121m}Sn	75 years	37.2
^{197}Au	^{197}Pt	20 hours	77.3
^{99}Ru	^{99}Rh	16 days	90.0
^{193}Ir	^{193}Os	32 hours	73.0
^{195}Pt	^{195}Au	192 days	98.8

5.3
Mössbauer Spectroscopy

Suppose that the atom of the absorbing nucleus has a different chemical environment than the emitting atom in the source. For example, the source is metallic iron and the absorber is an iron oxide. Because the nucleus is coupled to its environment through hyperfine interactions, the nuclear levels in the absorber have slightly different energies than in the emitter. Again, the Mössbauer effect will not be observed because the energy of the emitted γ-quantum does not match the energy difference between the levels in the absorber. Hence, we need to vary the energy of the photons. This can be done by using the Doppler effect: if we move the emitter towards the absorber at a velocity v, the energy of the photon becomes:

$$E(v) = E_o \left(1 + \frac{v}{c}\right) \tag{5-3}$$

where:
$E(v)$ is the energy of the γ-quantum emitted by the source;
v is the velocity of the source;

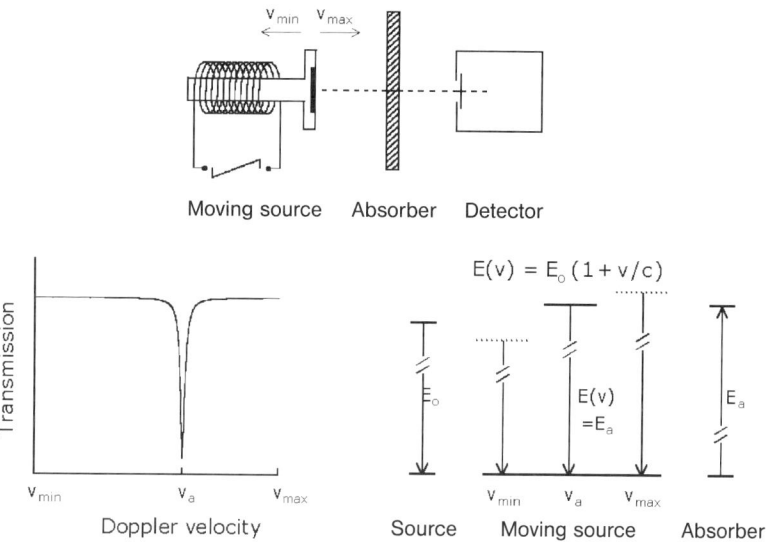

Fig. 5.4 In order to cover all possible transitions in the absorbing nucleus, the energy of the source radiation is modulated by using the Doppler effect. For ^{57}Fe the required velocities fall in the range between -1 and $+1$ cm s^{-1}. In Mössbauer emission spectroscopy, the sample under investigation is the source, and a single line absorber is used to scan the emission spectrum.

E_0 is the energy difference between the excited state and the ground state of the nucleus;
c   is the velocity of light.

Note that we ignore higher-order contributions such as $v^2/2c^2$ in Eq. (5-3); this has consequences, as we will see later when we discuss the second-order Doppler shift. In order to detect shifts and splitting in the nuclear levels due to hyperfine interactions in iron, one needs an energy range of at most $5 \cdot 10^{-8}$ eV around E_0, which is achieved with Doppler velocities in the range of -10 to $+10$ mm s^{-1}.

Figure 5.4 provides a schematic diagram of a Mössbauer experiment in transmission mode with a moving single-line source and the absorbing sample in fixed position. A Mössbauer spectrum is a plot of the γ-ray intensity transmitted by the sample, against the velocity v of the source. The latter is related to the

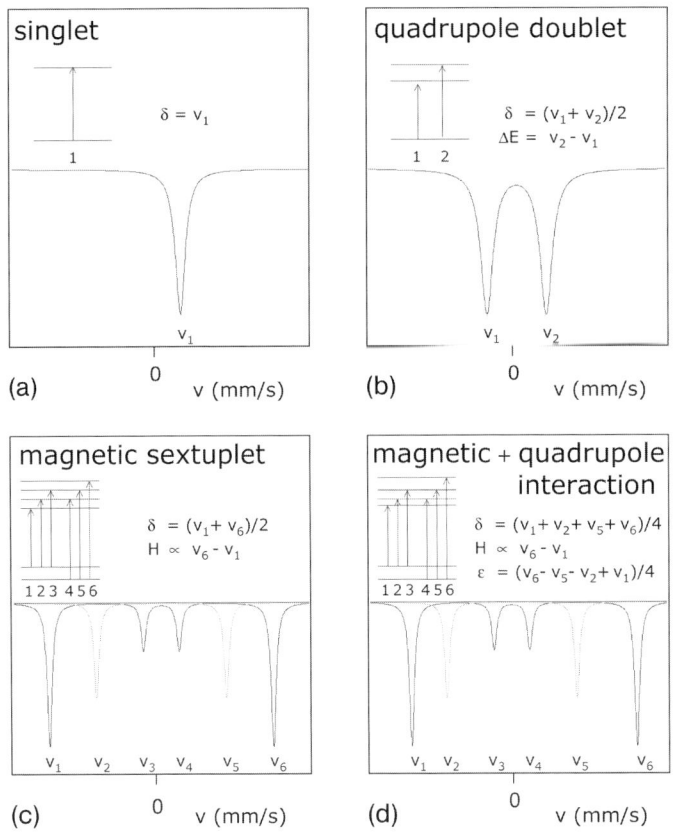

Fig. 5.5 (a–d) The four most common types of Mössbauer spectra observed in iron-containing catalysts, along with the corresponding nuclear transitions. Indicated is also how the Mössbauer parameters are derived from the spectra.

actual energy by Eq. (5-3). This is the common mode of operation, called Mössbauer absorption spectroscopy, sometimes abbreviated as MAS. It is also possible to fix the ^{57}Co-containing source and move the single-line ^{57}Fe absorber, in order to investigate cobalt-containing catalysts. This technique is called Mössbauer emission spectroscopy (MES). An application of MES is discussed in Chapter 9.

Hyperfine interactions couple the nucleus to its surroundings and make it a sensitive probe for the state of the absorber. Three interactions play a role, and these are illustrated in Figure 5.5.

5.3.1
Isomer Shift

The isomer shift, δ, is the consequence of the Coulomb interaction between the positively charged nucleus and the negatively charged s-electrons. Since the size of the nucleus in the excited state differs from that in the ground state, the Coulomb interaction energies are also different. The isomer shift therefore is a measure of the s-electron density at the nucleus, and yields useful information on the oxidation state of the iron in the absorber. Isomer shift values are expressed in velocity units (mm s^{-1}), and are usually given with respect to the peak position of a reference such as metallic iron. Some isomer shift values of common iron compounds are listed in Table 5.2.

The isomer shift contains a contribution from the thermal motion of the individual atoms in the absorber, the second-order Doppler shift, which makes the isomer shift temperature-dependent:

$$\delta = \delta_{T=0} - \frac{\langle v^2 \rangle}{2c} \tag{5-4}$$

where:
δ is the isomer shift as a function of temperature;
T   is the temperature;
$\langle v^2 \rangle$ is the mean-squared velocity of atoms due to lattice vibrations;
c   is the velocity of light.

The second term in Eq. (5-4) is the second-order Doppler shift. This is the higher-order term of the Taylor expansion that we ignored in Eq. (5-3). Like $\langle x^2 \rangle$, it can be calculated in the Debye model. Figure 5.6 shows plots of the second-order Doppler shift for the case of iron and for different values of the Debye temperature. Soft lattice vibrations are expected to decrease the isomer shift, although the effect becomes only significant at temperatures well above 80 K.

Table 5.2 Mössbauer parameters of common iron compounds.

Compound	δ [mm s^{-1}]	ΔE_Q [mm s^{-1}]	ε' [mm s^{-1}]	H [T]
α-Fe$_2$O$_3$	0.43	–	−0.10	51.5
α-FeOOH	0.35	–	0.13	38.4
Fe$_3$O$_4$	0.30	–		49.2
	0.63	–		45.5
FeO	1.08	0.55	–	–
FeS$_2$	0.28	0.60	–	–
α-Fe	0.00	–		33.0
θ-Fe$_3$C	0.19	–		20.8
SNP[a]	−0.26	1.70	–	–

[a] Sodium nitroprusside (sometimes used as calibration standard for the isomer shift, commonly metallic iron is used).

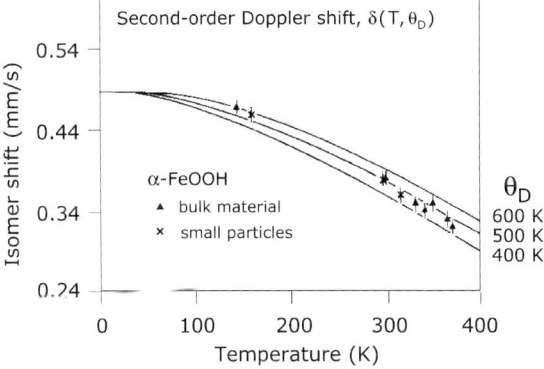

Fig. 5.6 Isomer shifts (relative to sodium nitroprusside) as a function of temperature in the Debye model, along with measurements on α-FeOOH. (From [12]).

5.3.2
Electric Quadrupole Splitting

The electric quadrupole splitting, ΔE_Q, is caused by the interaction of the electric quadrupole moment of the nucleus with an electric field gradient. The nucleus of iron in its ground state has a spherical charge distribution, and hence no quadrupole moment. When excited to the 14.4 keV level, however, the nucleus has the shape of an ellipsoid and possesses an electric quadrupole moment. The consequence is that the nucleus can orient itself in two ways in an electric field gradient with slightly different energies. Hence, two transitions from the ground state are now possible. The transition probabilities depend on the angle between the

Table 5.3 Relative peak areas in Mössbauer quadrupole doublets as a function of the angle φ between the radiation and the electric field gradient.

Line	Intensity	$\varphi = 90°$	$\varphi = 0°$	φ random	b_i
1	$3(1 + \cos^2 \varphi)$	3	3	1	0.5
2	$5 - 3\cos^2 \varphi$	5	1	1	0.5

electric field gradient and the direction of the γ-radiation (see Table 5.3), but for randomly oriented samples, such as catalyst powders, the angle-averaged transition probabilities are equal. Figure 5.5b shows what the effect is of an electric field gradient on the Mössbauer spectrum of iron (in the absence of magnetic interactions). The peak splits and forms a so-called quadrupole doublet, with the splitting being proportional to the magnitude of the electric field gradient at the nucleus.

The origin of the electric field gradient is twofold: it is caused by asymmetrically distributed electrons in incompletely filled shells of the atom itself, and also by charges on neighboring ions. The distinction is not always clear, because the lattice symmetry determines the direction of the bonding orbitals in which the valence electrons reside. If the symmetry of the electrons is cubic, the electric field gradient vanishes. Here, we consider two examples:

- In Fe^{3+} ions, the five 3d electrons occupy one atomic d-orbital each and form a half-closed shell. This arrangement has spherical symmetry and does not contribute to a field gradient. In this case the quadrupole splitting measures directly the lattice symmetry around the atom. For example, the quadrupole splitting of ions at the surface of small Fe_2O_3 particles, which are in a highly asymmetric environment, is about 0.9 mm s^{-1}, whereas for atoms in the interior, with a more symmetric environment, it is 0.5 mm s^{-1} [16]. In this way the quadrupole splitting provides direct information about the dispersion of freshly prepared supported catalysts, in which iron is usually in the 3+-oxidation state [17].

- In Fe^{2+}, the situation is more complicated. Here, the six 3d-electrons dominate the magnitude of the electric field gradient, albeit in a way that is determined by the lattice symmetry. If a Fe^{2+} ion enters a more asymmetric environment, the quadrupole splitting decreases in general, because the lattice contribution to the electric field gradient is smaller than the electronic contribution, and has the opposite sign.

Figure 5.7 demonstrates the relationship between isomer shift and quadrupole splitting for iron in various environments. In oxidic components of iron catalysts, one usually only encounters the high-spin configurations of Fe^{2+} and Fe^{3+} ions.

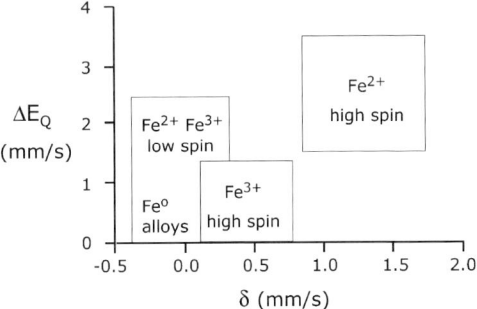

Fig. 5.7 The relationship between the isomer shift, the quadrupole splitting, and the oxidation state of iron.

5.3.3
Magnetic Hyperfine Splitting

Magnetic hyperfine splitting – the Zeeman effect – arises from the interaction between the nuclear magnetic dipole moment and the magnetic field H at the nucleus. This interaction gives rise to six transitions; the separation between the peaks in the spectrum is proportional to the magnetic field at the nucleus.

Magnetic fields at the nuclei of iron atoms are extremely high: 33 Tesla (T) in metallic iron, and over 50 T in Fe_2O_3. This is much higher than the fields of electromagnets (≤ 1 T) or even of superconducting magnets (≤ 10 T) that are commonly used in the laboratory. Hence, a few words on how the very high field at the nucleus arises are in order.

In the case of iron, magnetism is due to the unpaired electrons in the 3d-orbitals, which have all parallel spin. These electrons interact with all other electrons of the atom, including the s-electrons that have overlap with the nucleus. As the interaction between electrons with parallel spins is slightly less repulsive than between electrons with anti parallel spins, the s-electron cloud is polarized, which causes the large but also highly localized magnetic field at the nucleus. The field of any externally applied magnet adds vectorially to the internal magnetic field at the nucleus.

Figure 5.5c shows that a magnetic field removes all degeneration from the nuclear levels: the ground state (spin 1/2) splits in two and the excited state (spin 3/2) splits in four levels. Two of the eight conceivable transitions between these levels are forbidden, and the spectrum consists of six lines, often called the sextuplet or magnetic sextet. The relative intensities of the lines are listed in Table 5.4.

Clearly, all hyperfine interactions can occur simultaneously. In magnetically ordered compounds with a non-vanishing electric field gradient, the shape of the spectrum depends on the relative strengths of the magnetic and the electric quadrupole interaction. In catalysts, the usual situation is that the quadrupole interac-

Table 5.4 Relative peak areas in magnetically split Mössbauer spectra as a function of the angle φ between the γ-radiation and the magnetic field at the nucleus.

Line	Intensity	$\varphi = 90°$	$\varphi = 0°$	φ random	b_i
1	$3(1 + \cos^2 \varphi)$	3	3	3	0.25
2	$4 \sin^2 \varphi$	4	0	2	0.167
3	$1 + \cos^2 \varphi$	1	1	1	0.083
4	$1 + \cos^2 \varphi$	1	1	1	0.083
5	$4 \sin^2 \varphi$	4	0	2	0.167
6	$3(1 + \cos^2 \varphi)$	3	3	3	0.25

tion is much smaller than the magnetic interaction. In this case, all peaks shift by an energy ε' (again expressed in units of mm s^{-1}), although the outer lines shift in the opposite direction to the inner four lines (see Fig. 5.5d). Examples of this situation are observed in the Mössbauer spectra of Fe$_2$O$_3$ and FeOOH. As in the case of quadrupole splitting, ε' is proportional to the electric field gradient, but also depends on the orientation of the electric field gradient with respect to the magnetic field:

$$\varepsilon' = \frac{1}{2} \Delta E_Q \frac{3 \cos^2 \varphi - 1}{2} \tag{5-5}$$

where:
ε' is the quadrupole shift in a magnetic sextet;
ΔE_Q is the quadrupole splitting of a doublet (measurable above the Curie temperature where magnetic splitting is absent);
φ is the angle between the electric field gradient and the magnetization.

An interesting situation arises in the spectra of α-Fe$_2$O$_3$, where the angle φ switches from 0 to 90° at a temperature (T_M) of 260 K; this is termed the Morin transition. At temperatures below T_M, the parameter ε' equals +0.20 mm s^{-1}, whereas at the transition temperature ε' changes sign and magnitude to a value of -0.10 mm s^{-1} [16], in agreement with Eq. (5-5).

5.3.4
Intensity

The intensity of a Mössbauer spectrum depends not only on the recoil-free fractions of the source and the absorber and on the number of absorbing nuclei, but also on the linewidth of the absorption lines and on whether or not saturation

5.3 Mössbauer Spectroscopy

effects occur. The following approximate expression is valid for relatively thin absorbers [18]:

$$A_i = \frac{\pi}{2} f_s \Gamma_{nat} t_i \left(1 - \frac{\Gamma_{nat}}{\Gamma_i} \frac{t_i}{4}\right) \quad (5\text{-}6)$$

$$t_i = b_i \sigma_o n f_a \quad (5\text{-}7)$$

where:
- A_i is the absorption area of the ith line in a spectrum;
- f_s is the recoil-free fraction of the source ($f_s \approx 0.75$ for ^{57}Co sources at room temperature);
- f_a is the recoil-free fraction of the absorber (see Fig. 5.2);
- Γ_{nat} is the natural line width of the Mössbauer transition (≈ 0.1 mm s^{-1} for ^{57}Fe);
- Γ_a is the linewidth of the ith line in the absorber spectrum;
- t_i is the reduced thickness for line i of the absorber;
- b_i is the relative intensity of the ith line (see Tables 5.3 and 5.4);
- σ_o is the cross-section for resonant absorption;
- n is the number of Mössbauer nuclei per unit area in the absorber (for iron, the total number of iron atoms times the natural abundance of the ^{57}Fe isotope).

Equation (5-6) indicates that the intensity of a Mössbauer spectrum is only proportional to the concentration of Mössbauer atoms if the term $(1 - t\Gamma_{nat}/4\Gamma_a) \approx 1$. This is so when absorbers are thin, and when the lines are broader than the natural linewidth. For heavier samples, saturation effects come into play. In case of a sextet, the factor b_i in Eq. (5-7) forms the reason that the outer peaks are more affected than the inner peaks, with the result that the line intensity ratios become lower than the expected value of 3:1.

Thick samples also degrade the spectral resolution: according to Mørup and Both [18], the linewidth is to a good approximation given by:

$$\Gamma(t) = \Gamma_s + \Gamma_a + \tfrac{1}{4}\Gamma_{nat} t \quad (5\text{-}8)$$

where:
- $\Gamma(t)$ is the measured linewidth of a Mössbauer peak;
- Γ_s is the linewidth of the source;
- Γ_a is the linewidth of the infinitely thin absorber;
- Γ_{nat} is the natural linewidth; and
- t is the reduced thickness of the absorber given in Eq. (5-7).

Thus, preparing samples that are not too-thick helps to obtain sharper spectra and facilitates the quantitative interpretation. Finally – and particularly in the Mössbauer spectra of small catalyst particles – one should be aware of the temperature-dependence of the absorption area through the recoil-free fraction. If the spectrum contains contributions from surface and bulk phases, then the intensity of the former will be greatly underestimated if the spectrum is measured at room temperature. The only way to obtain reliable concentrations of surface and bulk phases is to determine their spectral contributions as a function of temperature and to make an extrapolation to zero Kelvin [14].

5.4
Mössbauer Spectroscopy in Catalyst Characterization

The application of Mössbauer spectroscopy to the investigation of catalysts began around 1970, and subsequently became quite popular, owing mainly to the importance of iron – with its rather rich chemistry – in many catalysts. By 1990 over 600 scientific reports had been published on the subject [19]. Most applications of Mössbauer spectroscopy to catalysts fall into one of the following categories:

- Identification of phases
- Determination of oxidation states
- Structure information
- Determination of particle size
- Kinetics of bulk transformations.

A typical example of how Mössbauer spectroscopy is used in the identification of oxidic, metallic and carbidic phases can be demonstrated by a study on titania-supported iron, prepared by impregnating the TiO_2 support with a solution of iron nitrate (see Fig. 5.8) [20]. In the figure, the top spectrum is that of a freshly impregnated and dried Fe/TiO_2 catalyst; this shows a doublet with an isomer shift of 0.37 mm s^{-1} and a quadrupole splitting of 0.82 mm s^{-1}. Comparison with the scheme in Figure 5.7 reveals that the iron is a high-spin Fe^{3+} species. It is difficult to draw conclusions on the type of compound that is present. Reference compounds of iron oxide, Fe_2O_3 or Fe_3O_4 possess magnetically split Mössbauer spectra (e.g., see the Mössbauer parameters of reference compounds in Table 5.2). The relatively high value of the quadrupole splitting points to a highly asymmetric environment, as surface atoms would have. Our chemical intuition – for what it is worth – suggests that we should expect a well-dispersed layer of iron oxide, or oxyhydroxide. This would be consistent with the spectrum.

After reduction in H_2 at 675 K, the catalyst consists mainly of metallic iron, as evidenced by the sextet ($\delta = 0.00$ mm s^{-1}, H = 331 kOe), along with some unreduced iron, which gives rise to two doublet contributions of Fe^{2+} and Fe^{3+} in the center. The overall degree of iron reduction, as reflected by the relative area under the bcc-ion sextet, is high. One should not consider the relative spectral contributions as concentrations, however, because the three types of iron species may have different recoil-less fractions.

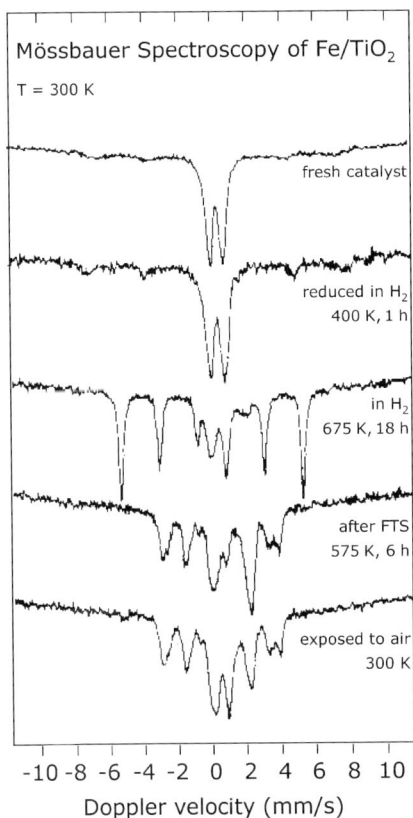

Fig. 5.8 Mössbauer spectra at room temperature gives detailed information on the state of iron in a TiO$_2$-supported iron catalyst after different treatments. Reduction times are shown in hours (h). FTS: Fischer–Tropsch synthesis. (From [20]).

When reduced Fe/TiO$_2$ is used as a catalyst for the reaction between CO and H$_2$ to hydrocarbons (the Fischer–Tropsch synthesis) the spectrum changes entirely. All metallic iron has been converted into a new phase, and the spectrum is that of a crystallographically well-defined iron carbide, namely the Hägg carbide, or χ-Fe$_5$C$_2$. Apparently, the strongly reducing atmosphere has also affected the unreduced iron, and all ions are now present as Fe^{2+}.

The bottom spectrum in Figure 5.8 has been recorded after exposing the used catalyst to air at room temperature. The spectrum clearly has changed; although most of the carbide phase is still present, some of the ferrous iron has been oxidized to ferric iron. Hence, it is essential that the catalyst be studied under *in-situ* conditions.

The conversion of iron catalysts into iron carbide under Fischer–Tropsch conditions is well known, and has been the subject of several studies [21–25]. A funda-

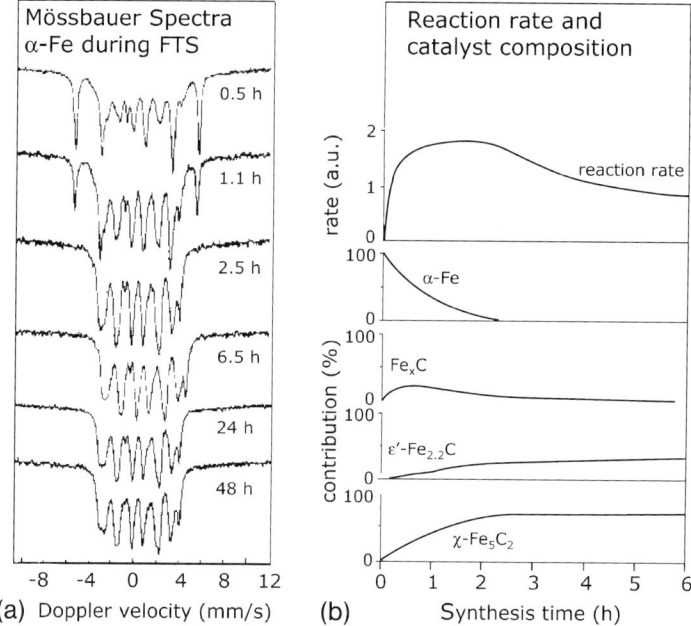

Fig. 5.9 (a) Mössbauer spectra of a metallic iron catalyst after different periods of Fischer–Tropsch synthesis in CO + H_2 at 240 °C, showing the conversion of metallic iron (visible by the outer two lines in the upper two spectra) into iron carbides; all spectra were recorded at room temperature. (b) Reaction rate of the Fischer–Tropsch synthesis (*upper curve*) and the relative contributions of metallic iron and various carbides to the Mössbauer spectra. (From [23]).

mentally intriguing question is why the active iron Fischer–Tropsch catalyst consists of iron carbide, while cobalt, nickel and ruthenium are active as a metal. Figure 5.9a shows how metallic iron particles convert to carbides in a mixture of CO and H_2 at 515 K. After 0.5 and 1.1 h of reaction, the sharp six-line pattern of metallic iron is still clearly visible, in addition to the complicated carbide spectra, but after 2.5 h the metallic iron has disappeared. At short reaction times, a rather broad spectral component appears (this is better visible in carburization experiments at lower temperatures), and is indicated as Fe_xC. The eventually remaining pattern can be understood as the combination of two different carbides: ε'-$Fe_{2.2}C$ and χ-Fe_5C_2.

The relationship between carbide composition and catalytic activity is unclear. Figure 5.9b compares the two properties, whereby the catalytic activity begins low, increases rapidly to a maximum, and decreases slowly thereafter. The increase in activity occurs simultaneously with the carburization, but without any direct correlation, the maximum in rate occurs when there is still metallic iron present. The interpretation given to the results in Figure 5.9 is as follows [23, 26].

CO dissociates readily on iron. The carbon atoms have three possibilities to react:

$$C + \alpha\text{-Fe} \rightarrow \text{carbides} \quad (1)$$
$$C + H_2 \rightarrow \text{hydrocarbons} \quad (2)$$
$$C \rightarrow \text{inactive carbon} \quad (3)$$

Reaction (1) is fast, as carbon diffusion in metallic iron has an activation energy on the order of only 60 kJ mol^{-1}. However, as soon as the iron lattice becomes significantly disturbed, the activation energy increases and the rate of diffusion slows to the values common in cobalt and nickel. Hence, in the early stage of the Fischer–Tropsch synthesis, step (1) consumes most of the carbon until the rate of carbon diffusion into the bulk of the catalyst slows down. Gradually, reaction (2) – the actual synthesis of hydrocarbons – takes over. When the catalyst becomes saturated with carbon, increasing amounts of carbon remain at the surface where it can either be hydrogenated or will form inactive carbon, as in step (3), a reaction which is responsible for the deactivation of the hydrocarbon formation. Thus, the time-dependent behavior of iron in the Fischer–Tropsch synthesis is initially governed by a competition between bulk carburization and hydrocarbon formation. For cobalt and nickel, the activation energy of carbon diffusion through the lattice is more than twofold higher than for iron, and consequently reaction (1) has hardly any influence.

These two examples illustrate how Mössbauer spectroscopy can reveal the identity of iron phases in a catalyst after different treatments. The examples are typical of many applications of the technique in catalysis – a catalyst is reduced, carburized, sulfided, or passivated and, after cooling down, its Mössbauer spectrum is monitored at room temperature. However, a complete characterization of phases in a catalyst sometimes requires that spectra are measured at cryogenic temperatures, in particular when catalysts are highly dispersed.

5.4.1
In-Situ Mössbauer Spectroscopy at Cryogenic Temperatures

Surface phases have low Debye temperatures, and as a result the recoil-free fraction may be low at room temperature (see Fig. 5.2). Thus, measuring at cryogenic temperatures will increase the Mössbauer intensity of such samples considerably. However, there are other reasons which call for low-temperature experiments.

Spectra of small particles – for example, those of supported iron particles with a diameter of a few nanometers [27] – often show poorly resolved contributions in the central region of the spectrum, instead of the magnetic sextuplet expected for metallic iron or iron oxide. The reason is that small particles become superparamagnetic, a phenomenon that occurs when thermal excitations of energy kT are energetic enough to decouple the magnetization from the lattice [28, 29]. As a result, the magnetization vector of each particle fluctuates rapidly over all directions and the Mössbauer transition, which takes place on a time scale of 10^{-8} s, has an average magnetization of zero. The lower spectra of Figure 5.10a demonstrates superparamagnetic behavior in a carbon-supported iron catalyst. In order to as-

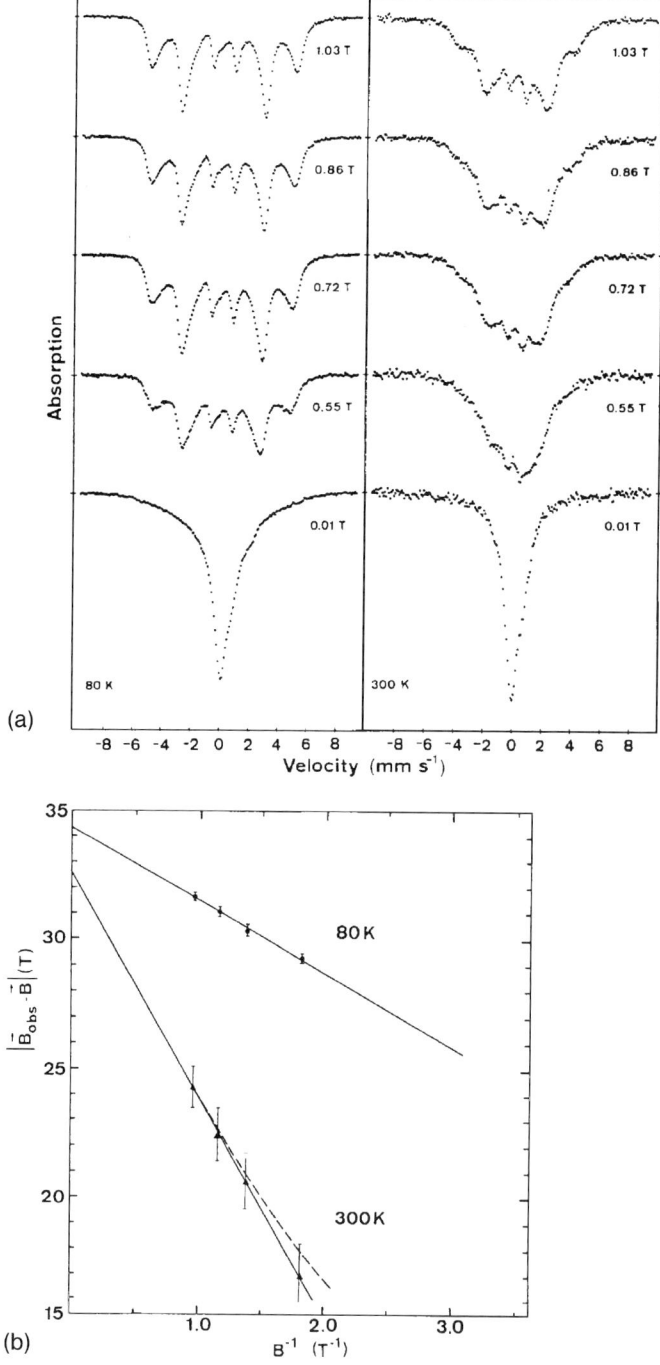

Fig. 5.10 (legend see p. 139)

sign such peaks correctly, one can either cool the sample below the superparamagnetic transition temperature [27, 30], or apply an external magnetic field, as in Figure 5.10 [31]. Because iron particles are highly susceptible to oxidation, it is essential that these spectra are taken *in situ*, with the catalyst under the gas atmosphere of interest. For recent examples, the reader is referred to the investigations of Mørup and co-workers [32].

5.4.2
Particle Size Determination

The occurrence of superparamagnetism allows the particle size to be determined if an external magnetic field is applied. Figure 5.10a illustrates how this is achieved on a Fe/C catalyst [31]. Here, the spectra without an external field show the single peak of superparamagnetic iron, but as soon as the magnetic field is applied then magnetic splitting sets in, the magnitude of which increases with increasing field strength. The external field orients the magnetization vector of the particles, but thermal excitations allow the magnetization vector to fluctuate around the direction of the applied field. Thus, one measures an average magnetic splitting, as given by the Langevin equation [29]:

$$\vec{H}_{obs} = \vec{H}_o \cdot L\left(\frac{\mu H}{kT}\right) + \vec{H} \tag{5-9}$$

or

$$|\vec{H}_{obs} - \vec{H}_{ext}| = H_o\left(1 - \frac{kT}{\mu H}\right) \quad \text{for} \quad \frac{\mu H}{kT} \geq 3 \tag{5-10}$$

where:
HH_{obs} is the observed magnetic splitting;
HH_{ext} is the externally applied field;
HH_o is the bulk magnetic field;
μ is the magnetic moment of the particle;
L is the Langevin function;
k is Boltzmann's constant;
T is the temperature.

Fig. 5.10 (a) Mössbauer spectra of a reduced carbon-supported iron catalyst at 80 and 300 K, obtained in different applied magnetic fields. The spectra at the bottom, measured without external field, consist mainly of a singlet due to superparamagnetic metallic iron. The application of magnetic fields induces magnetic splitting. (b) Langevin plots according to Eq. (5-10) for the spectra in (a). The lines extrapolate for $1/H \to 0$ to the magnetic splitting expected for single domain metallic iron particles; the slopes correspond to a particle diameter of 2.5 ± 0.2 nm. The dashed line is a plot according to Eq. (5-9), and confirms that the use of the high field approximation is justified. (From [31]).

Equation (5-10) is a simplification which is often applicable in practice.

Figure 5.10b illustrates the analysis of the magnetic hyperfine splitting with Eqs. (5-9) and (5-10). A plot of $|H_{obs} - H_{ext}|$ against $1/H_{ext}$ gives a straight line with a slope $H_o k T/\mu$, from which μ – the magnetic moment of the particle – follows. As the total magnetic moment is equal to the atomic moment (2.2 Bohr magneton) times the number of atoms in a particle, the latter can thus be calculated and converted into a diameter, if it is assumed that the particles have a spherical shape. In this way, a diameter of 2.5 ± 0.2 nm is determined for the iron particles of Figure 5.10 [31].

The intercepts of the plots in Figure 5.10b correspond to H_o, and equal 34.3 ± 0.5 and 32.7 ± 2.0 T at 77 and 300 K, respectively. These values are in agreement with the bulk fields of metallic iron, after corrections for the influence of the demagnetizing field (0.7 T) in isolated spherical single domain particles [30].

A final remark about the line intensities in Figure 5.10a: as the field is directed perpendicular to the γ-beam, the situation corresponding to $\varphi = 90°$ in Table 5.4 applies, and the line intensities from outward to inward are in the proportions of 3:4:1. One can also apply the external field parallel to the γ-beam, with the result that the second and fifth lines of the sextet disappear from the spectra ($\varphi = 0$ in Table 5.4). Bødker et al. [27] used this approach to simplify the spectrum of small iron particles, and in this way were able to analyze the shape of the outer lines in greater detail.

The usual techniques for determining the particle sizes of catalysts include electron microscopy, chemisorption, X-ray diffraction (XRD), line broadening or profile analysis, and magnetic measurements. The advantage of using Mössbauer spectroscopy for this purpose is that one simultaneously characterizes the state of the catalyst. As the state of supported iron catalysts depends often on subtleties in the reduction, the simultaneous determination of particle size and degree of reduction – as in the studies defined in Figure 5.10 – represents an important advantage for Mössbauer spectroscopy.

5.4.3
Kinetics of Solid-State Reactions from Single Velocity Experiments

In general, the recording of a complete Mössbauer spectrum for an iron catalyst takes several hours, which is too slow for following reactions in real time. Nevertheless, by measuring the intensity of a characteristic peak at constant velocity, it is possible to monitor processes that occur on a time scale of minutes to hours, such as the reduction of oxides. An elegant example has been reported by Hummel et al. [33], who studied the stability of the nitride ξ-Fe_2N in different gases (see Fig. 5.11). Complete Mössbauer spectra show that denitridation of Fe_2N in H_2 yields metallic iron, but in CO/H_2 a carbonitride forms. The constant velocity spectra of Figure 5.11 reflect the rate at which the nitride is converted, and also reveal the kinetics of these processes. Other examples of constant velocity experi-

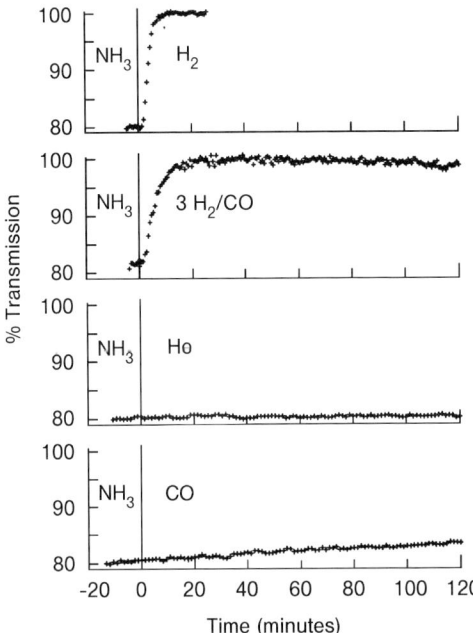

Fig. 5.11 A constant-velocity Mössbauer experiment reveals the kinetics of the denitridation of an iron nitride in different gases at 525 K. The negative part of the time scale gives the transmission of the most intense peak of the nitride; at time zero the gas atmosphere is changed to the desired gas. Denitridation occurs relatively rapidly in H_2, but is retarded by CO, whereas the nitride is stable in an inert gas such as helium. (From [33]).

ments have been reported by Raupp and Delgass [22], who studied the conversion of iron into carbides with this technique.

Alternative ways of studying the kinetics of bulk transformations would be to monitor changes in weight or in magnetization. Such methods, however, are less specific with regards to the initial and final state of the catalyst than is Mössbauer spectroscopy.

5.4.4
In-Situ Mössbauer Spectroscopy Under Reaction Conditions

In-situ characterization becomes an absolute necessity in cases where catalysts change their structure during the start-up of a catalytic reaction. Figure 5.12 shows the example of a bimetallic Fe–Ir catalyst during the synthesis of methanol from CO and H_2. Noble metals such as platinum and iridium are poor CO hydrogenation catalysts as they produce mainly methane. The addition of iron, however, increases the activity and shifts the product distribution towards methanol

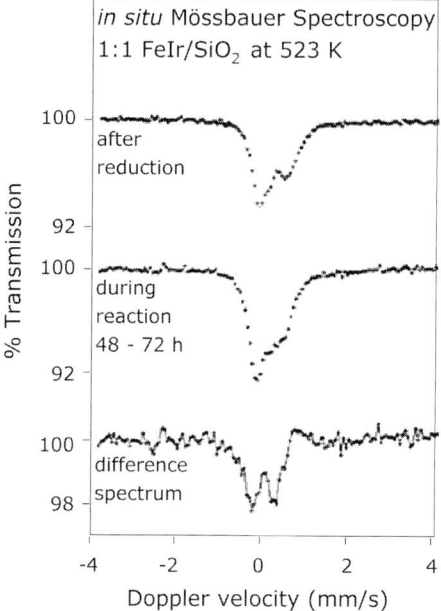

Fig. 5.12 *In-situ* Mössbauer spectra of a reduced FeIr/SiO₂ catalyst at a reaction temperature of 525 K and during CO hydrogenation when the catalyst is in its steady-state methanol-producing mode. The bottom spectrum represents the difference between the two upper spectra; it is characteristic for an iron carbide in superparamagnetic state. (Figure courtesy of Hyung Woo; adapted from [8]).

and higher oxygenates, particularly at pressures above 10 bar [34]. At steady state, methanol selectivities of up to 97% have been observed, albeit at activity levels below those of the commercially applied Cu/ZnO systems. However, a catalyst containing equal amounts of iron and iridium on silica begins to convert CO and H_2 almost exclusively into methane. Typically, it takes 10 to 50 hours for the product distribution to change gradually from methane to methanol [34]; the question to be asked, therefore, is what happens with the FeIr catalyst during this activation period?

Mössbauer spectra recorded *in situ* under high-pressure reaction conditions (Fig. 5.12) show that the initially reduced Fe–Ir catalyst, consisting of Fe–Ir particles and some unreduced iron believed to be in intimate contact with the support, has changed significantly when the catalyst has reached steady state [8]. The lower trace in Figure 5.12 represents the difference between the working and the initial catalyst. It is characteristic of an iron carbide in the superparamagnetic state. The total absorption area of the spectrum was also increased, indicating a loss of overall dispersion of the iron. Further characterization studies by Mössbauer spectroscopy and extended X-ray absorption fine structure (EXAFS) after cooling to ambient and cryogenic temperatures confirmed the interpretation, and also added much detail with regards to the composition of the surface [34]. In brief, during high-pressure CO hydrogenation, the active part of the Fe–Ir catalyst restructures from Fe–Ir alloy particles with a surface enriched in iron to an iridium-rich alloy accompanied by a (probably largely inactive) iron carbide phase. During the reconstruction, the chemical properties of the catalyst surface also change significantly [34]. The interesting point of these investigations is that the Fe–Ir catalyst restructures itself by allowing excess iron to segregate into carbide particles, leading to an Fe–Ir alloy that is well-tuned towards methanol formation. This process of self-assembly has been monitored successfully using *in-situ* Mössbauer spectroscopy.

5.4.5
Mössbauer Spectroscopy of Elements Other Than Iron

Although the majority of Mössbauer studies on catalysts are concerned with iron, other elements exhibiting the Mössbauer effect have also been used.

As explained in connection with Figure 5.4, cobalt catalysts enriched in the isotope ^{57}Co can be studied in emission mode. As an example, we refer to the studies of Van Berge et al. [35], who used Mössbauer emission spectroscopy to investigate cobalt Fischer–Tropsch catalysts with respect to stability versus oxidation in water. Oxidation had been suggested to cause deactivation, as it would imply a loss of reduced cobalt and hence of activity. These authors found that oxidation can indeed occur, depending on the partial pressure of water in the environment. The group also described an interesting *in-situ* cell for emission studies of cobalt catalysts under high pressures [36]. Mössbauer emission studies on cobalt-containing hydrodesulfurization catalysts are discussed in Chapter 9.

Almost all other studies on elements other than iron are performed in absorption mode. For example, Bussiere and co-workers used the Mössbauer effect to study the state of tin in supported Pt–Sn [37] and Ir–Sn [38] reforming catalysts, and of tin and antimony in mixed Sb–Sn oxides for the selective oxidation of propylene [39]. Also of note are Millet's investigations using the ^{125}Te isotope to characterize the state of the tellurium promoter in multicomponent ammoxidation catalysts [40].

Similarly well-known investigations were conducted by Clausen and Good on supported ruthenium catalysts, by means of the difficult ^{99}Ru isotope [41]. A more recent ^{99}Ru study was reported by the group of Wagner [42], who also used Ir Mössbauer spectroscopy to study supported Ir and Pt–Ir catalysts, used in the catalytic reformation of naphtha to hydrocarbon mixtures of higher octane numbers [43]. Such experiments are much less straightforward than those with iron: because the ^{192}Os source decays with a half-life of only 31 hours, very few experiments can be performed with one source, and consequently access to a nuclear reactor facility for reactivating the source is essential for carrying out Ir Mössbauer spectroscopy. Moreover, the high energy of the transition, 73 keV, implies that the recoil energy is high and the recoil-free fraction low [Eq. (5-1)]. Thus, both the absorber and the source need to be cooled to low temperatures, preferably to that of liquid helium.

Similar considerations hold for the Mössbauer spectroscopy of gold, although with the recently increased interest in small particles of gold in catalysis, several investigations have been undertaken; one such study is reviewed in the following section.

Supported gold particles of a few nanometers in size have been found to display remarkable activity in, for instance, selective and total oxidation reactions. The gold isotope ^{197}Au exhibits the Mössbauer effect and provides clear information on the oxidation state of the element. Kobayashi and co-workers [44] reported the evolution of gold species during the calcination of $Mg(OH)_2$-supported Au, prepared by soaking an MgO support with an aqueous solution of $HAuCl_4$. During the impregnation, the gold deposits as $Au(OH)_3$ and the MgO transforms to $Mg(OH)_2$. The corresponding Mössbauer spectrum in Figure 5.13 shows two doublets of Au^{III} species. Upon calcination, the Au^{III} converts gradually to Au^{I} and Au^0. This example nicely demonstrates the rich information content of the technique for characterizing gold catalysts. Another interesting example by Finch et al. [45] describes the application of both ^{57}Fe and ^{197}Au Mössbauer spectros-

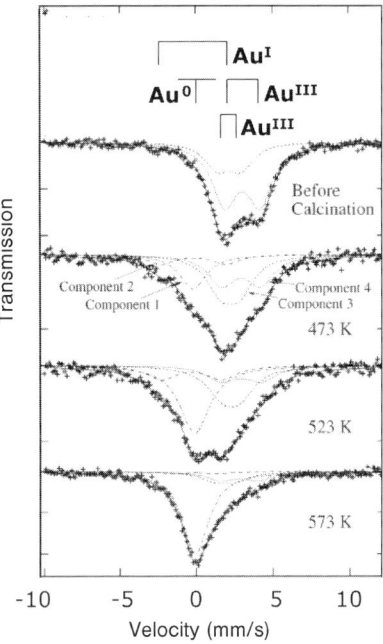

Fig. 5.13 ^{197}Au Mössbauer spectra of freshly prepared and dried Au catalysts supported on $Mg(OH)_2$, as well as the samples after calcination at the temperatures indicated. The spectra show that the initially present Au^{III} phases in the dried catalyst convert to Au^{I} and Au^0 upon calcination. (Adapted from [44]).

copy on Au/Fe$_2$O$_3$ catalysts, while the same group also reported Mössbauer spectra of Au/MgO oxidation catalysts [46].

5.5
Conclusion

Mössbauer spectroscopy has matured into one of the classical techniques for catalyst characterization, although its application is limited to a relatively small number of elements which exhibit the Mössbauer effect. The technique is used to identify phases, determine oxidation states, and to follow the kinetics of bulk reactions. Mössbauer spectra of super-paramagnetic iron particles in applied magnetic fields can be used to determine particle sizes. In favorable cases, the technique also provides information on the structure of catalysts. The great advantage of Mössbauer spectroscopy is that its high-energy photons can visualize the insides of reactors in order to reveal information on catalysts under *in-situ* conditions.

References

1 G.K. Wertheim, *Mössbauer Effect: Principles and Applications*. Academic Press, New York, 1964.
2 N.N. Greenwood and T.C. Gibb, *Mössbauer Spectroscopy*. Chapman & Hall, London, 1971.
3 T.E. Cranshaw, B.W. Dale, G.O. Longworth, and C.E. Johnson, *Mössbauer Spectroscopy and its Applications*. Cambridge University Press, Cambridge, 1985.
4 J.A. Dumesic and H. Topsøe, *Adv. Catal.* **26** (1977) 121.
5 H. Topsøe, J.A. Dumesic, and S. Mørup, in: *Applications of Mössbauer Spectroscopy*, R.L. Cohen (Ed.), Vol. II. Academic Press, New York, 1980, p. 55.
6 A.M. van der Kraan and J.W. Niemantsverdriet, in: *Industrial Applications of the Mossbauer Effect*, G.J. Long and J.G. Stevens (Eds.). Plenum, New York, 1985, p. 609.
7 F.J. Berry, in: *Spectroscopic Characterization of Heterogeneous Catalysts*, J.L.G. Fierro (Ed.), Part A. Elsevier, Amsterdam, 1990, p. A299.
8 J.W. Niemantsverdriet and W.N. Delgass, *Topics Catal.* **8** (1999) 133.
9 J.-M.M. Millet, *Adv. Catal.* **51** (2007) 309.
10 R.L. Mössbauer, *Z. Phys.* **151** (1958) 124; *Naturwissenschaften* **45** (1958) 538.
11 C. Kittel, *Quantum Theory of Solids*. Wiley, New York, 1963.
12 J.W. Niemantsverdriet, C.F.J. Flipse, B. Selman, J.J. van Loef, and A.M. van der Kraan, *Phys. Lett.* **100A** (1984) 445.
13 G.A. Somorjai, *Chemistry in Two Dimensions: Surfaces*. Cornell University Press, Ithaca, 1981.
14 J.W. Niemantsverdriet, A.M. van der Kraan, and W.N. Delgass, *J. Catal.* **89** (1984) 138.
15 J.W. Niemantsverdriet, C.F.J. Flipse, A.M. van der Kraan, and J.J. van Loef, *Appl. Surface Sci.* **10** (1982) 303.
16 A.M. van der Kraan, *Phys. Stat. Sol.* **18a** (1973) 215.
17 A.F.H. Wielers, A.J.H.M. Kock, C.E.C.A. Hop, J.W. Geus, and A.M. van der Kraan, *J. Catal.* **117** (1989) 1.
18 S. Mørup and E. Both, *Nucl. Instr. Meth.* **124** (1975) 445.
19 *Mössbauer Effect Reference and Data Journal*, J.G. Stevens (Ed.), Mössbauer Effect Data Center, Asheville, North Carolina.

20 A.M. van der Kraan, R.C.H. Nonnekens, F. Stoop, and J.W. Niemantsverdriet, *Appl. Catal.* **27** (1986) 285.

21 J.A. Amelse, J.B. Butt, and L.H. Schwartz, *J. Phys. Chem.* **82** (1978) 558.

22 G.B. Raupp and W.N. Delgass, *J. Catal.* **58** (1979) 337; *J. Catal.* **58** (1979) 348; *J. Catal.* **58** (1979) 361.

23 J.W. Niemantsverdriet, A.M. van der Kraan, W.L. van Dijk, and H.S. van der Baan, *J. Phys. Chem.* **84** (1980) 3363.

24 G. LeCaer, J.M. Dubois, M. Pijolat, V. Perrichon, and P. Bussiere, *J. Phys. Chem.* **86** (1982) 4799.

25 T.R. Motjope, H.T. Dlamini, G.R. Hearne, and N.J. Coville, *Catal. Today* **71** (2002) 335.

26 J.W. Niemantsverdriet and A.M. van der Kraan, *J. Catal.* **72** (1981) 385.

27 F. Bødker, S. Mørup, and J.W. Niemantsverdriet, *Catal. Lett.* **13** (1992) 195.

28 S. Mørup, J.A. Dumesic, and H. Topsøe, in: *Applications of Mössbauer Spectroscopy*, R.L. Cohen (Ed.), Vol. II, Academic Press, New York, 1980, p. 1.

29 P.W. Selwood, *Chemisorption and Magnetization.* Academic Press, New York, 1975.

30 J.W. Niemantsverdriet, A.M. van der Kraan, W.N. Delgass, and M.A. Vannice, *J. Phys. Chem.* **89** (1985) 67.

31 P.H. Christensen, S. Mørup, and J.W. Niemantsverdriet, *J. Phys. Chem.* **89** (1985) 4898.

32 F. Bødker, S. Mørup, and S. Linderoth, *Phys. Rev. Lett.* **72** (1994) 282.

33 A.A. Hummel, A.P. Wilson, and W.N. Delgass, *J. Catal.* **113** (1988) 236.

34 L.M.P. van Gruijthuijsen, G.J. Howsmon, W.N. Delgass, D.C. Koningsberger, R.A. van Santen, and J.W. Niemantsverdriet, *J. Catal.* **170** (1997) 331.

35 P.J. van Berge, J. van de Loosdrecht, S. Barradas, and A.M. van der Kraan, *Catal. Today* **58** (2000) 321.

36 M.W.J. Crajé, A.M. van der Kraan, J. van de Loosdrecht, and P.J. van Berge, *Catal. Today* **71** (2002) 369.

37 R. Bacaud, P. Bussiere, and F. Figueras, *J. Catal.* **69** (1981) 399.

38 K. Lazar, P. Bussiere, M. Guenin, and R. Frety, *Appl. Catal.* **38** (1988) 19.

39 B. Benaichouba, P. Bussiere, J.M. Friedt, and J.P. Sanchez, *Appl. Catal.* **8** (1983) 237.

40 J.-M.M. Millet, H. Roussel, A. Pigamo, J.L. Dubois, and J.C. Dumas, *Appl. Catal. A* **232** (2002) 77.

41 C.A. Clausen and M.L. Good, *J. Catal.* **38** (1975) 92; *J. Catal.* **46** (1977) 58.

42 L. Stievano, S. Calogero, F.E. Wagner, S. Galvagno, and C. Milone, *J. Phys. Chem. B* **103** (1999) 9545.

43 H. von Brandis, F.E. Wagner, J.A. Sawicki, K. Marcinkowska, and J.H. Rolston, *Hyperfine Interactions* **57** (1990) 2127.

44 Y. Kobayashi, S. Nasu, S. Tsubota, and M. Haruta, *Hyperfine Interactions* **126** (2000) 95.

45 R.M. Finch, N.A. Hodge, G.J. Hutchings, A. Meagher, Q.A. Pankhurst, M.R.H. Siddiqui, F.E. Wagner, and R. Whyman, *Phys. Chem. Chem. Phys.* **1** (1998) 485.

46 K. Blick, T.D. Mitrelias, J.S.J. Hargreaves, G.J. Hutchings, R.W. Joyner, C.J. Kiely, and F.E. Wagner, *Catal. Lett.* **50** (1998) 211.

6
Diffraction and Extended X-Ray Absorption Fine Structure (EXAFS)

Keywords

X-ray diffraction (XRD)
Low-energy electron diffraction (LEED)
X-ray absorption fine structure (XAFS)
Extended X-ray absorption fine structure (EXAFS)
X-ray absorption near edge spectroscopy (XANES)

6.1
Introduction

This chapter deals with the study of structural properties of catalysts and catalytic model surfaces by means of interference effects in scattered radiation. X-ray diffraction (XRD) is one of the oldest and most frequently applied techniques in catalyst characterization. It is used to identify crystalline phases inside catalysts by means of lattice structural parameters, and to obtain an indication of particle size. Low-energy electron diffraction (LEED) is the surface-sensitive analogue of XRD which, however, is only applicable to single crystal surfaces. LEED reveals the structure of surfaces and of ordered adsorbate layers. Both XRD and LEED depend on the constructive interference of radiation that is scattered by relatively large parts of the sample. As a consequence, these techniques require long-range order.

Extended X-ray absorption fine structure (EXAFS) on the other hand, is due to the interference of electron waves between atoms, and provides local structure information that is limited to a few interatomic distances. Here, we talk about the distance and the number of nearest and next-nearest neighbors of atoms in the catalyst. The more uniform the environment is through the catalyst, the more meaningful is the EXAFS information. Related to this method is X-ray absorption near edge spectroscopy (XANES), which deals with the detailed shape of the absorption edge, and yields important information on the chemical state of the absorbing atom. Commonly, one uses nowadays the acronym XAFS to include both EXAFS and XANES.

XRD and LEED are laboratory techniques, although synchrotrons offer advantages for XRD. XAFS, on the other hand, is conducted almost exclusively at synchrotrons. This – and the fact that EXAFS data analysis is complicated and not always without ambiguity – has inhibited the widespread use of the technique in catalysis. XANES, however, is becoming increasingly popular, as it may routinely yield similar information as X-ray photoelectron spectroscopy (XPS), but under *in-situ* conditions.

6.2
X-Ray Diffraction

X-rays have wavelengths in the Ångstrøm range, and are sufficiently energetic not only to penetrate solids but also to probe their internal structure. XRD is used to identify bulk phases, to monitor the kinetics of bulk transformations, and to estimate particle sizes. An attractive feature is that the technique can be applied *in situ*. We will first discuss XRD as conducted in the laboratory, and then describe some of the newer applications of XRD as are available by using synchrotron radiation. Typically, the theory of XRD is provided in textbooks of solid-state physics [1] and in other, more specialized, books [2–6].

A conventional X-ray source consists of a target that is bombarded with high-energy electrons. The emitted X-rays arise from two processes. Electrons slowed down by the target emit a continuous background spectrum of Bremsstrahlung. Superimposed on this are characteristic, narrow lines; the Cu Kα line, with an energy of 8.04 keV and a wavelength of 0.154 nm, arises because a primary electron creates a core hole in the K-shell, which is filled by an electron from the L-shell under emission of an X-ray quantum. Kβ radiation is emitted when the K-hole is filled from the M-shell, and so on. This process, which is called X-ray fluorescence, is the basis for X-ray sources and is also encountered in electron microscopy, EXAFS, and XPS.

X-ray diffraction is the elastic scattering of X-ray photons by atoms in a periodic lattice. The scattered monochromatic X-rays that are in phase produce constructive interference. Figure 6.1 illustrates how diffraction of X-rays by crystal planes allows one to derive lattice spacings by using the Bragg relationship:

$$n\lambda = 2d \sin \theta; \quad n = 1, 2, \ldots \tag{6-1}$$

where:
- λ is the wavelength of the X-rays;
- d is the distance between two lattice planes;
- θ is the angle between the incoming X-rays and the normal to the reflecting lattice plane;
- n is the integer called the order of the reflection.

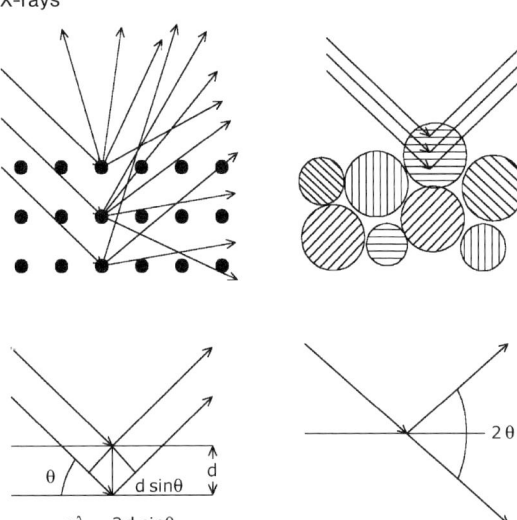

Fig. 6.1 X-rays scattered by atoms in an ordered lattice interfere constructively in directions given by Bragg's law. The angles of maximum intensity enable one to calculate the spacings between the lattice planes and allow furthermore for phase identification. Diffractograms are measured as a function of the angle 2θ. When the sample is a polycrystalline powder, the diffraction pattern is formed by a small fraction of the particles only. Rotation of the sample during measurement enhances the number of particles that contribute to diffraction.

If one measures the angles, 2θ, under which constructively interfering X-rays leave the crystal, the Bragg relationship [Eq. (6-1)] gives the corresponding lattice spacings, which are characteristic for a certain compound.

The XRD pattern of a powdered sample is measured with a stationary X-ray source (usually Cu Kα) and a movable detector, which scans the intensity of the diffracted radiation as a function of the angle 2θ between the incoming and the diffracted beams. When working with powdered samples, an image of diffraction lines occurs because a small fraction of the powder particles will be oriented such that by chance a certain crystal plane is at the correct angle θ with the incident beam for constructive interference (see Fig. 6.1). Rotating powders during measurement enhances the fraction of particles that contributes to the diffraction pattern.

In catalyst characterization, diffraction patterns are mainly used to identify the crystallographic phases that are present in the catalyst. Figure 6.2 illustrates this with an example from the studies of Kunimori et al. [7] on manganese-promoted rhodium catalysts. These authors prepared their catalysts in two-steps: by first impregnating the silica support with a solution of $RhCl_3$ in water, followed by a second impregnation with $Mn(NO_3)_2$ in water. Figure 6.2a shows XRD patterns

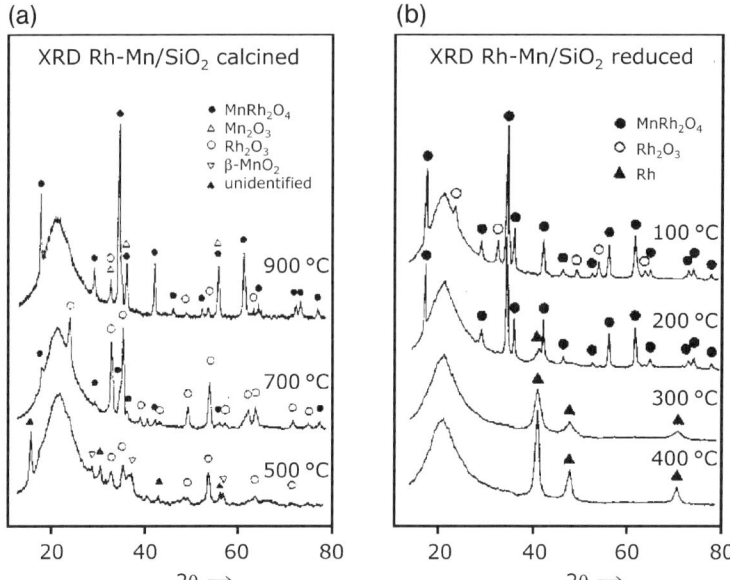

Fig. 6.2 X-ray diffraction identifies crystallographic phases in a rhodium-manganese catalyst on SiO$_2$. (a) Catalyst with atomic ratio Rh/Mn = 1 after calcination in air at the indicated temperatures. (b) Calcined Rh–Mn catalyst (atomic ratio Rh/Mn = 2) after reduction in H$_2$ at the indicated temperature. (From [7]).

of the catalysts after calcination in air. Heating in air to 500 °C produces a mixture of Rh$_2$O$_3$ and MnO$_2$ and an unidentified phase, which the authors interpret as a mixed oxide of Rh and Mn. The broad background peak around 22° is due to poorly crystallized SiO$_2$. Upon heating to 700 °C, the peaks of the Rh$_2$O$_3$ become more intense and the unidentified phase disappears. After calcination in air at 900 °C, peaks due to the mixed Rh–Mn oxide MnRh$_2$O$_4$ dominate the XRD pattern, although some Rh$_2$O$_3$ is still detectable. Note how the peaks sharpen up after treatment at progressively higher temperatures, indicating that particles become crystallographically better defined.

Figure 6.2b indicates what happens if a calcined catalyst (which according to XRD consists mainly of 27-nm MnRh$_2$O$_4$ particles) is reduced in hydrogen. Reduction at 100 °C reveals no changes that are detectable by XRD, but after reduction at 200 °C, the minor constituent Rh$_2$O$_3$ has disappeared and a weak signal due to metallic Rh becomes visible. Reduction at 300 °C decomposes all large MnRh$_2$O$_4$ particles to metallic Rh and X-ray amorphous (i.e., invisible) manganese and perhaps also rhodium phases. The width of the Rh peaks corresponds to a diameter of approximately 6 nm. Reduction at 500 °C leads to sharper diffraction lines corresponding to Rh particles with a mean diameter of 12 nm, but still

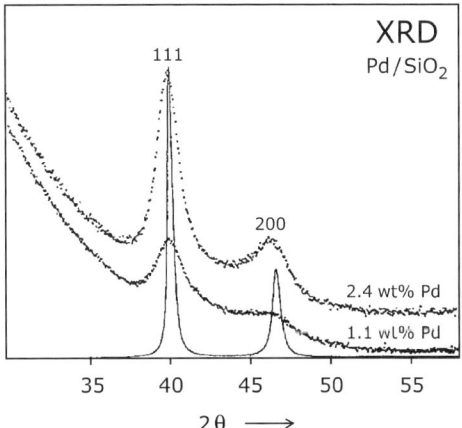

Fig. 6.3 X-ray diffraction (XRD) pattern showing the (111) and (200) reflections of Pd in two silica-supported palladium catalysts and of a Pd reference sample. The reader may use the Bragg equation [Eq. (6-1)] to verify that the Pd (111) and (200) reflections are expected at angles 2θ of 40.2 and 46.8° with Cu Kα radiation (lattice constant of Pd is 0.389 nm, $d_{111} = 0.225$ nm, $d_{200} = 0.194$ nm, $\lambda = 0.154$ nm. (From [8]).

there is no indication of any Mn-containing phase in the XRD pattern. The catalytic properties of the sample differ from those of a pure Rh catalyst, which led the authors to believe that the surface of rhodium is affected by manganese.

XRD has one important limitation, namely that clear diffraction peaks are only observed when the sample possesses sufficient long-range order. The advantage of this limitation is that the width (or rather the shape) of diffraction peaks carries information on the dimensions of the reflecting planes. Diffraction lines from perfect crystals are very narrow, see for example the (111) and (200) reflections of large palladium particles in Figure 6.3. For crystallite sizes below 100 nm, however, line broadening occurs due to incomplete destructive interference in scattering directions where the X-rays are out of phase. The two XRD patterns of supported Pd catalysts in Figure 6.3 show that the reflections of palladium are much broader than those of the reference [8].

The Scherrer formula relates crystal size to line width:

$$\langle L \rangle = \frac{K\lambda}{\beta \cos \theta} \tag{6-2}$$

where:
$\langle L \rangle$ is a measure for the dimension of the particle in the direction perpendicular to the reflecting plane;

λ   is the X-ray wavelength;
β   is the peak width;
θ   is the angle between the beam and the normal on the reflecting plane;
K   is a constant (often taken as 1).

When applied to the XRD patterns of Figure 6.3, average diameters of 4.2 and 2.5 nm are found for the catalysts with 2.4 and 1.1 wt% Pd, respectively [8].

The Scherrer equation [Eq. (6-2)] indicates that measuring at smaller wavelengths gives sharper peaks, not only because λ becomes smaller but also because the diffraction lines shift to lower angles, which decreases the $1/\cos\theta$ term in Eq. (6-2). Both effects help to reduce line broadening. Thus, by using Mo Kα X-rays (17.44 keV; $\lambda \approx 0.07$ nm), one can obtain diffraction patterns from smaller particles than with Cu Kα radiation (8.04 keV; 0.15 nm).

X-ray line broadening provides a quick – but not always reliable – estimate of the particle size. As Cohen [9] points out, the size thus determined is merely a ratio of two moments in the particle size distribution, equal to $\langle L^2 \rangle / \langle L \rangle$. Both averages are weighted by the volume of the particles, and not by number or by surface area, as would be more meaningful for a surface phenomenon such as catalysis. Internal strain and instrumental factors also contribute to broadening.

Better procedures to determine particle sizes from XRD are based on line-profile analyses with Fourier transform methods. The average size is obtained from the first derivative of the cosine coefficients, and the distribution of particle sizes from the second derivative. When used in this way, XRD offers a fundamental advantage over electron microscopy, because it samples a much larger portion of the catalyst. The reader is referred to publications by Cohen and co-workers for more details and examples [3, 10, 11].

The examples in Figures 6.2 and 6.3 demonstrate the strengths and the weaknesses of XRD for investigating catalysts. The technique provides clear and unequivocal information on particles that are sufficiently large, but it does not see particles that are either too small or amorphous. Hence, one can never be sure that no other phases are present than those detected with XRD. In particular, the surface – where the catalytic activity resides – is invisible in XRD.

6.2.1
In-Situ XRD: Kinetics of Solid-State Reactions

One great advantage of X-rays is that they have considerable penetrating power, such that XRD can be used to study catalysts under *in-situ* conditions. Canton et al. [12] have described a number of different cells for this purpose. Figure 6.4 shows a set-up for *in-situ* XRD studies as published by Jung and Thomson [13]. This consists of a cell with windows which are largely transparent to X-rays, and a heating stage covered by a platinum foil, which serves as a thermally conducting sample holder for the powdered catalyst. These authors used this cell to follow the reduction of supported iron oxide catalysts, and obtained detailed insight in the kinetics of the reduction process. During reduction the initial Fe_2O_3 is rapidly

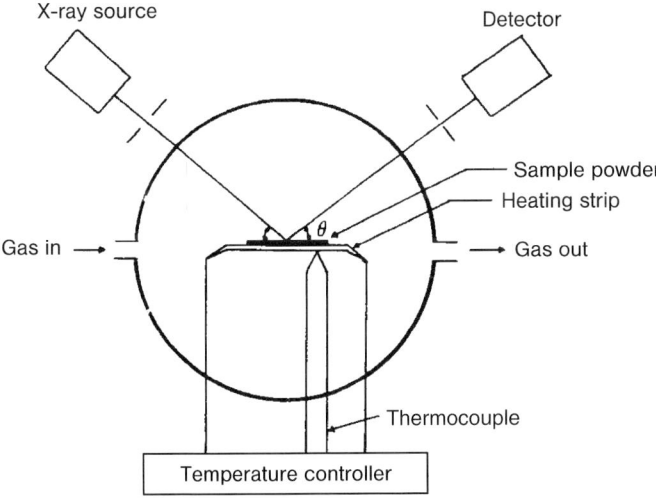

Fig. 6.4 Experimental set-up for *in-situ* or dynamic X-ray diffraction studies of catalysts. (From [13]).

converted to Fe_3O_4, which in turn converts at a slower rate to metallic iron, in agreement with the phase diagram in Figure 2.2.

Monitoring solid-state reactions that play a role in catalyst activation forms a useful application of XRD. In this regard, Figure 6.5 illustrates an elegant example of the reduction of copper oxide powder to metallic copper, although this was measured at a synchrotron [14].

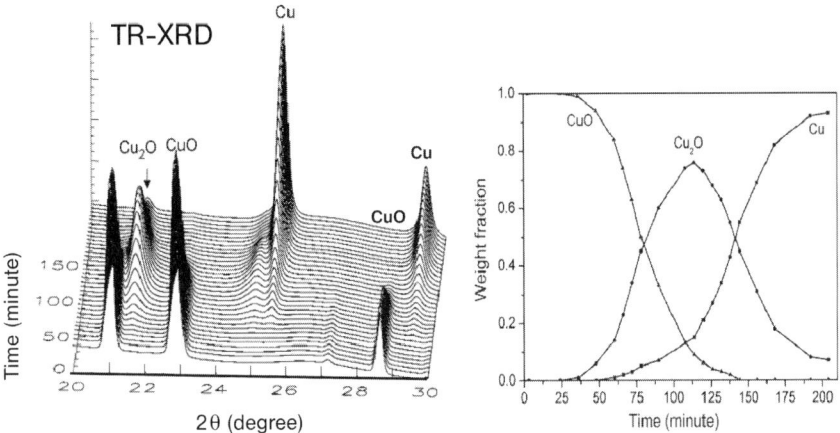

Fig. 6.5 Time-resolved (TR) XRD patterns measured at a synchrotron for the isothermal reduction of CuO powder in CO at about 500 K, along with weight fractions of CuO, Cu_2O and Cu. (Adapted from [14]).

Using synchrotron radiation as a source for such studies has considerable advantages [15].

- First, the high intensity of the radiation produces data of much better signal-to-noise ratio, and as a consequence patterns with broad peaks from small particles can be determined with much better accuracy. The collection times for a diffractogram are significantly shorter, which is an advantage for *in-situ* studies during a reaction. (An example of the temperature-programmed reduction of supported copper oxide, which was followed *in situ* simultaneously by XRD and XAFS, is provided in Figure 6.17.)

- Second, one can easily vary the wavelength of the X-rays. This offers several attractive opportunities. As mentioned earlier, measuring at higher energy reduces line broadening, enabling one to observe smaller particles. Another advantage is that one can use the phenomenon that the scattering efficiency of an element decreases if the energy of the radiation is close to an absorption edge. This so-called *anomalous scattering* can be used to distinguish the diffraction pattern of the supported particles from that of the support. Cohen and co-workers successfully applied this elegant procedure to platinum particles with diameters as small as 1.5 nm [16].

In addition to diffraction, one can perform small-angle scattering, where the scattering centers are the interfaces between particle and support, particle and gas phase, and support and gas phase. Examples can be found in the studies of Goodisman et al. [17] on supported catalysts, and of Beelen and colleagues on catalyst supports [18].

6.2.2
Concluding Remarks

The examples illustrate the strong points of XRD for catalyst studies: XRD identifies crystallographic phases, if desired under *in-situ* conditions, and can also be used to monitor the kinetics of solid-state reactions such as reduction, oxidation, sulfidation, carburization or nitridation, that are used in the activation of catalysts. In addition, careful analysis of diffraction line shapes or – more common but less accurate – a simple determination of the line broadening yields information on particle size.

However, XRD has some serious disadvantages. Because it depends on interference between X-rays reflecting from lattice planes, the technique requires samples that possess sufficient long-range order. Amorphous phases and small particles give either broad and weak diffraction lines or no diffraction at all, with the consequence that if catalysts contain particles with a size distribution, XRD may detect only the larger ones. XRD at synchrotrons greatly improves the possibilities to study small particles. Finally, although the catalytic activity resides at the surface region, this part of the catalyst is virtually invisible to XRD.

6.3
Low-Energy Electron Diffraction (LEED)

LEED is used to determine the surface structure of single crystal surfaces and the structure of ordered adsorbate layers [19–22]. The principle of LEED is illustrated in Figure 6.6: a beam of mono energetic low-energy electrons (50–200 eV, minimum mean free path) falls on a surface, whereupon electrons are scattered elastically in all directions. The electrons can be considered as a wave with wavelength

$$\lambda = \frac{h}{\sqrt{2m_e E_{kin}}} \tag{6-4}$$

where:
λ is the wavelength of the electrons;
h   is Planck's constant;
m_e is the mass of the electron;
E_{kin} is the kinetic energy of the electron.

Hence, scattered electrons will exhibit an interference pattern with constructive interference in directions with

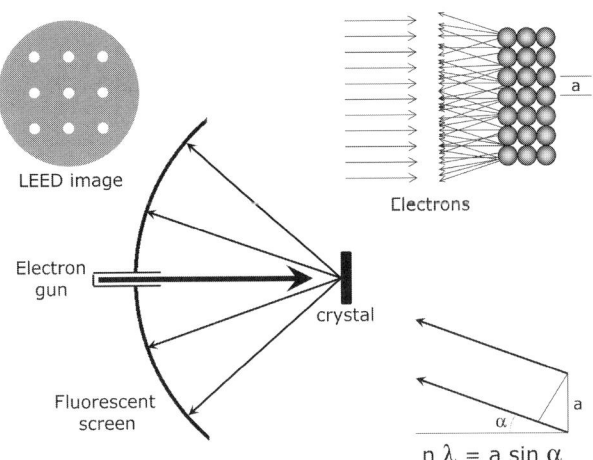

Fig. 6.6 The principle of low-energy electron diffraction (LEED) is that a beam of monoenergetic electrons scatters elastically from a surface. Due to the periodic order of the surface atoms, electrons show constructive interference in directions for which the path lengths of the electrons differ by an integral number times the electron wavelength. Directions of constructive interference are made visible by collecting the scattered electrons on a fluorescent screen.

$$\sin \alpha = \frac{n\lambda}{a} = \frac{nh}{a\sqrt{2m_e E_{kin}}}, \quad n = 0, 1, 2, \ldots \tag{6-5}$$

where:
- α is the angle between scattered electrons and the surface normal;
- n is the order of the diffraction;
- a is the distance between two atoms in the surface;

and the other symbols as defined after Eq. (6-4) above. Hence, if the scattered electrons are collected with a fluorescent screen, one observes a pattern of spots, each of which corresponds to a direction in which constructive interference takes place. The set-up is shown schematically in Figure 6.6.

Because of the inverse relationship between interatomic distances and the directions in which constructive interference between the scattered electrons occurs, the separation between LEED spots is large when interatomic distances are small, and vice versa. The LEED pattern has the same form as the so-called "reciprocal lattice". This concept plays an important role in the interpretation of diffraction experiments, as well as in understanding the electronic or vibrational band structure of solids. In two dimensions the construction of the reciprocal lattice is simple. If a surface lattice is characterized by two base vectors a_1 and a_2, the reciprocal lattice follows from the definition of the reciprocal lattice vectors $a_1{}^*$ and $a_2{}^*$:

$$\vec{a}_i \cdot \vec{a}_j{}^* = \delta_{ij} \tag{6-6}$$

where:
- a_i are the base vectors of the real lattice ($i = 1, 2$);
- $a_j{}^*$ are the base vectors of the reciprocal lattice ($j = 1, 2$);
- δ_{ij} is the Kronecker delta, $\delta_{11} = \delta_{22} = 1$, $\delta_{12} = \delta_{21} = 0$.

The most important properties of the reciprocal lattice are summarized in Figure 6.7. It is important that the base vectors of the surface lattice form the smallest parallelogram from which the lattice may be constructed through translations. Figure 6.8 shows the five possible surface lattices and their corresponding reciprocal lattices, which are equivalent to the shape of the respective LEED patterns. The unit cells of both the real and the reciprocal lattices are indicated. Note that the actual dimensions of the reciprocal unit cell are irrelevant, only the shape is important.

In case adsorbed gases form ordered layers with periodical structures, the unit cell of the overlayers can readily be determined with LEED. The exact position of the adsorbate atoms with respect to the substrate, however, remains undetermined, because the LEED pattern reflects only the periodicity in the layer.

Figure 6.9 illustrates LEED patterns of the clean Rh(111) surface, and the surface after adsorption of 0.25 monolayer of NH_3 [23]. The latter orders in the primitive (2 × 2) overlayer structure (see Appendix for the Wood's notation). In

Definition: $\vec{a}_i \cdot \vec{a}_j^* = \delta_{ij}$

Directions: $\vec{a}_1^* \perp \vec{a}_2 \, ; \, \vec{a}_2^* \perp \vec{a}_1$

Length: $|\vec{a}_i^*| = \dfrac{1}{|\vec{a}_i| \cdot \sin \gamma}$

Angle: $\angle (\vec{a}_1^*, \vec{a}_2^*) = 180^\circ - \gamma$

Area unit cell: $A^* = |\vec{a}_1^* \times \vec{a}_2^*|$
$= \dfrac{1}{|\vec{a}_1 \times \vec{a}_2|} = \dfrac{1}{A}$

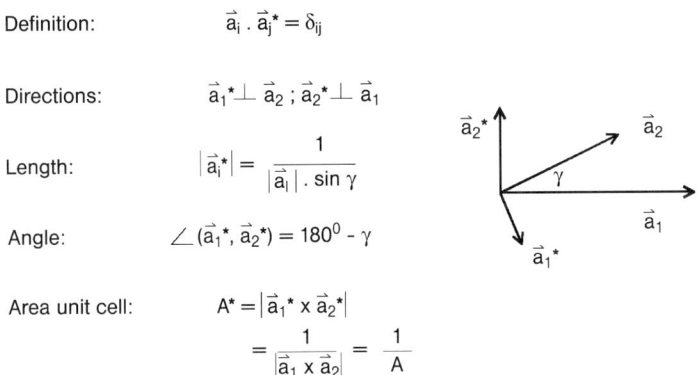

Fig. 6.7 Definition and properties of the two-dimensional reciprocal lattice; a_1 and a_2 are base vectors of the surface lattice, a_1^* and a_2^* are the base vectors of the reciprocal lattice. The latter is equivalent to the LEED pattern.

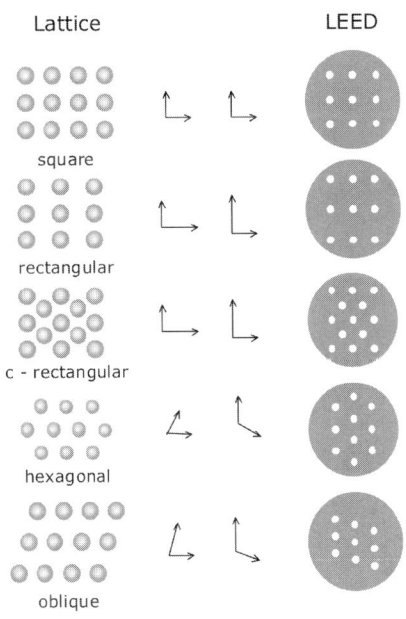

Fig. 6.8 The five different surface lattices, base vectors of the real and reciprocal lattices, and the corresponding LEED patterns.

the (2 × 2) overlayer, a new unit cell exists on the surface with twice the dimensions of the substrate unit cell. Hence, the reciprocal unit cell of the adsorbate has half the size of that of the substrate, and the LEED pattern shows four times as many spots. Calculations by Frechard et al. [24] on the $NH_3/Rh(111)$ have con-

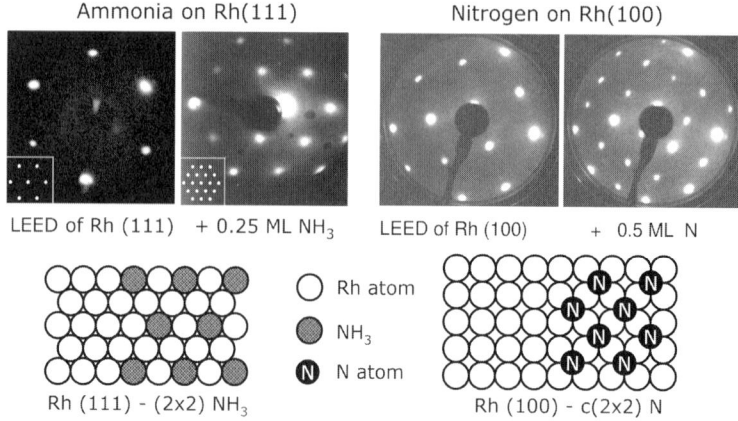

Fig. 6.9 LEED images of Rh(111) and (100) with ordered layers of ammonia and nitrogen. The insets show the idealized patterns. In the left image, the middle spot in the Rh(111) image is missing because it is covered by the electron gun (see the LEED apparatus in Fig. 6.6). In the second image of Rh(111) with NH$_3$, the sample has been rotated slightly away from the gun, which is visible as the black shadow in the center. Note also that the crystal has been rotated over 30° in this image. The ammonia orders in a primitive (2 × 2) structure, corresponding to a coverage of 0.25 monolayer (ML). (Adapted from [23]). N-atoms order in a c(2 × 2) structure on Rh(100), corresponding to 0.5 ML coverage. (Adapted from [25]).

firmed that the molecule adsorbs on top of the rhodium atoms, as indicated in the schematic structure of Figure 6.9.

Also shown in Figure 6.9 are the LEED patterns of the square Rh(100) and an ordered layer of nitrogen atoms in a c(2 × 2) structure [25]. Actually, c(2 × 2) is a convenient – but incorrect – way to indicate this pattern, based on a unit cell that is larger than needed to describe the structure. The correct unit cell – this is important because it is needed to interpret the diffraction pattern – is $(\sqrt{2} \times \sqrt{2})R45°$ with respect to the (1 × 1) unit cell of the substrate, and the right way to describe the structure would be $\{Rh(100) - (\sqrt{2} \times \sqrt{2})R45° \ N\}$ (which hardly anyone uses). Calculations indicate that the N-atoms occupy the fourfold hollow positions between the rhodium atoms [26]. In Chapter 8, we discuss the LEED patterns of CO adsorbed on Rh(111), which gives two more examples of ordered adsorbate structures and how they appear in LEED experiments (see Fig. 8.15).

So far, we have discussed LEED patterns entirely in terms of two-dimensional lattices. However, the third dimension is also involved, because electrons can be scattered from deeper layers. For each layer spacing there are combinations of electron energies and angles of diffraction for which a Bragg condition in the vertical direction is satisfied. In such situations one observes very intense spots. Plots of the spot intensity as a function of the electron energy are called I–V plots; these contain, in combination with the LEED pattern itself, three-dimensional structure information. In this way it is possible to determine the exact location of adsorbed atoms. Although the measurement of I–V curves is rather straightforward, their interpretation requires a comparison with model calculations based on the full diffraction process, including multiple scattering contributions [19–22]. Automated procedures for optimizing the structures corresponding to a given set of LEED intensities have recently been developed [27].

The most recent advances in structure determination by LEED make use of holographic effects. In short, adsorbed atoms in an ordered superstructure on the surface act as beam splitters, reflecting a reference wave and transmitting a wave that reflects from the surface as the object wave. Both waves together constitute the holographic image, from which the adsorption geometry can, in principle, be reconstructed [28].

LEED intensities also depend on the lattice vibrations at the surface of the crystal. A high Debye temperature of the surface results in intense LEED spots. In fact, measurement of spot intensities as a function of primary electron energy provides a way to determine the surface Debye temperature [21].

In summary, LEED is most often used to verify the structure and quality of single crystal surfaces, to study the structure of ordered adsorbates, and to study surface reconstructions. In more sophisticated uses of LEED, one determines also exact positions of atoms, the nature of defects, and the morphology of steps as well as Debye temperatures of the surface.

6.4
X-Ray Absorption Fine Structure (XAFS)

XAFS provides detailed information on chemical composition and local structure [29, 30]. The technique is based on the absorption of X-rays and the creation of photoelectrons, which are either excited to holes in the valence levels, or to unbound states and scattered by nearby atoms in a lattice. The situation is illustrated in Figure 6.10. First, we consider the X-ray absorption spectrum of a free atom which has an electron with binding energy E_b. If we irradiate this atom with X-rays of energy $h\nu$, absorption takes place when $h\nu \geq E_b$ and the electron leaves the atom with a kinetic energy $E_k = h\nu - E_b$. The X-ray absorption spectrum shows edges corresponding to the binding energy of all electron core levels in the atom, but contains no further structure. If, however, the atom is bound in a lattice, the absorption is modulated due to its local coordination, and fine structure arises, which contains highly valuable information, as will be seen later.

X-ray absorption near edge spectroscopy (XANES) focuses on the shape of the absorption edge, and is highly sensitive for the valence state of the atom and its bonding geometry. Extended X-ray absorption fine structure (EXAFS) deals with the interference effects visible in the absorption spectrum beyond the edge, and provides detailed information on the distance, number, and type of neighbors of the absorbing atom.

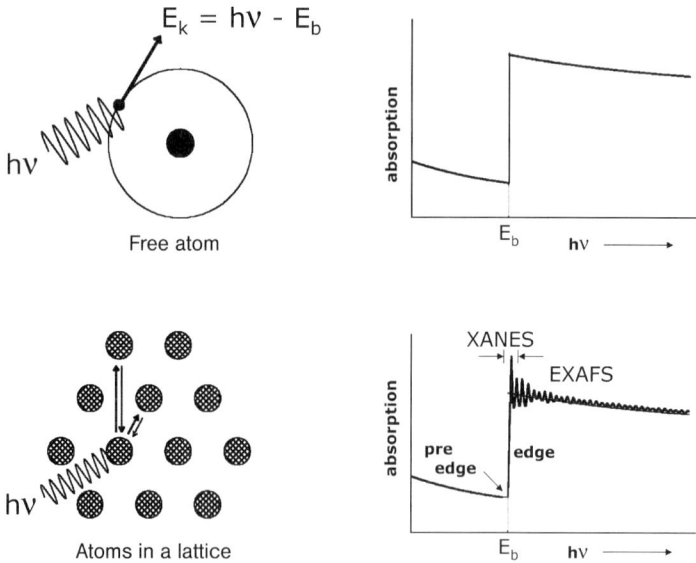

Fig. 6.10 Absorption of X-rays as a function of photon energy $E = h\nu$ by a free atom and by atoms in a lattice. The fine structure represents the extended X-ray absorption fine structure (EXAFS) function. XANES: X-ray absorption near edge spectroscopy.

6.4.1
EXAFS

Although the EXAFS phenomenon was first recognized during the 1920s [31, 32], it was not exploited as an analytical tool until the bright, tunable sources of X-rays at synchrotrons became available during the 1970s. Subsequently, laboratory systems have been built [29], and although these spectrometers may be quite successful for applications in the low-energy range, they have in general been abandoned in favor of synchrotrons. Important points in the progress of EXAFS were the theoretical developments by Sayers et al. [33], who showed how local structure information is extracted from EXAFS measurements.

At the time that EXAFS was introduced into catalysis, around 1975, the technique was considered to be one of the most promising tools for investigating catalysts. Unfortunately, these high expectations have not quite been fulfilled, mainly because the analysis of data for EXAFS is highly complicated and not always possible without ambiguity. Several successful applications, however, have proven that EXAFS – when applied with care on optimized catalysts – can be a very powerful tool in catalysis [30, 34, 35].

As illustrated in Figure 6.10, fine structure arises due to neighbors that surround the absorbing atom. The photoelectron, having both particle and wave character, can be scattered back from a neighboring atom. Because of its wave character, the outgoing and the backscattered electrons interfere. Depending on the wavelength of the electron, the distance between the emitting and scattering atom, and also the shift in phase caused by the scattering event, the two waves either enhance or destroy each other. As a result, the cross-section for X-ray absorption is modulated by the interference between the photoelectron waves, such that it is enhanced at energies where constructive interference occurs. As indicated in Figure 6.10, the X-ray absorption spectrum exhibits fine structure, which extends to several hundred eV above the absorption edge. The absorption around the edge arises from electrons with low kinetic energies which interact with valence electrons. This part of the spectrum is often referred to as the NEXAFS or XANES [36].

Before we discuss the mathematics behind EXAFS, it is good to have a feeling for how distances, coordination numbers and concentrations affect the EXAFS spectrum. The intensity of the wiggles goes up if the number of neighbors increases, the number of oscillations depends inversely on interatomic distances (as in any scattering or diffraction experiment), and the step height of the edge is proportional to the concentration of atoms in the sample.

The EXAFS function, $\chi(k)$, is extracted from the X-ray absorption spectrum in Figure 6.10 by removing first the approximately parabolic background and next the step – that is, the spectrum of the free atom in Figure 6.10. As in any scattering experiment, it is customary to express the signal as a function of the wave number, k, rather than of energy. The relationship between k and the kinetic energy of the photoelectron is:

$$k = \frac{2\pi}{h}\sqrt{2m_e E_k} = \frac{2\pi}{h}\sqrt{2m_e(h\nu - E_b)} \tag{6-7}$$

where:
- k is the wavenumber of the photoelectron;
- h is Planck's constant;
- m_e is the mass of an electron;
- E_k is the kinetic energy of the photoelectron;
- ν is the X-ray frequency;
- E_b is the binding energy of the photoemitted electron.

For those unfamiliar with wave vectors, $hk/2\pi$ is the momentum of a wave quantum, whereas $\sqrt{(2m_e E_{kin})} = m_e v$ is the classical momentum of the electron when considered as a particle.

It is impossible to explain EXAFS properly without theoretical formulae. In a mono atomic solid, the EXAFS function $\chi(k)$ is the sum of the scattering contributions of all atoms in neighboring coordination shells:

$$\chi(k) = \sum_j A_j(k) \sin(2kr_j + \phi_j(k)) \tag{6-8}$$

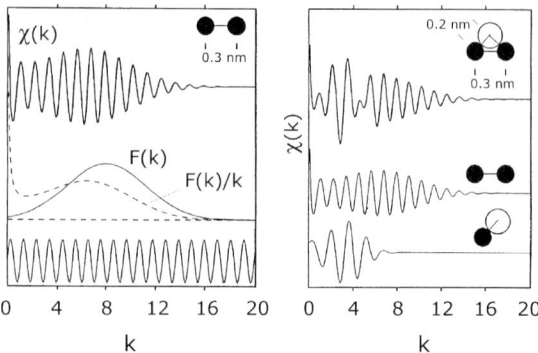

Fig. 6.11 Left: Simulated EXAFS spectrum of a dimer such as Cu_2, showing that the EXAFS signal is the product of a sine function and a backscattering amplitude $F(k)$ divided by k, as expressed by Eqs. (6-8) and (6-9). Note that $F(k)/k$ remains visible as the envelope around the EXAFS signal $\chi(k)$. Right: The Cu EXAFS spectrum of a cluster such as Cu_2O is the sum of a Cu–Cu and a Cu–O contribution. Fourier analysis is the mathematical tool to decompose the spectrum in the individual Cu–Cu and Cu–O contributions. Note the different backscattering properties of Cu and O, revealed in the envelope of the individual EXAFS contributions. For simplicity, phase shifts have been ignored in the simulations.

where:
$\chi(k)$ is the EXAFS function, with the wave number as defined in Eq. (6-7);
j is the label of the coordination shells around the electron-emitting atom;
$A_j(k)$ is the amplitude, the scattering intensity due to the jth coordination shell;
r_j is the distance between the central atom and atoms in the jth shell;
$\phi(k)$ is the total phase shift, equal to the phase shift of the backscattering atom plus twice that of the absorbing atom.

Thus, each coordination shell contributes a sine function multiplied by an amplitude, as illustrated in Figure 6.11 for the simple case of a dimer. EXAFS analysis boils down to recognizing all sine contributions in $\chi(k)$. The obvious mathematical tool to achieve this is Fourier analysis.

The argument of each sine contribution in Eq. (6-8) depends on k (which is known), on r (which is to be determined), and on the phase shift $\phi(k)$. The latter needs to be known before r can be determined. The phase shift is a characteristic property of the scattering atom in a certain environment, and is best derived from the EXAFS spectrum of a reference compound, for which all distances are known. For example, the phase shift for zero-valent rhodium atoms in the EXAFS spectrum of a supported rhodium catalyst is best determined from a spectrum of pure rhodium metal (as in Fig. 6.12), while Rh_2O_3 may provide a reference for the scattering contribution from oxygen neighbors in the metal support interface.

The amplitude $A_j(k)$ of each scattering contribution contains the number of neighbors in a coordination shell as the most desirable information:

$$A_j(k) = N_j \frac{e^{-2r_j/\lambda(k)}}{kr_j^2} S_o^2(k) F_j(k) e^{-2k^2\sigma_j^2} \qquad (6\text{-}9)$$

where:
N_j is the coordination number of atoms in the jth shell;
S_o is the correction for relaxation effects in the emitting atom;
F_j is the backscattering factor of atoms in the jth shell;
λ is the inelastic mean free path of the electron (see Fig. 3.1);
σ^2 is the mean-squared displacement of atoms in the sample;

and all other symbols as defined with Eqs. (6-7) and (6-8). All of these contributions to $A_j(k)$ will be discussed in the order that they appear in the formula.

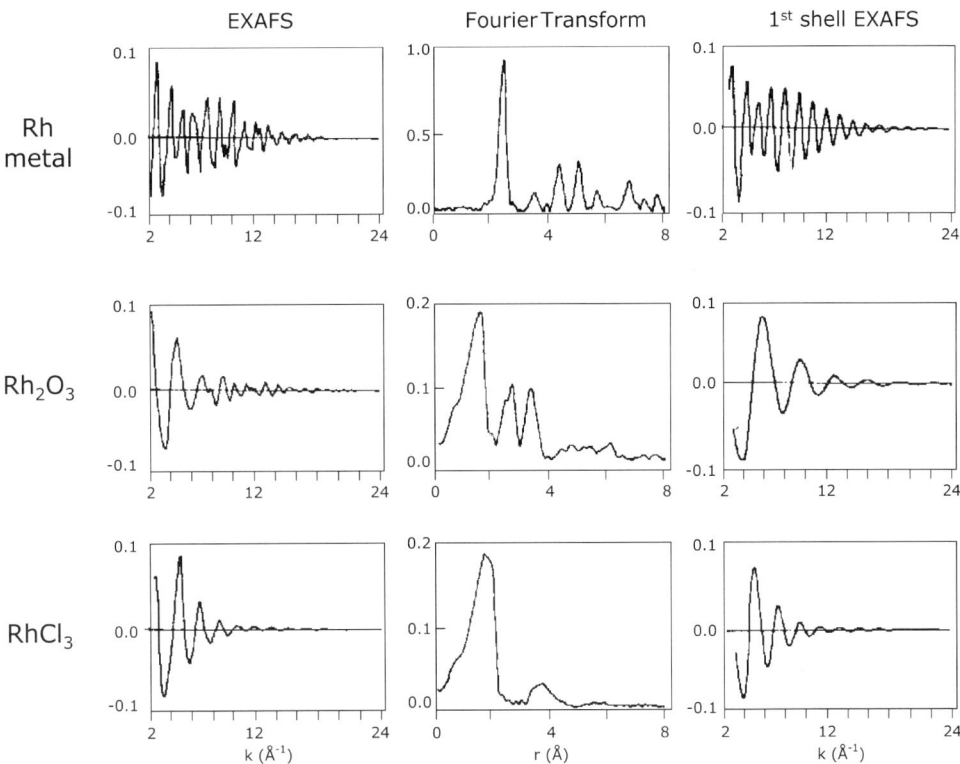

Fig. 6.12 Rh K-edge EXAFS spectra, uncorrected Fourier transforms according to Eq. (6-10) and isolated EXAFS contribution from the first neighbor shell of Rh metal (top), Rh_2O_3 (middle) and $RhCl_3$ (bottom). The first shell contributions clearly reflect the different backscattering properties of Rh, O and Cl atoms. Note the high number of coordination shells that are visible in Rh bulk metal. (From [38]).

N_j is the coordination number, equal to the number of neighbors in the jth coordination shell. For an fcc metal such as rhodium, we expect 12 neighbors in the first shell. If a particle becomes small, the average coordination number decreases. Note that unless the sample is that of a single element, N is a fractional coordination number – that is, the product of the real coordination number and the concentration of the element involved.

The second term in Eq. (6-9) expresses that nearest and next-nearest neighbors dominate the scattering contributions to the EXAFS signal, while contributions from distant shells are weak. The dependence of the amplitude on $1/r^2$ reflects that the outgoing electron is a spherical wave, the intensity of which decreases with the distance squared. The term $\exp(-2r/\lambda)$ represents the exponential attenuation of the electron when it travels through the solid, similarly as in the electron spectroscopies described in Chapter 3. The factor 2 is there because the electron must make a round trip between the emitting and the scattering atom in order to cause interference.

The term $S_o^2(k)$ in Eq. (6-9) is a correction for relaxation or final state effects in the emitting atom, such as the shake-up, shake-off and plasmon excitations discussed in Chapter 3. The result of these processes is that not all absorbed X-ray quanta of energy $h\nu$ are converted to photoelectrons of kinetic energy $h\nu - E_b$, but also to electrons with lower kinetic energy.

The backscattering factor $F_j(k)$ represents the scattering performance of the jth neighbor. Figure 6.11 illustrates how the backscattering factor determines the shape of an EXAFS signal if there is only one type of neighbor at a single distance. In this case, $F_j(k)$ is recognizable in the envelope of the oscillating EXAFS function. The dependence of the backscattering factor on energy is characteristic for an element. Hence, via $F_j(k)$ one can often identify the scattering atom. For example, one easily distinguishes between the scattering contributions of O and Rh neighbors in the EXAFS spectrum of Rh_2O_3, because the backscattering function of oxygen neighbors, recognizable as the envelope of the EXAFS signal corresponding to the first shell in Figure 6.12, falls off much more rapidly with energy than that of rhodium.

The term $\exp(-2k^2\sigma^2)$ in Eq. (6-9) accounts for the disorder of the solid. Static disorder arises if atoms of the same coordination shell have slightly different distances to the central atom. Amorphous solids, for instance, possess large static disorder. Dynamic disorder, on the other hand, is caused by lattice vibrations of the atoms, as explained in Appendix 1. Dynamic disorder becomes much less important at lower temperatures, and it is therefore an important advantage to measure spectra at cryogenic temperatures, especially if a sample consists of highly dispersed particles. The same argument holds in X-ray and electron diffraction, as well as in Mössbauer spectroscopy.

The EXAFS function becomes understandable if we examine the Fourier transform of $\chi(k)$, which resembles a radial distribution function:

$$\theta_n(r) = \frac{1}{\sqrt{2\pi}} \int_{k_{min}}^{k_{max}} k^n \chi(k) e^{2ikr} \, dk \qquad (6\text{-}10)$$

where:
$\theta(r)$ is the Fourier transform of the EXAFS signal;
n is an integer, usually chosen as 1, 2, or 3;

and all other symbols as defined in connection with Eqs. (6-8) and (6-9). The function $\theta_n(r)$ represents the probability of finding an atom at a distance r, modified by the two r-dependent terms in the amplitude, which progressively decrease the intensity of distant shells. The transform is often weighted with either k or k^3, to emphasize the role of light or heavy atoms, respectively. Vaarkamp, however, has argued that there is little mathematical justification for doing so, and advocates the use of weight factors based on the statistical errors in the EXAFS data [37]. In principle, the Fourier transform becomes more accurate when the k-interval is larger, but in practice the signal-to-noise ratio of the spectrum sets the limit for k.

A straightforward Fourier transform of the EXAFS signal does not yield the true radial distribution function. First, the phase shift causes each coordination shell to peak at the incorrect distance; second, due to the element specific back-scattering amplitude, the intensity may not be correct. The appropriate corrections can be made, however, when phase shift and amplitude functions are derived from reference samples or from theoretical calculations. The phase- and amplitude-corrected Fourier transform becomes:

$$\theta_n(r) = \frac{1}{\sqrt{2\pi}} \int_{k_{min}}^{k_{max}} k\chi(k) \frac{e^{-i\phi(k)}}{F_j(k)} e^{2ikr} \, dk \tag{6-11}$$

with all symbols as defined above. Figure 6.13 illustrates the effect of phase and amplitude correction on the EXAFS of a Rh foil [39]. Indicated is the magnitude of the Fourier transform, plotted twice – both positively and negatively – together with its imaginary part, which is the oscillating function in Figure 6.13. The latter is a sensitive function of distances, and has important diagnostic value.

Fourier transforms, as shown in Figures 6.12 and 6.13, usually show intensity at distances that are too small to correspond to neighbor atoms. Although artifacts induced by the limited k-range over which the Fourier transform has been taken may occur, part of the intensity at low values of r is due to scattering of the electrons inside the atom where they originate from [40, 41].

The analysis of an EXAFS spectrum is a time-consuming and expertise-demanding venture, for which the following are required:

- Extract the EXAFS function $\chi(k)$ from measured data.
- Select a k-interval and compute the k- or k^3-weighted Fourier transform.
- Identify individual contributions to the Fourier transform and use these to construct a set of parameters that gives acceptable fits to the spectrum $\chi(k)$, the magnitude, and the imaginary part of the Fourier transforms.

In order to do this, one requires phases and backscattering amplitudes, preferably from spectra of reference compounds, although calculated data are also used.

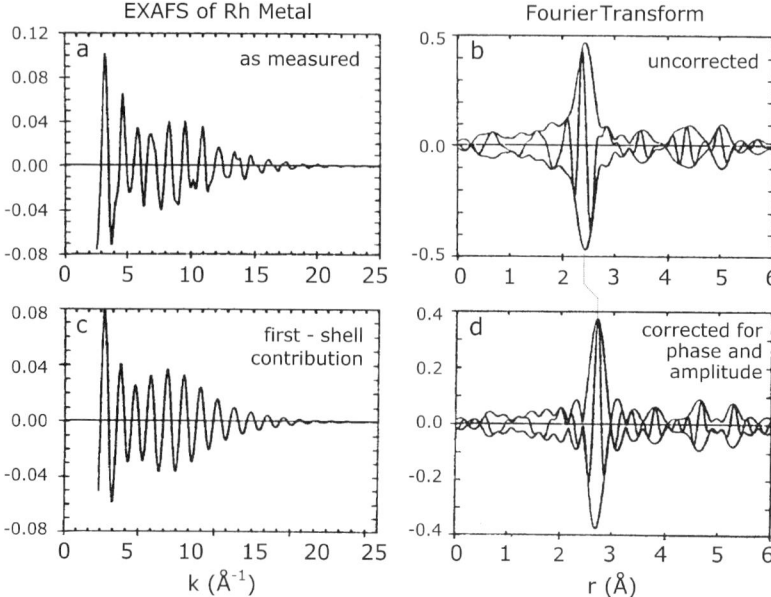

Fig. 6.13 EXAFS and Fourier Transform of rhodium metal, showing: (a) the measured EXAFS spectrum; (b) the uncorrected Fourier Transform according to Eq. (6-10); (c) the first Rh–Rh shell contribution being the inverse of the main peak in the Fourier transform; and (d) the phase- and amplitude-corrected Fourier transform according to Eq. (6-11). The Fourier transform is a complex function, and hence the transforms give the magnitude of the transform (the positive and the negative curve are equivalent) as well as the imaginary part, which oscillates between the magnitude curves. (From [39]).

We illustrate EXAFS with a study performed by Sinfelt et al. [42] on bimetallic Ru–Cu/SiO$_2$. Although this example has found its way into several reviews [34, 35], we repeat part of it here because the investigations nicely illustrate the information that EXAFS provides.

Ruthenium and copper are not miscible; hence, homogeneous alloy particles will not be formed in supported Ru–Cu catalysts. As copper has a smaller surface free energy than ruthenium, we expect that if the two metals are present in one particle, copper will be at the surface and ruthenium in the interior (see also Appendix). This is indeed what chemisorption experiments and catalytic tests suggest [43]. EXAFS, being a probe for local structure, is of particular interest here because it investigates the environment of both Ru and Cu in the catalysts.

Let us first consider what EXAFS might tell us in the case of bimetallic particles that are not too small, say a few nanometers in diameter. For a truly homogeneous alloy with a 50–50% composition, EXAFS should see a coordination shell of nearest neighbors with 50% Cu and 50% Ru around both ruthenium and copper atoms. If, on the other hand, the particle consists of a Ru core surrounded by a Cu shell of monatomic thickness, we expect that the Ru EXAFS shows Ru as the dominant neighbor, because only Ru atoms in the layer directly below the surface are in contact with Cu. The Cu EXAFS should see both Cu neighbors in the surface and Ru neighbors from the layer underneath, with a total coordination number smaller than that of the Ru atoms. The latter situation is indeed observed in Ru–Cu/SiO$_2$ catalysts.

Quantitative analysis of the data revealed that the first coordination shell of the average Ru atom in the bimetallic catalyst contains 90% Ru and 10% Cu, whereas the first coordination shell of the average Cu atom contains 50% Cu and 50% Ru. This, together with the overall lower coordination of copper, is in agreement with particles consisting of a Ru core and an outer shell of Cu atoms.

Exposing a bimetallic catalyst to oxygen, to see which of the constituent elements is affected, offers an additional method to investigate which element enriches at the surface. Figure 6.14 shows the Fourier transforms of the catalysts before and after exposure to oxygen. It is important to realize that the transforms have been taken over a range from 3 to 15 Å^{-1} in k-space. As the backscattering function of oxygen has its maximum well below 3 Å^{-1} and falls rapidly off with increasing k (see Fig. 6.12), the peaks of Ru–O and Cu–O are largely suppressed. Hence, Figure 6.14 shows the influence of oxygen adsorption on the first coordination shell of the metal neighbors, which appears quite informative.

The Fourier transform of the Ru EXAFS of the Ru/SiO$_2$ catalyst under O$_2$ is approximately 30% lower than that of the reduced catalyst, indicating that ruthenium atoms in the outer layer(s) are oxidized. The Fourier transform of Ru in the EXAFS of Ru–Cu/SiO$_2$, however, is not affected by oxygen, indicating that copper at the outside prevents the oxidation of ruthenium in the interior of the particles. The 50% decrease in the Fourier transform of the Cu EXAFS for Cu/SiO$_2$ under O$_2$ implies that the copper catalyst is substantially oxidized, but not entirely. By far the largest effect occurs in the Cu EXAFS data of the bimetallic catalysts: the Fourier transform decreases in intensity and changes its shape completely after exposure of the Ru–Cu/SiO$_2$ catalyst to oxygen. Thus, oxygen at room temperature oxidizes the copper, but leaves ruthenium largely unaffected. These results again confirm that copper is present in a shell which surrounds a core consisting mainly of ruthenium.

The EXAFS study on Ru–Cu/SiO$_2$ of Sinfelt et al. [42] is one of the few examples where conclusions can be drawn on a qualitative base, without the need to construct computer fits to the data. The following example illustrates EXAFS as a quantitative tool in a study of particle sizes.

Selective chemisorption of hydrogen is a common technique for measuring the dispersion – that is, the fraction of surface atoms, of a supported metal catalyst. One must, however, make an assumption about the hydrogen-to-metal stoichiometry, the usual choice being one H atom per surface atom (H/M = 1). Unfortunately, this is not always the case, and in practice H/M ratios as high as 3 have been measured for highly dispersed Ir catalysts [44]. Spillover of hydrogen to the support, or solution of hydrogen inside the particles, have been quoted as possible explanations for H/M ratios in excess of unity. In any case, in order to use hydrogen adsorption as an indicator of particle size, the H/M scale must be cali-

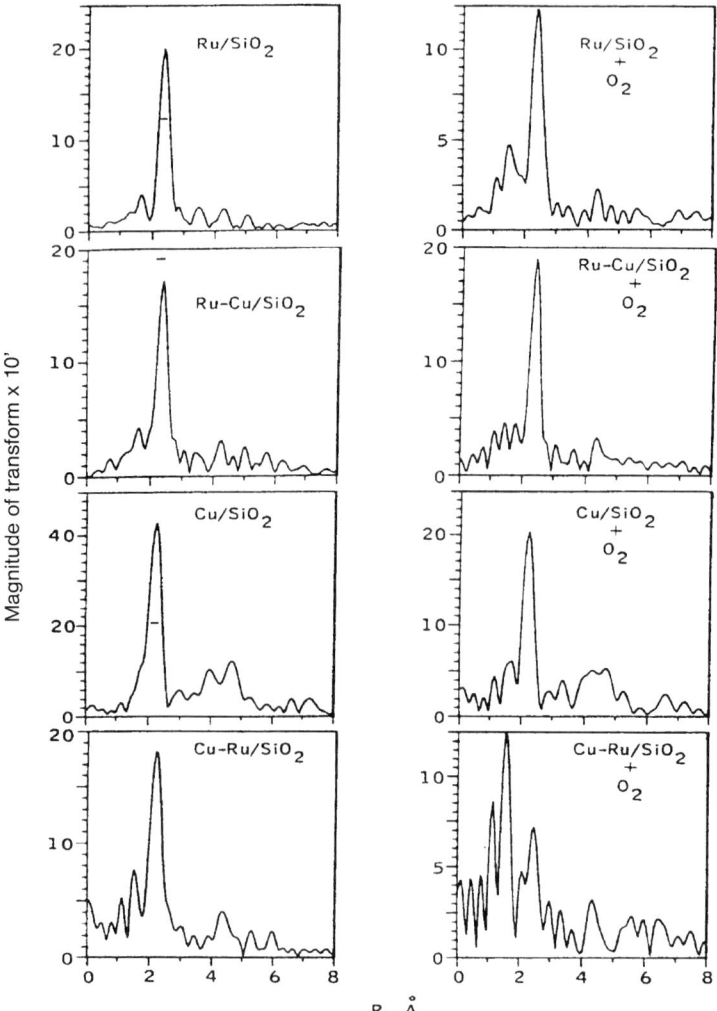

Fig. 6.14 Fourier transforms of Ru and Cu K-edge EXAFS spectra of Ru/SiO$_2$, Cu/SiO$_2$ and Ru–Cu/SiO$_2$ catalysts before and after exposure to oxygen at room temperature. The data show that almost all Cu in the bimetallic Ru–Cu catalyst is oxidized, while Ru is hardly affected. The monometallic Ru and Cu catalysts are oxidized only to a limited extent. (From [42]).

brated. Kip et al. [44] and Wijnen et al. [45] each used coordination numbers determined with EXAFS for this purpose.

Table 6.1 incorporates the results of chemisorption and EXAFS measurements on a large number of supported metal catalysts. These data provide the average number of metal neighbors in the first coordination shell. All metals have the

Table 6.1 Hydrogen chemisorption and EXAFS results of Rh, Ir, and Pt catalysts [44].

Catalyst	Reduction temperature [K]	H/M ratio	N
4.2% Pt/Al_2O_3	1100	0.23	10.0
4.2% Pt/Al_2O_3	1058	0.43	10.2
4.2% Pt/Al_2O_3	573	0.77	7.6
1.06% Pt/Al_2O_3	673	1.14	5.2
2.00% Rh/Al_2O_3	673	1.2	6.6
2.4% Rh/Al_2O_3	473	1.2	6.3
1.04% Rh/Al_2O_3	773	1.65	5.8
0.47% Rh/Al_2O_3	773	1.7	5.1
0.57% Rh/Al_2O_3	573	1.98	3.8
7.0% Ir/SiO_2	773	0.43	11.1
15% Ir/SiO_2	773	0.83	11.0
5.3% Ir/SiO_2	773	1.24	8.6
1.5% Ir/SiO_2	773	1.70	8.6
2.4% Ir/Al_2O_3	773	1.96	7.7
1.5% Ir/Al_2O_3	773	2.40	7.3
0.8% Ir/Al_2O_3	773	2.68	6.0

N: Coordination number.

fcc structure, and hence the bulk value of the coordination number is 12 (see the Appendix). In general, distances can be determined with high accuracy, because the fits are quite restrictive (in particular the fit to the imaginary Fourier transform). Coordination numbers, on the other hand, are accurate to within 5–10% at best.

The interesting information is the correlation between first shell coordination numbers from EXAFS and H/M ratios from chemisorption, shown in Figure 6.15. The correlation is as expected: high dispersions correspond to low coordination numbers. Of course, what we really need is a relationship between particle size and H/M ratios. The right-hand panel of Figure 6.15 translates the experimentally determined H/M ratios of the catalysts to the diameter of particles with a half-spherical shape. Similar calibrations can be made for spherical particles, or for particles of any other desired shape. Such calibration curves form indispensable information for the correct interpretation of chemisorption data. For example, a H/M ratio of 1 measured for nickel catalysts corresponds to particles as small as three atoms in diameter. For a rhodium catalyst, however, a H/M ratio of one translates to particles about twice as large. Kip et al. [44] attribute a hydrogen-to-metal stoichiometry greater than 1 to multiple adsorption of hydrogen atoms, in analogy with metal polyhydride complexes, where several H atoms can be coordinated to the same metal atom.

In principle, transmission electron microscopy may also be used for determining the particle size of the catalysts. However, electron microscopy requires a vac-

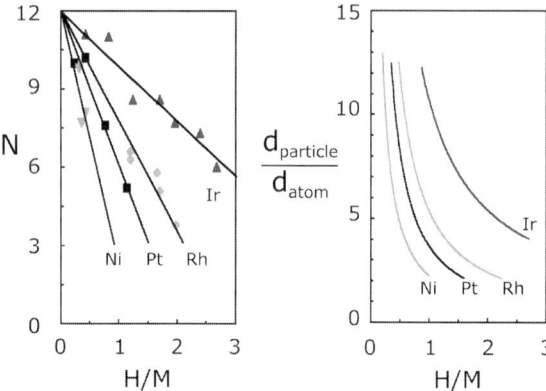

Fig. 6.15 Left: First shell coordination numbers from EXAFS versus H/M ratios from selective hydrogen chemisorption for a number of supported Ni, Rh, Ir, and Pt catalysts. Right: Relative diameter of half-spherical metal particles as a function of H/M. The curves correspond to the straight lines in the left part of the figure. (Adapted from [44, 45]).

uum, which may affect the degree of metal reduction. In addition, the highly energetic electron beam may alter the morphology of the particles. The advantage of using EXAFS for this purpose is that measurements can be made *in situ* under hydrogen.

6.4.2
Quick EXAFS for Time-Resolved Studies

The high penetrating power of X-rays makes them eminently suitable for investigations of a catalyst under operating conditions inside a reactor. Figure 6.16 illustrates, schematically, a catalytic reactor that Clausen and co-workers used to measure XRD and EXAFS data at synchrotrons. One particularly attractive point about this reactor is that all gas flows through the catalyst bed, and hence it essentially operates as a plugged-flow reactor, enabling kinetic measurements. The use of a stream of air for heating and cooling ensures that the reactor responds almost instantaneously to changes in temperature, which is not easily achieved when conventional heating elements are used.

Over the past few years the time needed to collect EXAFS spectra has been reduced by an order of magnitude, owing to improvements in detectors, data management in the computers, and monochromator operation. Instead of increasing the wavelength of the X-rays in stepwise fashion, implying dead time during changing the settings and mechanical stabilization of the monochromator after each step, the latter is now changed continuously and therefore data are available without interruption. As a result, the collection time of an entire spectrum is a few seconds, allowing time-resolved studies to be conducted, for example during

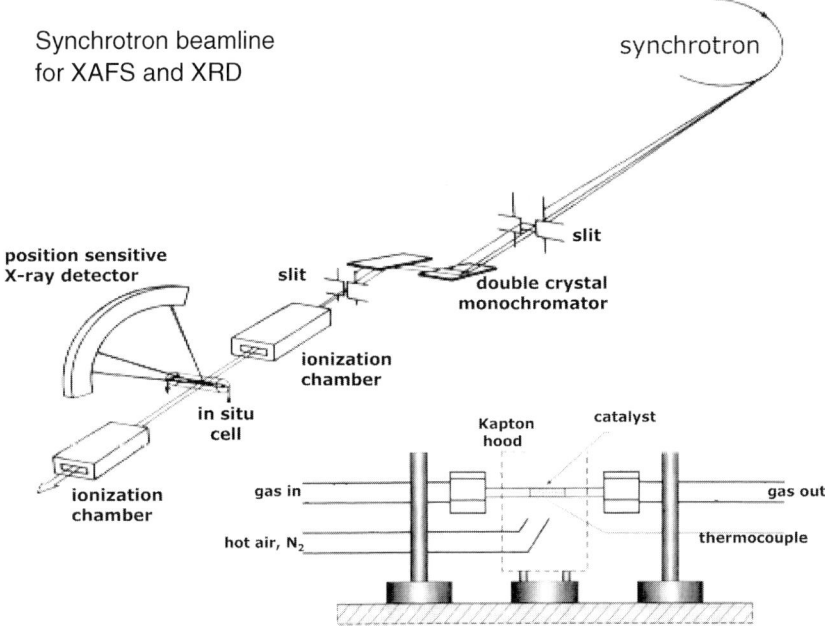

Fig. 6.16 Catalytic *in-situ* reactor made of a quartz capillary, suitable for use at synchrotrons for the collection of EXAFS and XRD data. The scheme of the synchrotron beamline shows the positions of the monochromator, the ion chambers which measure the intensity of the X-rays before and after the sample, and the position-sensitive X-ray detector which records the XRD diffractogram. (Adapted from [47]).

a temperature-programmed reduction. This rapid mode of the technique is referred to as Quick EXAFS, or QEXAFS.

Figure 6.17 shows an example in which QEXAFS has been used in combination with XRD to study the temperature-programmed reduction of copper oxide in a $Cu/ZnO/Al_2O_3$ catalyst for the synthesis of methanol [46, 47]. Reduction to copper metal takes place in a narrow temperature window of 430 to 440 K, and is clearly revealed by both the EXAFS pattern and the appearance of the (111) reflection of metallic copper in the XRD spectra. Note that the QEXAFS detects the metallic copper at a slightly lower temperature than the XRD does, indicating that the first copper metal particles to form are too small to be detected by XRD, which requires a certain extent of long-range order [46, 47].

In conclusion, EXAFS has provided important structural information on several supported catalyst systems. Two more applications – on the structure of the metal support interface and on the structure of metal sulfide catalysts – are discussed in Chapter 9. An important requirement for meaningful EXAFS data on catalysts is that the particles are monodisperse, such that the average environment, which determines the EXAFS signal, is the same throughout the entire cat-

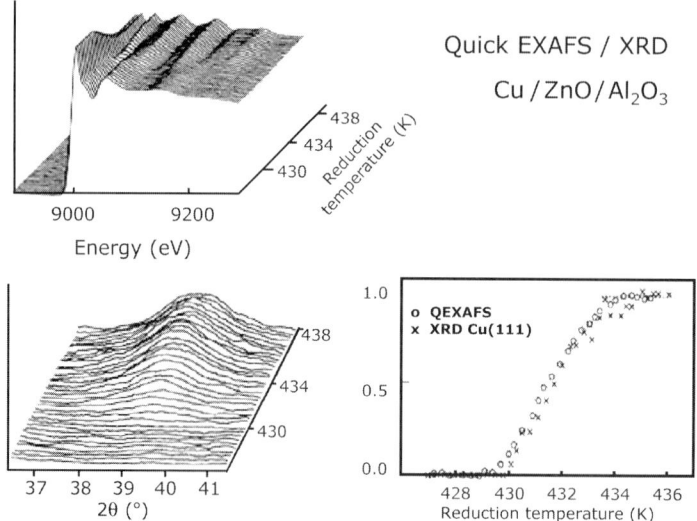

Fig. 6.17 Quick EXAFS and XRD measurements recorded during the temperature-programmed reduction of copper in a Cu/ZnO/Al$_2$O$_3$ methanol synthesis catalyst. The disappearance and appearance of peaks with increasing temperature in the series of EXAFS spectra corresponds to the conversion of oxidic to metallic copper. The intensity of the relatively sharp peak around 9040 eV, indicative of Cu metal, clearly illustrates the kinetics of the reduction process, as does the intensity of the (111) reflection of Cu metal in the XRD spectra. (Adapted from [47]).

alyst. For multicomponent catalysts, the technique has the drawback that data analysis becomes progressively more complicated and time-consuming, with an increasing number of constituent atoms, and that considerable expertise is required to avoid ambiguities. However, for optimized and monodisperse catalysts, for which the data analysis is carried out with care, EXAFS provides structure information on the scale of interatomic distances!

6.4.3
X-Ray Absorption Near Edge Spectroscopy

XANES might easily be the most popular characterization technique for determining catalyst composition, were it not that a synchrotron is needed in order to perform the experiments. Apart from this drawback, the technique has much to offer: chemical specificity, oxidation states, sometimes coordination round the absorbing atom, and, based on highly penetrating X-rays, all of this under *in-situ* conditions. As compared to XPS, the technique lacks surface sensitivity, but for supported catalysts where one is interested in the composition of the entire particle, this is not a particular disadvantage. XANES spectra are recorded in the same set-up as used for EXAFS, the main difference being that the former is limited to a precise measurement of typically a 50-eV range around the absorption edge.

The technique is also referred to as NEXAFS, although this acronym appears to be used increasingly for X-ray absorption at low energies and in light elements [36, 48]. Like QEXAFS, XANES spectra can be recorded in fast mode, to monitor the kinetics of solid-state reactions in real time [49].

In theory, to describe a XANES spectrum is not an easy task. The equations discussed earlier in this chapter for EXAFS are not valid at low k-values (i.e., energies close to that of the edge), and instead the X-ray absorption will have to be calculated from first principles, which is a specialism in itself. Fortunately, XANES spectra can usually be very well interpreted with the help of reference spectra of known compounds, and constructing linear combinations of references to fit the spectrum of the catalyst often works well to obtain quantitative information on composition.

XANES is based on the absorption of an X-ray photon which excites an electron from a core level (e.g., 1s or 2p3/2) to an unoccupied state close to the Fermi level. For final states below the Fermi level, selection rules generally require $\Delta l = \pm 1$; $\Delta j = \pm 1$; $\Delta s = 0$, implying that the predominant transitions around the K-edge are from s to p states, and around the L_{II}, L_{III} edges from p to d states. However, the final states are not necessarily purely atomic, and may have hybrid character. This causes the pre-edge region to display interesting effects due to coordination and oxidation state. For example, tetrahedrally coordinated Cr^{6+} ions feature a distinct pre-edge peak that is absent in XANES spectra of chromium in lower oxidation states. Such sharp pre-edge features as in chromium correspond to the transition of an s-electron to a bound p-state which is hybridized with the d-electron states near the Fermi level. Obviously, such pre-edge peaks have great diagnostic value. In general, one may say that purely octahedrally coordinated ions – where mixing of p-d states is not allowed – show no pre-edge peak whereas ions in a distorted octahedral environment – where p-d mixing is allowed – do show absorption in the pre-edge region. Purely tetrahedrally coordinated ions exhibit the largest pre-edge intensity, as for instance in the Cr^{6+} ion in a calcined Phillips catalyst [50].

The edge position corresponds to the ionization threshold, and thus reflects the oxidation state of the atom, similarly as in XPS. Several correlations between the edge position and the oxidation or charge state of the absorbing atom have been reported in the literature, both for transition metals [51–53] and other elements such as sulfur. For example, George and Gorbaty [54] report sulfur K-edge positions from 2469.1 eV for elemental sulfur to 2478.5 eV for S^{6+} in K_2SO_4, representing a shift of more than 9 eV.

Another indicator of the valence state is the white line – the intense absorption at the edge that extends above the intensity of the EXAFS region – or, in simple terms, the intensity overshoot. The terminology is derived from the times when photographic plates were used; strong X-ray absorption gave rise to unexposed regions that showed up as white bands in the photographic negative. For L-edges, where transitions are from p to d-states, the white line intensity is a measure for the number of empty d-states. For example, the s-metals silver and gold have filled d-states, and do not exhibit white lines in their metallic state, but moving to the left in the Periodic Table a clear white line develops. White lines are particularly prominent in XANES and EXAFS spectra of transition metal ions with high oxidation states.

At energies above the edge, core electrons are excited to continuum states, and the spectrum becomes dominated by scattering phenomena, similarly as in EXAFS. Hence, in a qualitative sense, the shape of the XANES spectrum depends straightforwardly on the states that are available to the excited electron. These depend on several factors, such as the coordination of the atom (octahedral, tetrahedral, etc.), its mode of bonding, oxidation state, the density of states in the electron band, etc., which give XANES its great chemical sensitivity.

Figure 6.18 illustrates the technique with a study on a proprietary cobalt on alumina Tropsch catalyst for Fischer–Tropsch synthesis (the reaction of synthesis gas, $CO + H_2$, to hydrocarbon fuels) [55]. Trace amounts of platinum help to obtain an appreciable degree of reduction for the cobalt (similarly as in the temperature-programmed reduction of bimetallic Fe–Rh catalysts in Fig. 2.4). The left part of Figure 6.18 shows Co K-edge XANES of metal and oxide reference compounds, and illustrates the strong intensity of the white line region for ionic cobalt compounds. The XANES spectrum of the calcined $CoPt/Al_2O_3$ catalyst re-

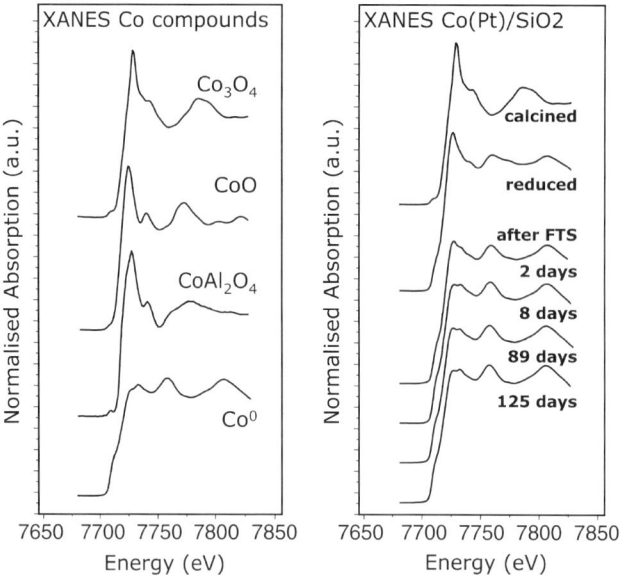

Fig. 6.18 XANES spectra of cobalt reference compounds and of a CoPt/Al$_2$O$_3$ Fischer–Tropsch catalyst in different stages of its working life. Samples labeled "after FTS" were taken from a pilot plant operated at commercially relevant conditions (220 °C, 20 bar, relatively high conversion of 50–70%). Although these catalysts were exposed to significant partial pressures of the byproduct water, the XANES indicate further reduction during usage. (Adapted from [55]).

sembles that of Co_3O_4, whereas the spectrum of the reduced catalyst is largely a mixture of cobalt metal (53%) and CoO (47%). Samples taken from the slurry reactor after various periods of high-pressure Fischer–Tropsch synthesis and stored and measured in waxes to preserve their oxidation state, indicate that the catalyst reduces further in the process, from about 53% metal at the start to 89% after 125 days on stream. These results are remarkable, because it has long been assumed that cobalt oxidizes during Fischer–Tropsch synthesis at high conversions, due to the water that is formed as a byproduct [56–58]. The measurements of Figure 6.18 disprove this hypothesis, at least for particles with a diameter on the order of 6 nm and larger [55].

We conclude this section with a comparison of EXAFS and XANES:

- XANES, including the absorption edge, offers a much stronger signal than EXAFS, and consequently measurements can be made with the absorbing element at lower concentrations.

- XANES spectra can be used as a fingerprint technique, provided that the spectra of suitable reference compounds are available. If this is the case, XANES spectra are much easier to interpret than EXAFS spectra.

- EXAFS spectra can, in principle, be well described on the basis of scattering theory (although data analysis is time-consuming and demand considerable expertise). XANES spectra can only be predicted on the basis of first-principle calculations. As noted above, this is generally not necessary because in practice XANES data are interpreted with the help of reference spectra.

- The beauty of XANES for catalyst characterization is that it readily yields oxidation states and average compositions of supported particles, whereas the strong point of EXAFS is that it yields structural information on the atomic scale.

- Both techniques have in common that they can be applied *in situ*, in the same experimental set-up, and that the most meaningful information is derived from samples with a uniform composition.

References

1 N.W. Ashcroft and N.D. Mermin, *Solid State Physics*. Holt-Saunders, Philadelphia, 1976.
2 B.D. Cullity, *Elements of X-ray Diffraction*. Addison-Wesley, Reading, 1978.
3 J.B. Cohen and L.H. Schwartz, *Diffraction from Materials*. Springer-Verlag, New York, 1987.
4 C. Suryanarayana and M. Grant Norton, *X-ray Diffraction: A Practical Approach*. Plenum, New York, 1998.
5 C. Hammond, *The Basics of Crystallography and Diffraction*. Oxford University Press, Oxford, 1998.
6 J. Als-Nielsen and D. McMorrow, *Elements of Modern X-ray Physics*. Wiley, Chichester, 2001.
7 K. Kunimori, T. Wakasugi, Z. Hu, H. Oyanagi, M. Imal, H. Asano, and T. Uchijima, *Catal. Lett.* 7 (1990) 337.
8 G. Fagherazzi, A. Benedetti, A. Martorana, S. Giuliano, D. Duca,

and G. Deganello, *Catal. Lett.* **6** (1990) 263.
9 J.B. Cohen, *Ultramicroscopy* **34** (1990) 41.
10 S.R. Sashital, J.B. Cohen, R.L. Burwell, and J.B. Butt, *J. Catal.* **50** (1977) 479.
11 J. Pielaszek, J.B. Cohen, R.L. Burwell, and J.B. Butt, *J. Catal.* **80** (1983) 479.
12 P. Canton, C. Meneghini, P. Riello, and A. Benedetti, in: *In Situ Characterization of Catalytic Materials*, B.M. Weckhuysen (Ed.). American Scientific Publishers, 2004, p. 293.
13 H. Jung and W.J. Thomson, *J. Catal.* **128** (1991) 218.
14 X. Wang, J.C. Hanson, A.I. Frenkel, J.-Y. Kim, and J.A. Rodriguez, *J. Phys. Chem. B* **108** (2004) 13667.
15 B.S. Clausen, G. Steffensen, B. Fabius, J. Villadsen, R. Feidenhans'l, and H. Topsøe, *J. Catal.* **132** (1991) 524.
16 P. Georgopoulos and J.B. Cohen, *J. Catal.* **92** (1985) 211.
17 J. Goodisman, H. Brumberger, and R. Cupello, *J. Appl. Crystallogr.* **14** (1981) 305.
18 P.W.J.G. Wijnen, T.P.M. Beelen, C.P.J. Rummens, H.C.P.L. Saeijs, and R.A. van Santen, *J. Appl. Crystallogr.* **24** (1991) 759.
19 J.B. Pendry, *Low Energy Electron Diffraction*. Academic Press, New York, 1974.
20 K. Heinz and K. Müller, in: *Structural Studies of Surfaces, Springer Tracts in Modern Physics*, Vol. 91. Springer, Berlin, 1982, p. 1.
21 G. Ertl and J. Küppers, *Low Energy Electrons and Surface Chemistry*. VCH, Weinheim, 1985.
22 M.A. van Hove, W.H. Weinberg, and C.-M. Chan, *Low Energy Electron Diffraction*. Springer Verlag, New York, 1986.
23 R.M. van Hardeveld, R.A. van Santen, and J.W. Niemantsverdriet, *Surface Sci.* **369** (1996) 23.
24 F. Frechard, R.A. van Santen, A. Siokou, J.W. Niemantsverdriet, and J. Hafner, *J. Chem. Phys.* **111** (1999) 8124.
25 A.P. van Bavel, M.J.P. Hopstaken, D. Curulla, J.W. Niemantsverdriet, J.J. Lukkien, and P.A.J. Hilbers, *J. Chem. Phys.* **119** (2003) 523.
26 D. Curulla, A.P. van Bavel, and J.W. Niemantsverdriet, *ChemPhysChem* **6** (2005) 473.
27 M. Kottcke and K. Heinz, *Surface Sci.* **376** (1997) 352.
28 K. Reuter, J. Bernhardt, H. Wedler, J. Schardt, U. Starke, and K. Heinz, *Phys. Rev. Lett.* **79** (1997) 4818.
29 D.C. Koningsberger and R. Prins (Eds.), *X-ray Absorption*. Wiley, New York, 1987.
30 Y. Iwasawa, *X-ray Absorption Fine Structure (XAFS) for Catalysts and Surfaces*. World Scientific, Singapore, 1996.
31 W. Kossel, *Z. Phys.* **1** (1920) 119.
32 R. de L. Kronig, *Z. Phys.* **70** (1931) 317; *Z. Phys.* **75** (1932) 191; *Z. Phys.* **75** (1932) 468.
33 D.E. Sayers, F.W. Lytle, and E.A. Stern, *Phys. Rev. Lett.* **27** (1971) 1204.
34 J.H. Sinfelt, G.H. Via, and F.W. Lytle, *Catal. Rev. – Sci. Eng.* **26** (1984) 81.
35 R. Prins and D.C. Koningsberger, in: D.C. Koningsberger and R. Prins (Eds.), *X-ray Absorption*. Wiley, New York, 1987, p. 321.
36 J. Stöhr, *NEXAFS Spectroscopy*. Springer Verlag, New York, 1996.
37 M. Vaarkamp, *Catal. Today* **39** (1998) 271.
38 J.B.A.D. van Zon, D.C. Koningsberger, H.F.J. van't Blik, and D.E. Sayers, *J. Chem. Phys.* **82** (1985) 5742.
39 J.H.A. Martens, PhD. Thesis, Eindhoven University of Technology, 1988.
40 J.J. Rehr, C.H. Booth, F. Bridges, and S.I. Zabinsky, *Phys. Rev. B* **49** (1994) 12347.
41 D.E. Ramaker, B.L. Mojet, D.C. Koningsberger, and W.E. O'Grady, *J. Phys. Condens. Matter* **10** (1998) 8753.
42 J.H. Sinfelt, G.H. Via, and F.W. Lytle, *J. Chem. Phys.* **72** (1980) 4832.
43 J.H. Sinfelt, *Acc. Chem. Res.* **10** (1977) 15.
44 B.J. Kip, F.B.M. Duivenvoorden, D.C. Koningsberger, and R. Prins, *J. Catal.* **105** (1987) 26.
45 P.W.J.G. Wijnen, F.B.M. van Zon, and D.C. Koningsberger, *J. Catal.* **114** (1988) 463.
46 B.S. Clausen, L. Gråbæk, G. Steffensen, P.L. Hansen, and H. Topsøe, *Catal. Lett.* **20** (1993) 23.
47 J.-D. Grunwaldt and B.S. Clausen, *Topics Catal.* **18** (2002) 37.
48 F.M.F. de Groot, A. Knop-Gericke, T. Ressler, and J.A. van Bokhoven, in: *In Situ Characterization of Catalytic Materials*, B.M. Weckhuysen (Ed.).

American Scientific Publishers, 2004, p. 107.
49 T. Ressler, J. Wienold, R.E. Jentoft, T. Neisius, and M. Günter, *Topics Catal.* **18** (2002) 45.
50 E. Groppo, C. Prestipino, F. Cesano, F. Bonino, S. Bordiga, C. Lamberti, P.C. Thüne, J.W. Niemantsverdriet, and A. Zecchina, *J. Catal.* **230** (2005) 98.
51 S.P. Cramer, T.K. Eccles, F.W. Kutzler, K.O. Hodgson, *J. Am. Chem. Soc.* **98** (1976) 1287.
52 J. Wong, F.W. Lytle, R.P. Messmer, and D.H. Maylotte, *Phys. Rev. B* **30** (1984) 5596.
53 T. Ressler, J. Wienold, R.E. Jentoft, and T. Neisius, *J. Catal.* **210** (2002) 67.
54 G.N. George and M.L. Gorbaty, *J. Am. Chem. Soc.* **111** (1989) 3182.
55 A.M. Saib, A. Borgna, J. van de Loosdrecht, P.J. van Berge, and J.W. Niemantsverdriet, *Appl. Catal. A* **312** (2006) 12.
56 E. Iglesia, *Appl. Catal. A* **161** (1997) 59.
57 P.J. van Berge, J. van de Loosdrecht, S. Barradas, and A.M. van der Kraan, *Catal. Today* **58** (2000) 321.
58 G. Jacobs, P.M. Patterson, Y. Zhang, T.K. Das, J. Li, and B.H. Davis, *Appl. Catal. A* **233** (2002) 215.

7
Microscopy and Imaging

Keywords

Transmission electron microscopy (TEM)
Scanning electron microscopy (SEM)
Electron microprobe analysis (EMA, EPMA)
Energy dispersive X-ray analysis (EDX, EDAX)
Field emission microscopy (FEM)
Field ion microscopy (FIM)
Atomic force microscopy (AFM)
Scanning tunneling microscopy (STM)
Photoemission electron microscopy (PEEM)

7.1
Introduction

Seeing the surface of a catalyst, preferably in atomic detail, is the ideal of every catalytic chemist. Unfortunately, optical microscopy is of no use to this end, simply because the rather long wavelength of visible light (a few hundred nanometers) does not enable the detection of features smaller than about 1 µm. Electron beams offer better opportunities, and development over the past 40 years has resulted in electron microscopes which routinely achieve magnifications on the order of one million times, thereby revealing details with a resolution of about 0.1 nm [1, 2]. The technique is very popular in catalysis; for the interested reader, several excellent reviews are available providing a good overview of what electron microscopy is, and what the related techniques tell us about a catalyst [3–7].

Electron microscopy routinely yields pictures (see Fig. 7.1) which stimulate our imagination of what a reacting molecule "sees" on approaching a catalyst from the gas phase. However, since the wavelength associated with high-energy electrons is on the order of interatomic distances or even smaller, diffraction effects come into play with the result that an image does not necessarily represent the shape of a particle as it would in optical microscopy. Thus, the fact that magnifi-

Spectroscopy in Catalysis: An Introduction, Third Edition
J. W. Niemantsverdriet
Copyright © 2007 WILEY-VCH Verlag GmbH & Co. KGaA, Weinheim
ISBN: 978-3-527-31651-9

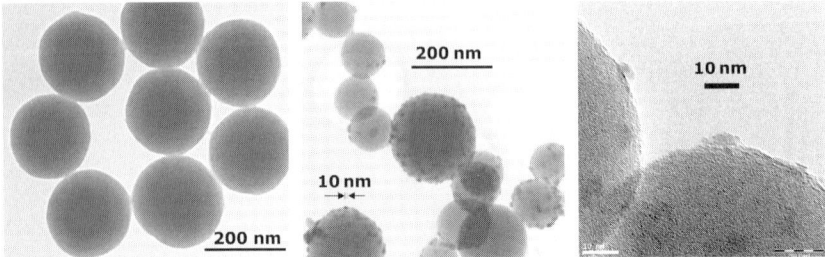

Fig. 7.1 Transmission electron microscopy images of a spherical model for a silica supported model catalysts for Fischer–Tropsch synthesis. Left: the silica spheres. Center and right: the reduced Co/SiO$_2$ catalyst. (Adapted from [8]).

cations of 10^6 are readily obtained does not imply that all catalysts can routinely be imaged at atomic resolution.

The more recent scanning probe microscopies [e.g., scanning electron microscopy (STM), developed in 1982, and atomic force microscopy (AFM), in 1986] each take a different approach to "see" surfaces: When a sharp tip is scanned over the surface at a few Angstroms distance, the tunneling current or the forces between the tip and atoms in the sample produce an image of the surface, sometimes even at atomic resolution. While STM is limited to (semi) conducting samples, AFM can be used on almost all surfaces, which makes it a promising tool for investigating catalysts.

Less generally applicable than electron or scanning probe microscopy – but still capable of revealing great detail – are field emission microscopy (FEM) and field ion microscopy (FIM). These techniques are limited to the investigation of sharp metallic tips, however, with the attractive feature that the facets of such tips exhibit a variety of crystallographically different surface orientations, which can be studied simultaneously, for example in gas adsorption and reaction studies.

Imaging on larger scale – and in particular when large numbers of adsorbate molecules organize themselves into patterns of micrometer size – is possible by using recently developed methods based on photoemission microscopy (PEEM) and ellipsometry. These techniques have provided us with spectacular movies of reaction wave fronts moving over surfaces. These phenomena are characteristic of oscillating surface reactions, a subject that has fascinated many of those who have investigated catalysis and surface science [9].

7.2
Electron Microscopy

Electron microscopy is a rather straightforward technique that can be used to determine the size and shape of supported particles. It can also reveal information on the composition and internal structure of the particles, for example by detect-

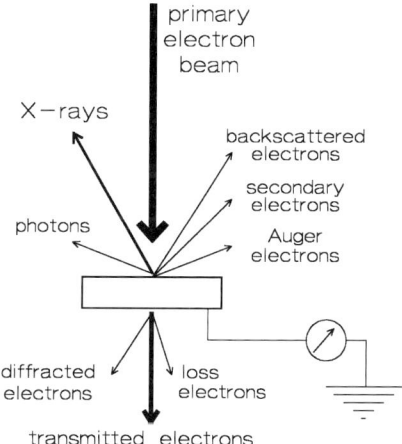

Fig. 7.2 The interaction between the primary electron beam and the sample in an electron microscope leads to a number of detectable signals.

ing the characteristic X-rays that are produced by the interaction of the electrons with matter, or by analyzing how the electrons are diffracted [1, 2].

Electrons have characteristic wavelengths of less than an Ångstrøm, and come close to seeing atomic detail. What happens to the electrons when the primary beam of energy between 100 and 400 keV hits the sample is shown schematically in Figure 7.2.

- Depending on the sample thickness, a fraction of the electrons passes through the sample without suffering energy loss. As the attenuation of the beam depends on density and thickness, the transmitted electrons form a two-dimensional projection of the sample.
- Electrons are diffracted by particles if these are favorably oriented towards the beam, enabling one to obtain dark-field images as well as crystallographic information.
- Electrons can collide with atoms in the sample and be scattered back; backscattering is more effective when the mass of the atom increases. If a region of the sample contains heavier atoms (e.g., Pt) than the surroundings, it can be distinguished due to a higher yield of backscattered electrons.
- Auger electrons and X-rays are formed in the relaxation of core-ionized atoms, as discussed in Chapter 3 for AES and XPS.
- Electrons excite characteristic vibrations in the sample that can be studied by analyzing the energy loss suffered by the primary electrons (see Chapter 8, Vibrational spectroscopy).
- Many electrons lose energy in a cascade of consecutive inelastic collisions. Most of the secondary electrons emitted by the sample had their last loss process in the surface region.

- The emission of a range of photons from ultraviolet (UV) to infrared, referred to as *cathodoluminescence*, is mainly caused by recombination of electron-hole pairs in the sample.

Thus, the interaction of the primary beam with the sample provides a wealth of information on morphology, crystallography, and chemical composition.

7.2.1
Transmission Electron Microscopy

The operational modes of three types of electron microscope are shown schematically in Figure 7.3.

In transmission electron microscopy (TEM), which uses transmitted and diffracted electrons, the instrument is in a sense similar to an optical microscope, if one replaces optical lenses with electromagnetic lenses. In TEM, a primary electron beam of high energy and high intensity passes through a condenser to produce parallel rays, which impinge on the sample. As the attenuation of the beam depends on the density and the thickness, the transmitted electrons form a two-dimensional projection of the sample mass, which is subsequently magnified by the electron optics to produce a so-called bright-field image. The dark-field

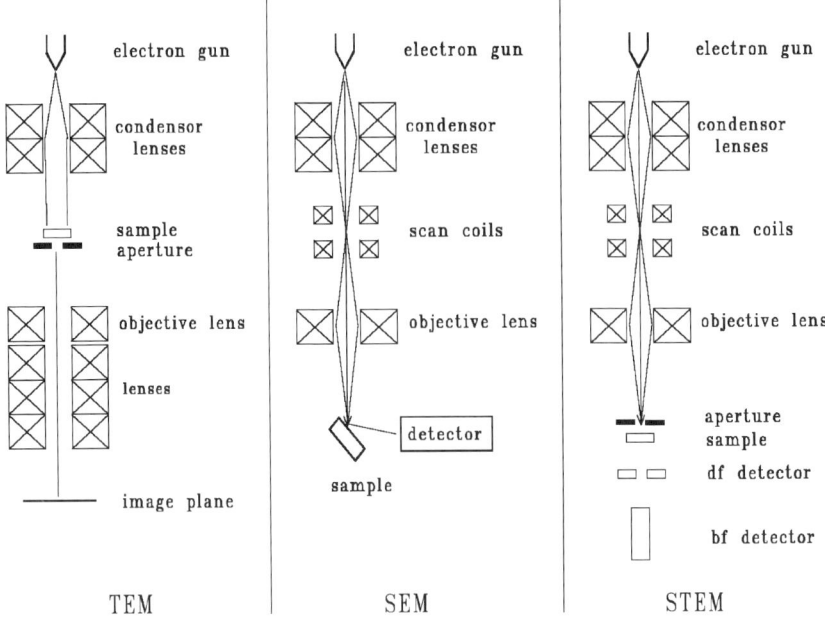

Fig. 7.3 Schematic set-up of an electron microscope in the transmission (TEM), scanning (SEM), and combined (STEM) modes. (After [10]).

image is obtained from the diffracted electron beams, which are slightly off-angle from the transmitted beam. Typical operating conditions of a TEM instrument are 100 to 200 keV electrons, 10^{-6} mbar vacuum, 0.3 nm resolution, and a magnification of $3 \cdot 10^5$ to 10^6.

For studying supported catalysts, TEM is the commonly applied form of electron microscopy, and today images as shown in Figure 7.1 are obtained routinely. In general, the detection of supported particles is possible provided that there is sufficient contrast between the particles and the support. This may impede applications of TEM on well-dispersed supported oxides [11]. Contrast in the transmission mode is caused not only by the attenuation of electrons due to density and thickness variations over the sample, but also by diffraction and interference. For example, a particle in a TEM image may show less contrast than other identical

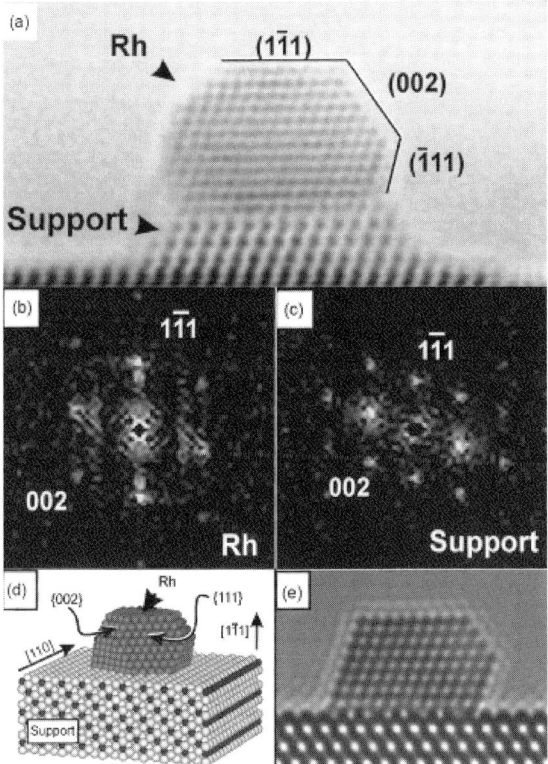

Fig. 7.4 High-resolution transmission electron microscopy (TEM) image of a rhodium on a modified cerium oxide ($Ce_{0.8}Tb_{0.2}O_{2-x}$) after reduction, showing how the metal is almost epitaxially attached to the support. (a) TEM image; (b,c) digital diffraction patterns obtained by Fourier transformation from the images; (d) structure model; (e) simulated TEM image belonging to the model in (d). (Reproduced from [13]).

particles because it is favorably oriented for Bragg diffraction by its lattice planes, such that the diffracted beam does not contribute to the image (amplitude or diffraction contrast). Tilting the sample changes the orientation of the particle, and this is used to recognize amplitude contrast. Another possibility is that the diffracted beam does contribute to the image, where it interferes destructively with the undeflected beam (phase contrast). Both effects are a possible source for misinterpretation of TEM images [12].

The current literature on TEM in catalysis is vast, and forms a rich source of inspiration for many investigators in the field [3–7]. One noteworthy line of research is the application of TEM to investigate particle morphology in relation to metal support interaction. Figure 7.4 shows a side view of a rhodium particle on a modified ceria support ($Ce_{0.8}Tb_{0.2}O_{2-x}$), taken from studies of Bernal et al. [13]. This picture shows beautifully the shape of the particle, together with the lattice fringes characteristic of certain orientations of the particles and the support. Figure 7.4b and c shows digital diffraction patterns calculated from a Fourier transform of the image. These diffraction patterns provide evidence that lattice planes in the rhodium run parallel to lattice planes with the same indexes in the support. The structure model in Figure 7.4d is based on this epitaxial relationship. This model, finally, is taken as input for a simulation of how the particle and support would appear in a TEM image [13]. Other elegant examples of such studies on metal-support epitaxy may be found in the investigations conducted by Henry and co-workers [14].

7.2.2
Scanning Electron Microscopy

Scanning electron microscopy (SEM) is carried out by rastering a narrow electron beam over the surface, and detecting the yield of either secondary or backscattered electrons as a function of the position of the primary beam [15]. Contrast is caused by the orientation: parts of the surface facing the detector appear brighter than parts of the surface with their surface normal pointing away from the detector. The secondary electrons have mostly low energies ($\approx$ 5 to 50 eV), and originate from the surface region of the sample. Backscattered electrons come from deeper and carry information on the composition of the sample, because heavy elements are more efficient scatterers and appear brighter in the image.

Dedicated SEM instruments have resolutions of about 3 to 10 nm. Simple versions of SEM with micron resolution are often available on Auger electron spectrometers, for the purpose of sample positioning. The main difference between SEM and TEM is that SEM sees contrast due to the topology and composition of a surface, whereas the electron beam in TEM projects all information on the mass it encounters in a two-dimensional image which, however, is of subnanometer resolution.

Charging of insulating samples under the electron beam may be a problem in SEM, although coating the sample with a thin film of gold or carbon alleviates the problem [15]. The use of an environmental SEM (ESEM) offers a convenient way to avoid these problems. Whereas electron microscopes normally operate under high vacuum in order to avoid scattering of the electrons, ESEM operates with the sample under less-stringent vacuum conditions. The advantage here is that secondary electrons emitted by the sample ionize the residual gas molecules and create a cascade of additional secondary electrons, which are then drawn towards a detector. In fact, the cascade magnifies the signal from the secondary electrons originating from the sample. At the same time, the positive ions move to the sample, where they are neutralized by the negative charge that builds up on the surface of non-conductive materials. In this way, sample charging is largely avoided, and samples can in general be measured in their as received state.

The lower resolution of SEM compared to TEM is the main reason why SEM is less often used in catalysis. In cases where phenomena occur at length scales exceeding 5 to 10 nm, however, the technique is indispensable. Figure 7.5 shows ESEM images of catalytically formed polyethylene on chromium catalysts (usually referred to as the Phillips catalyst [16]; see also Chapter 9). The polymer is non-conductive and would normally build up charge rapidly but, as discussed above, this is not a problem in ESEM. The inset shows the polymer formed around a spherical Cr/SiO$_2$ catalyst particle. In fact, due to polymer formation inside the pores, the particle has broken up, and is present in finely divided form in the polymer. The larger image shows a thick layer of polymer that formed on the surface of a flat model catalyst, consisting of a submonolayer of chromium ions on

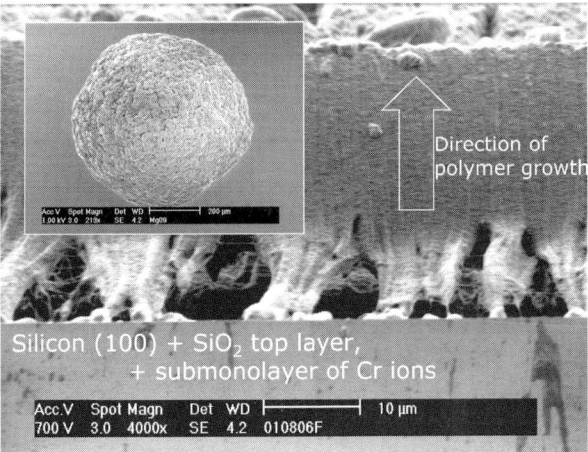

Fig. 7.5 Environmental scanning electron microscopy (ESEM) images of polyethylene formed on Cr/SiO$_2$ catalysts. The inset shows polymer grown on a commercial catalyst particle; the larger image shows a polymer film on a planar model catalyst. (Courtesy of P.C. Thüne, Eindhoven).

an $SiO_2/Si(100)$ support. The ESEM image reveals that polymer growth starts locally, and organizes into larger crystals that together form a dense tapestry which eventually covers the catalyst entirely. Note that a tiny amount of Cr ions (<10% of a monolayer) gives rise to a polyethylene film of 25 μm thickness in about 1 hour. The amazingly active Phillips catalyst is discussed as a case study in Chapter 9.

7.2.3
Scanning Transmission Electron Microscopy

Scanning transmission electron microscopy (STEM) combines the TEM and SEM modes of operation. Here, the scanning coils are used to illuminate a small area of the sample, from which either bright- or dark-field images are obtained. If the primary beam is generated with a field emission gun, the resolution is comparable to that in TEM, with the advantage that selected regions of the sample (e.g., a supported catalyst particle) can be investigated separately.

STEM offers a wealth of information about catalysts if all the opportunities in Figure 7.2 are utilized [7]. The contrast of imaging supported metal particles can be greatly enhanced by collecting images in dark-field mode, where the image is composed by electrons that are diffracted by the metal particles. Figure 7.6 indicates the arrangement of the ring-shaped detectors for such annular dark-field (ADF) measurements. At even higher angles (larger than those under which Bragg diffraction occurs), the image is formed by electrons scattered by heavy elements. This high-angle annular dark field (HAADF) detection mode is also called Z-contrast microscopy. The following example illustrates the capabilities of this technique.

The activity of a catalyst often depends on the type of sites that are present on the catalytically active particles. Hence, ideally one wants to know how atoms are distributed over terraces, edges, corners, etc. – a question that has long occupied the thoughts of many investigators of catalysis. Van Hardeveld and Hartog [17] addressed this problem by analyzing the different atom environments on a number of different crystal geometries, such as cubes and truncated octahedrons, as a function of their size, and their report from 1969 has subsequently become a true citation classic.

Modern electron microscopes are very well capable of imaging individual particles, but of course it is impossible to do so, even for a representative fraction of particles in a supported catalyst. Carlsson et al. [18] described an interesting method to obtain the particle geometry distribution, such that the fraction of edge and corner sites in a supported catalyst can be estimated. An assumption must be made on the shape of the particles, for which these authors used the truncated octahedron, and were able to demonstrate the procedure for gold particles on three different supports.

Figure 7.7 shows the STEM images obtained in HAADF mode. In this mode, the image formed is due to that part of the electron beam that is diffracted by

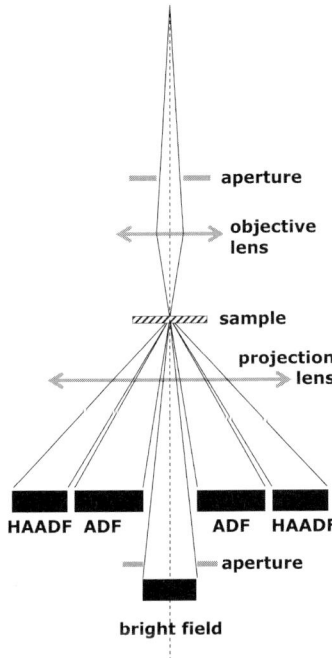

Fig. 7.6 Scheme of a scanning transmission electron microscope with bright- and dark-field detectors. ADF: annular dark field; HAADF: high-angle annular dark field. (Adapted from [7]).

the gold particles, and hence these appear bright. In fact, the brightness of an imaged particle is proportional to the number of atoms it contains, and consequently the images contain information on the volume distribution of the gold particles, be it that the absolute value of the volumes is not known, and requires some form of calibration. The second and third rows in Figure 7.7 show the particle size distribution, as well as graphs of how the intensity of the particles (which are proportional to the number of atoms) depends on the diameter, plotted in a double logarithmic mode. The slope of such a plot corresponds to the dimensionality of the particles. Interestingly, small particles with a diameter below about 2.5–3 nm appear to grow mainly in two dimensions, while the larger particles grow in three dimensions. The alumina-supported gold particles are also, on average, smaller and flatter than the gold particles on the titania and Mg–Al spinel support.

In order to arrive at the distribution of different sites on the particles, one must adopt a certain shape. Carlsson et al. assumed that the particles are truncated octahedrons (see Fig. 7.8), where the dark part corresponds to a supported particle of fcc structure, with edges of four atoms long, and five atom layers thick. This

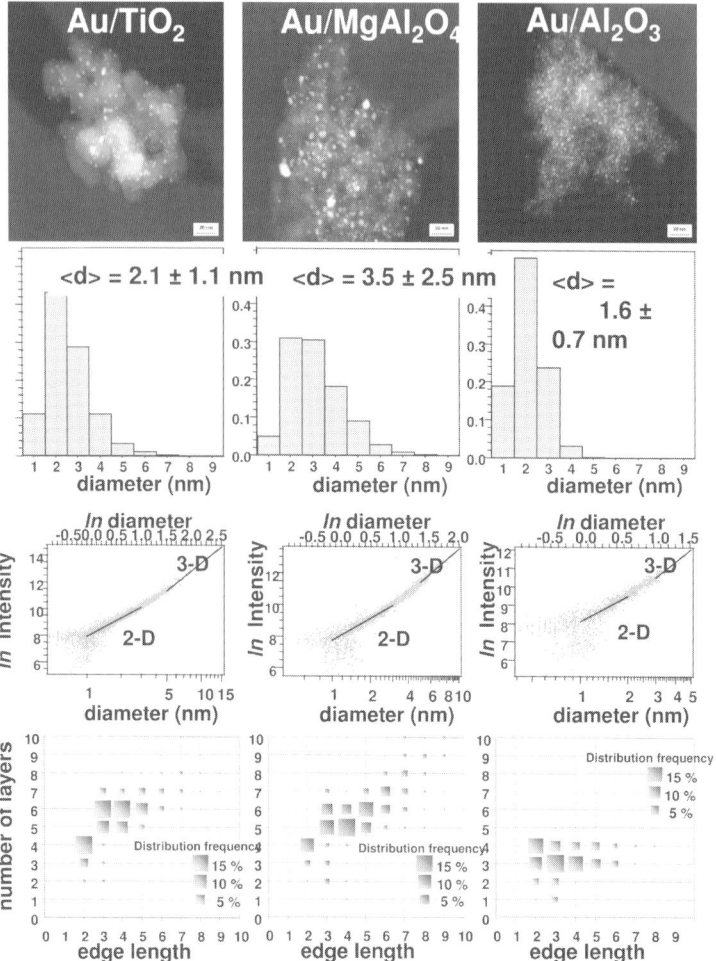

Fig. 7.7 Scanning transmission electron microscopy (STEM) images of supported gold catalysts, along with particle diameter distributions, double-logarithmic plots showing how particle volume (proportional to intensity) depends on particle size, and geometric distributions of truncated octahedrons with certain edge lengths and thickness, as indicated in Figure 7.8. (Adapted from [18]).

particle exhibits fcc (111) and (100) facets, with edge sites between the facets, and corner atoms at the end of the edges. The interface with the support is a (111) facet.

The distribution of diameters and volumes must now be translated into a distribution of the corresponding edge lengths and thicknesses of truncated octahedrons. This, however, requires an implicit calibration of the particle intensities in the HAADF-STEM image. Carlsson et al. used the average Au–Au coordination

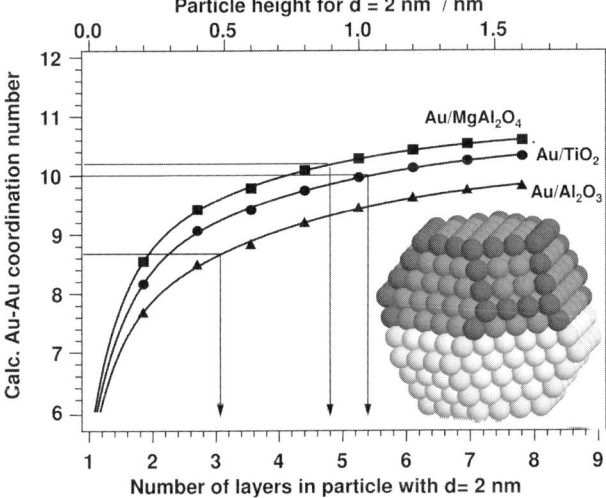

Fig. 7.8 Calculated Au–Au coordination numbers for truncated octahedrons of diameter corresponding to the average size of the three catalysts in Figure 7.7 for different thicknesses. (Adapted from [18]).

number from extended X-ray absorption fine structure (EXAFS) for this purpose. Alternatively, one could use the specific surface area from chemisorption measurements. Figure 7.8 illustrates how the calibration is achieved. Due to the particle size distributions, all samples contain at least some particles with a diameter of 2 nm, and these are used for calibration. The Au–Au coordination number from EXAFS is 10. The curve labeled Au/TiO$_2$ in Figure 7.8 corresponds to the coordination numbers of Au octahedrons with diameter 2 nm and thicknesses between one and eight atomic layers. A 2-nm particle of five atomic layers thickness matches the experimental model best. In this way, the STEM-HAADF intensity distribution can now be converted to a calibrated volume distribution. Each combination of diameter and volume corresponds to a truncated octahedron with a certain edge length and thickness. The distribution of corresponding particles is shown in the lower graphs of Figure 7.7.

Although the procedure necessarily rests on a number of assumptions on particle geometry, it undoubtedly represents a highly original and useful approach to quantify the distribution of sites in supported particles. In comparing the geometry distributions of gold on titania and gold on alumina, one also sees that the particles in the latter are flatter, indicating that the gold particles bind more strongly to the alumina than to the titania support. Hence, the method appears attractive for systematic studies of the often elusive metal–support interaction in supported catalysts.

In conclusion, TEM belongs to the most often-used techniques for the characterization of catalysts. The determination of particle sizes or of distributions therein has become a matter of routine, although the results rest of course on the assumption that the size of the imaged particle is truly proportional to the size of the actual particle, and that the detection probability is the same for all particles, independent of their dimensions. *In-situ* studies of catalysts are of special interest, and are rendered possible by coupling the instrument to a reactor [5, 6, 19, 20]. Numerous applications of electron microscopy in catalysis have been described in the literature, and several excellent reviews are available for the interested reader [3–7, 13, 19, 20].

7.2.4
Element Analysis in the Electron Microscope

As indicated in Figure 7.2, X-rays are among the byproducts of electron microscopy. Although the finding that matter emits X-rays when bombarded with electrons had already been recognized during the early part of the 20th century, the explanation of this phenomenon was only realized with the development of quantum mechanics. Today, it forms the basis for determining composition on a submicron scale and, with still increasing spatial resolution, in a technique referred to variously as electron microprobe analysis (EMA), electron probe microanalysis (EPMA) or energy-dispersive analysis of X-rays (EDAX, EDX) [15, 21, 22].

The interaction of an electron with an atom gives rise to two types of X-ray: characteristic emission lines and Bremsstrahlung. The atom emits element-characteristic X-rays when the incident electron ejects a bound electron from an atomic orbital. The core-ionized atom is highly unstable and has two possibilities for decay: X-ray fluorescence and Auger decay. The first is the basis for electron microprobe analysis; the second forms the basis of Auger electron spectroscopy (see Chapter 3).

X-ray fluorescence occurs when the initial core-hole in, for example the K-shell, is filled with an electron from a higher shell, say the L_{III} or L_{II} shell. The energy gain in this process, $(E_K - E_L)$, is used to emit a Kα X-ray photon, which is characteristic of the emitting atom. X-ray fluorescence competes with Auger decay, in which the energy gain, $(E_K - E_L)$, is taken up by a bound electron in the L shell (or a higher shell), which leaves the atom with kinetic energy $(E_K - E_L - E_L)$ as a so-called KLL Auger electron. Auger decay dominates in light elements, whereas X-ray fluorescence becomes significant for elements heavier than magnesium ($Z = 12$) and dominates in the heavy elements (see also Fig. 3.25). For example, the probability of X-ray fluorescence is 80% for zirconium ($Z = 40$), but is as low as 0.1% for carbon ($Z = 6$) [21].

Bremsstrahlung is emitted by an electron that is accelerated in the attractive force field of the positively charged nucleus. The electron follows a hyperbolic path with a curvature depending on its distance to the nucleus. During the acceleration it emits radiation called Bremsstrahlung (literal translation "braking radiation"), which produces a structure-less continuum background in the X-ray spectrum.

Thus, X-ray fluorescence offers the possibility of determining the composition of a sample. In SEM, where the beam can be positioned on desired positions of the sample, one can thus obtain the local composition on a scale of about 5 to 10 nm.

The most convenient means of analyzing the emitted X-rays is with an energy-dispersive X-ray detector located at a fixed position with respect to the sample. This detector is a liquid nitrogen-cooled solid-state device consisting of lithium-doped silicon, and functions as follows. An incident X-ray photon is converted by means of the photoelectric effect into an electron with kinetic energy ($E_{X\text{-ray}} - E_b$), in which E_b is the binding energy of the photoelectron. The latter dissipates its kinetic energy by creating electron-hole pairs in the semiconductor, at the cost of 3.8 eV per pair. The number of electron-hole pairs is reflected in the pulse height of the current that is measured if a voltage is applied over the detector. Thus, the pulse height is a measure of the kinetic energy of the photoelectron and hence of the energy of the incident X-ray photon. The energy resolution of this detector is limited to about 150 eV, implying that closely spaced X-ray lines cannot always be resolved. Wavelength dispersive spectrometers have much better resolution (on the order of 5 eV) but are slower, more expensive, and are restrictive on sample preparation.

Quantitation is usually achieved by comparing the X-ray yields from the sample with yields obtained from standards. The ease with which measurements can be interpreted quantitatively depends on the sample. As illustrated in Figure 7.9, the volume that is activated by the 10- to 100-keV electron beam has the shape of a

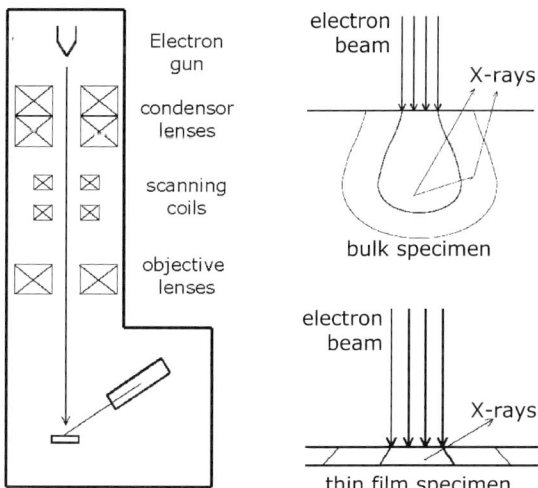

Fig. 7.9 Schematic set-up for measuring X-ray fluorescence with an energy-dispersive detector, as in energy dispersive X-ray analysis (EDX). Irradiation of a bulk sample activates a pear-shaped volume from which X-rays are emitted. The chance of secondary processes is considerable and requires correction of the measured X-ray yields; secondary effects are much less important if the sample is a thin film.

pear, with typical dimensions of a few micrometers. As a consequence, X-rays formed in the interior may be absorbed on their way out, and stimulate the emission of photoelectrons, Auger electrons and, again, X-rays. The latter process – termed *secondary fluorescence* – can lead to an overestimate of the concentrations. For example, if the specimen is a bulk Fe–Ni alloy, Ni Kα radiation is adsorbed by iron and causes photoemission of the K-electron of iron, which provides an additional possibility for Fe Kα radiation. Hence, the concentration of Ni in the alloy is underestimated by a high absorption of Ni Kα radiation, whereas the iron concentration is overestimated due to secondary fluorescence [21].

In bulk samples, X-ray yields need to be adjusted by the so-called "ZAF" correction. Here, Z represents the element number (heavier elements reduce the electron beam intensity more than lighter elements, because they are more efficient backscatterers), A indicates absorption (different elements have different cross-sections for X-ray absorption), and F indicates secondary fluorescence (the effect described above). Corrections are much less important when the sample is a film with a thickness of 1 μm or less, because secondary effects are largely reduced. The detection limit is set by the accuracy with which a signal can be distinguished from the Bremsstrahlung background. In practice, this corresponds to about 100 ppm for elements heavier than Mg.

Energy-dispersive X-ray analysis (EDX) has, for example, been used to determine the composition of individual particles in bimetallic catalysts, and here the studies of Wang et al. [23] on Pd–Au on carbon provide a good example. The fcc metals Pd and Au are easily miscible, and form alloys over almost the entire composition range. Figure 7.10 shows a TEM image of a reduced catalyst along with the particle size distribution. The right-hand part of the figure is an overall EDX spectrum, showing the X-ray emission lines of the carbon support, and the metals

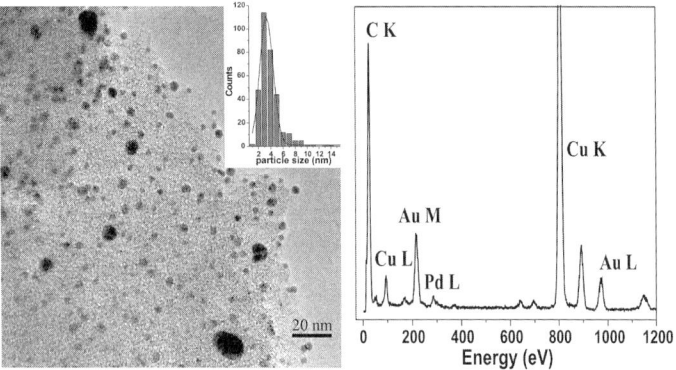

Fig. 7.10 TEM image of a bimetallic PdAu/C catalyst, along with particle size distribution and energy dispersive X-ray analysis of the composition, showing the X-ray emission lines of the carbon support, and the metals Au and Pd. The Cu lines are from the sample holder. (Adapted from [23]).

Au and Pd. The Cu signal is due to the TEM grid. In order to prove that the particles are truly bimetallic, the authors took EDX images of 50 single particles, which appeared to have a similar composition as the overall sample. Hence, the catalyst contains Pd–Au particles with a distribution of particle sizes, but a rather uniform composition.

7.3
Field Emission Microscopy and Ion Microscopy

Field emission microscopy (FEM) was the first technique capable of imaging surfaces at a resolution close to atomic dimensions. The pioneer in this area was E.W. Müller, who published details of the field emission microscope in 1936, and of the field ion microscope some 15 years later, in 1951 [24]. Both techniques are limited to sharp tips of high-melting metals (tungsten, rhenium, rhodium, iridium, and platinum), but have been extremely useful in exploring and understanding the properties of metal surfaces. Here we should mention the structure of clean metal surfaces, defects, ordering/disordering phenomena, diffusion of adatoms and interaction between adsorbed atoms. The techniques have, incidentally, been applied to catalytic problems. The theory behind field microscopy [25, 26], together with the details of a few examples, are briefly described in the following section.

7.3.1
Theory of FEM and FIM

The basic set-up for FEM and FIM is shown schematically in Figure 7.11. The sample is an electrochemically etched, needle-shaped single crystal with an approximately spherical end (radius ca. 100 nm) to which a high potential (5–10 kV) is applied. The tip faces an electrically conducting, fluorescing screen which is grounded. Figure 7.12 illustrates a potential energy diagram for the situ-

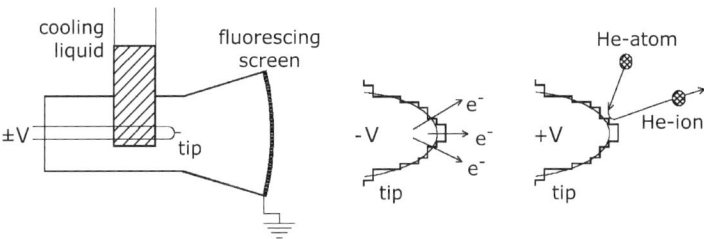

Fig. 7.11 The set-up for field emission microscopy and field ionization microscopy. Both techniques produce an image of the concave end of a single crystal tip on the fluorescent screen (see text for explanation). The tip exposes many facets of different crystallographic orientations.

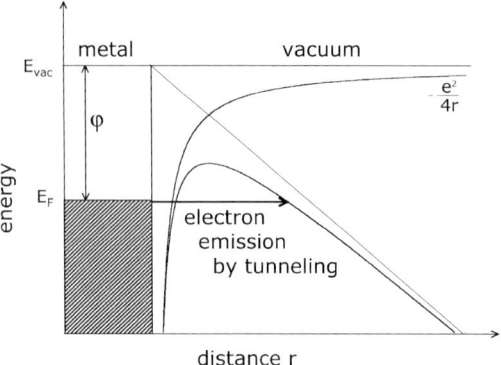

Fig. 7.12 Potential energy diagram for electrons in and near a metal on which a high negative potential is applied. Electrons in the valence band of the metal see an attractive potential equal to $-eFr$ (F is the applied field in V cm^{-1}) outside the metal, behind a barrier formed by the applied field and the image potential.

ation that the tip potential is negative. An electron inside the metal sees, as usual, a repulsive potential just outside the metal, but an attractive potential further away, due to the electric field between the tip and the screen. If the field is sufficiently high, then electrons from the metal are able to tunnel through the potential barrier, with a current density given by the Fowler–Nordheim equation:

$$j = \frac{c_1 V^2}{\varphi} \exp\left(-\frac{c_2 \varphi^{3/2}}{V}\right) \tag{7-1}$$

where:
- j is the current density;
- V is the potential of the tip;
- φ is the work function;
- $c_{1,2}$ are constants.

This entirely quantum mechanical phenomenon is called *field emission*. Equation (7-1) indicates that the current increases with increasing electric field and decreasing work function, in agreement with the potential energy diagram of Figure 7.12.

A spherical tip exposes many facets with different crystallographic orientations. The more open, high-index planes have a lower work function than the close-packed, low-index planes, and thus emit more electrons. The electrons travel in straight lines to the fluorescent screen, where they produce an image of the tip, in which the intensity of each facet is determined by its local work function. Bright and dark areas in the image are identified by comparing the pattern with a stereographic projection map, constructed by intersecting a sphere through the

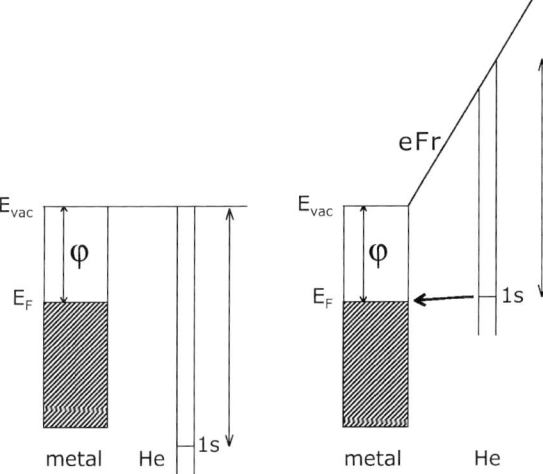

Fig. 7.13 The principle of field ionization. Left: the potential for a helium atom near a metal without a field. Right: in the presence of an electric field of strength F (V cm^{-1}). Field ionization by electron tunneling becomes possible when the He 1s level (ionization potential I) is above the Fermi level of the metal. Tunneling increases when the He atom is closer to the surface. This, however, requires high local fields, which are present at the edges of crystal facets, or at adsorbed atoms.

lattice and projecting the facets thus obtained onto a plane. The image resolution obtainable in field emission microscopy is about 2 nm.

In FIM, the ions of a gas such as hydrogen (or one of the rare gases) image the tip; the principle is shown in Figure 7.13 for helium. A high positive potential on the tip creates the situation that at some critical distance r_c from the surface, the He 1s electron lines up with the Fermi level of the metal. The electron may now tunnel to the metal, leaving a helium ion, which is strongly repelled by the positive tip and flies in a straight line to the fluorescent screen. The distance at which field ionization occurs follows approximately from the condition:

$$eFr_c = I - \varphi \rightarrow r_c = \frac{I - \varphi}{eF} \tag{7-2}$$

where:
r_c is the critical distance at which field ionization becomes possible;
I is the ionization potential of the imaging gas;
φ is the local work function of the site where ionization occurs;
e is the electron charge;
F is the electric field strength.

Ionization of the image gas by tunneling is most likely to occur where critical distances are small; this corresponds to sites where the local field strength F is

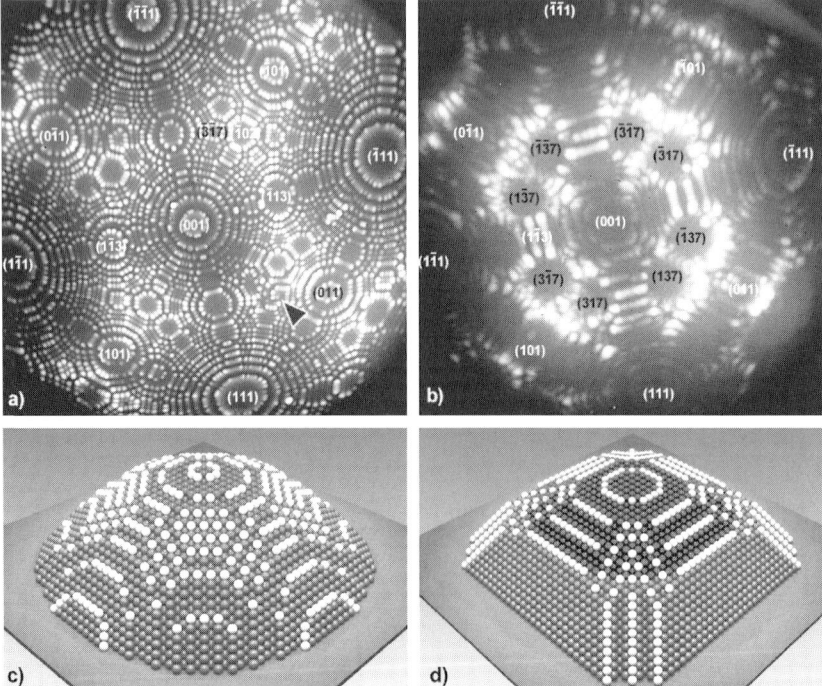

Fig. 7.14 (a) Field ion microscopy (FIM) image of a rhodium tip. As Rh has the fcc structure, the fourfold symmetry of the picture implies that the center corresponds to a (100) facet. Dark areas near the four edges of the figure are (111) facets. The arrowhead shows a small defect on the (135) plane. (b) The same Rh tip after adsorption of oxygen followed by field evaporation, leaving oxygen dissolved in the bulk and a short flash to desorb the surface oxygen. (c,d) Ball models of the tip morphologies. FIM and ball models demonstrate the oxygen-induced shape transformation from nearly hemispherical to polyhedral along with local missing-row restructuring of individual {011} and {113} planes. Oxygen is dissolved in the bulk, possibly leading to a surface Rh-oxide. (Courtesy of Norbert Kruse [27]).

high. Hence, adatoms, and atoms at edge positions on the tip, have the highest probability to cause field ionization and appear as bright features in the image. FIM is capable of imaging single, adsorbed atoms.

Figure 7.14 shows the – very beautiful – FIM image of a rhodium tip oriented in the (001) direction, prepared by Kruse [27]. The image possesses fourfold symmetry, implying that the center region corresponds to a (100) face. With the help of a stereographic projection map one can identify the other planes. The large dark areas near the outside of the image correspond to surfaces with a high work function, and are due to (111) facets. The many bright circles correspond to the edges of crystal facets. The right-hand part of Figure 7.14 is the same surface exposed to O_2 at elevated temperature, but after removal of the adsorbed oxygen; this shows that the rhodium tip is significantly restructured by dissolved oxygen. While the original rhodium tip resembles an almost perfect hemisphere, the restructured tip resembles a truncated polyhedron. This coarsening of the morphology becomes immediately evident if one draws a line from the central (001) facet to for example the (111) facet below: the path is a smooth trajectory on the original surface, but a stepped one on the reconstructed tip.

FEM and FIM are both real-time imaging techniques, implying that time-dependent phenomena can be followed. The FIM studies of Ehrlich and co-workers [28, 29] on surface diffusion of single adatoms, and of Bassett and Parseley [30] on the interaction between adsorbed atoms, form impressive examples. With respect to catalysis, the elegant field emission studies by van Tol et al. are also of note [31, 32]. These authors used FEM to follow the oscillating reaction between NO and H_2 over a rhodium tip. The application proved to be successful, because NO-covered rhodium has a higher work function than clean rhodium, or rhodium covered by NH_x fragments. Thus, the conversion of adsorbed NO gives rise to enhanced field emission from the substrate, which is observable in FEM. At the start of the reaction, the entire tip is covered by NO, and initiation of the reaction between NO and H_2 is observed at a high index surface; the reaction is subsequently seen to spread across the entire tip in a period of seconds. At the end of the reaction, the empty surface is refilled with NO and the cycle starts over again [32].

Atom probe microscopy is a variation on FIM in which either the field-ionized atoms or evaporated atoms from the tip are detected with a mass spectrometer, placed behind an aperture in the imaging screen. This allows one to identify the desorbing ions. If the tip is mounted on a manipulator, it is possible to zoom in on a desired surface plane. This technique has been used to study the composition of alloy surfaces, and the interested reader is referred to Tsong for some excellent reviews [33, 34].

Block and co-workers [35] modified the atom probe to develop a method called pulsed-field desorption mass spectrometry (PFDMS), whereby a short high-voltage pulse desorbs all species present on the tip during a catalytic reaction. The repetition frequency of the field pulse controls the time for which the reaction is allowed to proceed. Hence, by varying the repetition frequency between desorption pulses in a systematic way, one can study the kinetics of a surface reaction [35]. In fact, this type of experiment – where one focuses on a facet of desired structure, which may include steps and defects – comes close to one of the fundamental goals of catalyst characterization, namely studying a catalytic reaction on substrates of atomically resolved structure with high time resolution.

7.4
Scanning Probe Microscopy: AFM and STM

The most recent developments in determining the surface structure are the scanning tunneling microscope (STM) and the scanning or atomic force microscope (SFM or AFM) [36, 37]. These techniques are capable of imaging the local surface

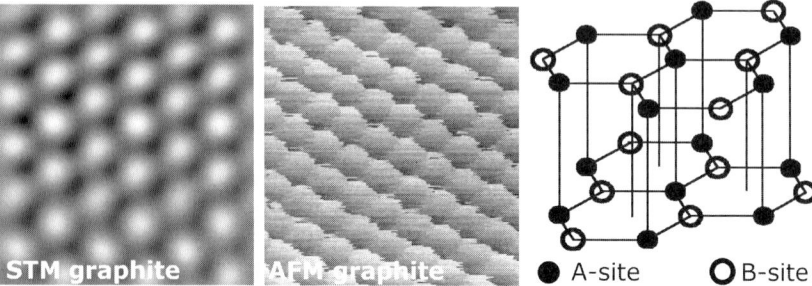

Fig. 7.15 Images at atomic resolution of graphite obtained with scanning tunneling microscopy (STM; *left*) and atomic force microscopy (AFM; *center*). The graphite lattice contains two types of site: A-sites with a carbon atom neighbor in the second layer; and B-sites without a neighbor in the next layer. STM detects the B-sites, whereas the A-sites show up better in AFM. (STM image courtesy of TopoMetrix; AFM image courtesy of M.W.G.M. Verhoeven, Eindhoven).

topography with a resolution determined by the geometric shape of a tip and the nature of the interaction between the tip and the surface. Under favorable conditions, atomic resolution is achievable.

The general concept behind scanning probe microscopy is that a sharp tip is rastered across a surface by piezoelectric translators, while a certain property reflecting the interaction between the tip and the surface is monitored. As a result, scanning probe microscopy yields local information. STM is based on the tunneling current that flows between a tip and a surface at different electric potentials, and was discovered by Binnig and Rohrer at IBM in Switzerland in 1981 [38]. As the tunneling between two surfaces is entirely dominated by the shortest distance between two atoms on either side of the gap, this technique offers atomic resolution. A few years later, in 1986, the same group reported the more generally applicable AFM, based on the forces between a tip and the surface [39]. Here, the interaction between the tip and the surface has contributions from short and longer range forces. Although in favorable cases AFM may image surfaces in atomic detail, it more commonly reveals the surface morphology on a larger scale (of nanometers); consequently, the technique is often referred to as scanning force microscopy (SFM). The AFM and STM images of graphite are shown in Figure 7.15; the separate atoms can be clearly discerned, although in STM only three of every six atoms in the graphite rings is seen. In contrast, when using AFM half of the atoms can be seen much more clearly better than the other half (we will return to this point later).

7.4.1
AFM and SFM

AFM (or SFM), which is the most generally applicable member of the scanning probe family, is based on the minute but detectable forces (of the order of nano

Newtons) between a sharp tip and atoms in the surface [40]. The tip is mounted on a flexible arm, called the cantilever, and is positioned at subnanometer distance from the surface. If the sample is scanned under the tip in the x-y plane, it feels the attractive or repulsive force from the surface atoms and hence it is deflected in the z-direction. Various methods exist to measure these deflections [37, 40–42]. Before we describe the equipment and applications to catalysts, we will briefly examine the theoretical aspects of AFM, which can be applied in either the contact mode or the non-contact mode.

7.4.1.1 Contact Mode AFM

In the contact mode, the tip is located within a few Ångstrøms of the surface, and the interaction between them is determined by the interactions between the individual atoms in the tip and in the surface. The description of atomic force interactions in the contact mode is highly complex. It requires a molecular dynamics simulation of the Coulomb interaction between charges or charge distribution, polarization due to induced dipole moments, and quantum mechanical forces when electron orbitals start to interact, for each pair of atoms from the tip and the surface [36, 37, 40–42]. Let us simplify the situation to that indicated in Figure 7.16a, where we consider the interaction between the atom at the apex of a tip and the atom directly underneath in the surface in terms of a Lennard–Jones potential. The force between the two atoms is given by the negative derivative of the potential. As indicated in the figure, one can work with either an attractive or a repulsive force within the contact mode. The contact mode is the usual choice of operation if one is interested in surface morphology, but is the only choice if one

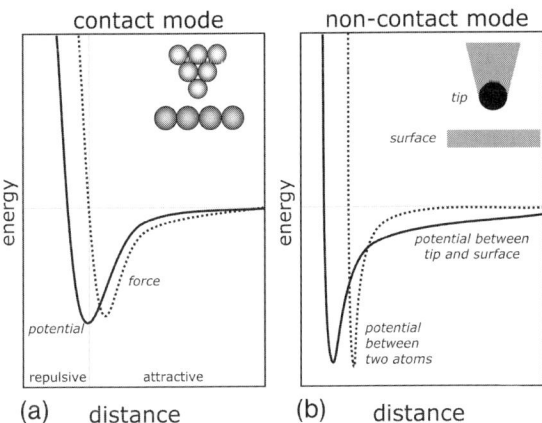

Fig. 7.16 (a) Interaction potential and force between an atom at the apex of the tip and an atom in the surface. Tip–surface interactions can be described by a summation of these potentials over all combinations of atoms from the tip and the surface.
(b) Interaction potential between the tip, approximated as a sphere, and a plane surface, valid in the non-contact mode of force microscopy. To stress the long-range character of the non contact-potential, the Lennard–Jones interaction potential between two atoms has also been included (dotted line).

requires atomic resolution. Forces in this mode are considerable, however (on the order of nano Newtons), and it should be borne in mind that the surface might also be affected.

7.4.1.2 Non-Contact Mode AFM

The second mode of operation is the non-contact mode, in which the distance between tip and sample is much larger, between 2 and 30 nm [42]. In this case, one describes the forces in terms of the macroscopic interaction between bodies. Figure 7.16b provides an example in which the sample has a flat surface and the tip is a spherical particle. Several forces may play a role here: electrostatic in case of a potential difference between the tip and the sample; magnetostatic if the sample is magnetic; and the always-present van der Waals forces due to fluctuating or induced dipoles in the tip and the sample. Forces in the non-contact mode are typically two to four orders of magnitude smaller than in the contact mode. It will be clear that details on the subnanometer scale are not obtained because the interaction is now between larger portions of the tip and the sample. The non-contact mode is of particular interest for imaging magnetic domains or electronic devices.

7.4.1.3 Tapping Mode AFM

A third mode which has recently become the standard for work on surfaces that are easily damaged, is in essence a hybrid between contact and non-contact mode, called tapping mode [36, 37, 42]. In this case, the cantilever is brought into oscillation such that the tip just touches the surface at the maximum deflection towards the sample. When the oscillating cantilever approaches the surface, it begins to feel the surface and the oscillation becomes damped; this damping is detected electronically and used as the basis for monitoring the topography when the sample is scanned. In tapping mode, shear forces due to dragging the tip horizontally along the surface ("scratching") are avoided, while forces in a perpendicular direction are greatly reduced. Recently, tapping mode has become the favored way of imaging small particles on planar substrates used as models for catalysts.

7.4.2
AFM Equipment

Atomic force microscopes have been built in many different versions, with at least six different ways of measuring the deflection of the cantilever [36, 37, 40–42]. The commercially available AFM systems use the double photo detector system shown in Figure 7.17 and described by Meyer and Amer [44]. Here, a lens focuses a laser beam on the end of the cantilever, which reflects the beam onto two photo detectors which measure intensities I_1 and I_2. When the cantilever bends towards the surface, detector 2 receives more light and the difference $(I_2 - I_1)$ becomes larger. If the tip is scanned over the sample by means of the x- and y-components of the piezo crystal, the difference signal $(I_1 - I_2)/(I_1 + I_2)$

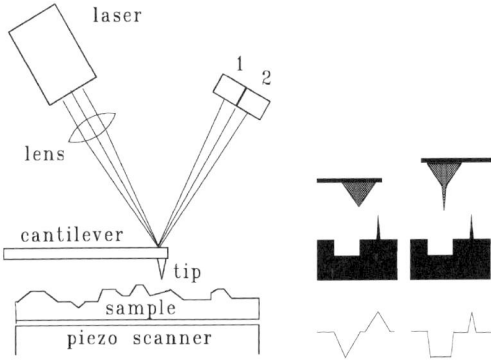

Fig. 7.17 Experimental set-up for atomic force microscopy (AFM). The sample is mounted on a piezoelectric scanner and can be positioned with a precision better than 0.01 nm in the x-, y-, and z-directions. The tip is mounted on a flexible arm, the cantilever. When the tip is attracted or repelled by the sample, the deflection of the cantilever/tip assembly is measured as follows. A laser beam is focused at the end of the cantilever and reflected to two photodiodes, numbered 1 and 2. If the tip bends towards the surface, photodiode 2 receives more light than 1, and the difference in intensity between 1 and 2 is a measure for the deflection of the cantilever and thus for the force between the sample and the tip. With four photodiodes, one can also measure the sideways deflection of the tip, for example at an edge on the sample surface. Scanning probe images represent a convolution between the morphology of the sample surface and the shape of the tip. Features on the samples that are sharper than the tip yield an image of the tip. In such cases so-called "supertips" improve the image greatly. Note, however, that normal tips are perfectly suitable to image flat surfaces in atomic detail.

provides a sensitive measure for the deflection of the tip and describes the topography of the surface.

The positioning of the sample at better than atomic precision is carried out with piezo crystals, which are ceramic electromechanical transducers that distort when a voltage is applied. As the distortion is proportional to the applied voltage, one can displace a sample by means of piezos at any desired precision, provided that the electronics are sufficiently accurate and stable. The rapid development of the scanning probe microscope would not have been possible without the availability of modern, stable electronics. In order to obtain optimum atomic resolution, one usually selects a tubular piezo element with a small scan range (ca. 1 µm); for larger scan ranges, the scanner is a tripod with a separate piezo element for each dimension.

In addition to piezo scanners, the atomic force microscope may contain a step motor for coarse x–y positioning of the sample. Most instruments also possess a built-in camera for selecting the desired area of the sample and for positioning the tip at a few micrometers distance from the surface. The final approach of the tip towards the surface is performed automatically.

Images can be made in either variable or constant force mode. In the latter case, the difference signal from the photo detectors is used to compensate the distance between tip and surface, such that the force between the two – and thus the deflection of the cantilever – remains constant. An important advantage of working in constant force mode is that the overall orientation of the surface with respect to the z-direction is less critical, because the z-piezo compensates for any inclination of the sample.

The image is always a convolution of the topography of the surface and that of the tip, and the one with the least steep feature determines the image (see Fig. 7.17). Flat surfaces are scanned with the conventional pyramidal tips, which have a wide opening angle and are relatively blunt. If surfaces contain features that are sharper than the tip, one images the tip shape rather than the surface topography. Figure 7.18 shows the AFM image of copper particles on a planar silica substrate. The particles, with average diameters of only a few nanometers, appear larger because the size of the tip determines the resolution of the image. Nevertheless, the correct information is obtained on the height of the particles; as long as the tip can reach the substrate, the correct height difference between particle apex and substrate level can be measured.

Of course, sharper tips produce sharper images. Commercially available super-tips have a diameter in the 0.1 µm range, and a characteristic radius at the tip of about 20 nm. Such tips can be made by focusing an intense electron beam of a scanning electron microscope in the presence of a low pressure of hydrocarbons

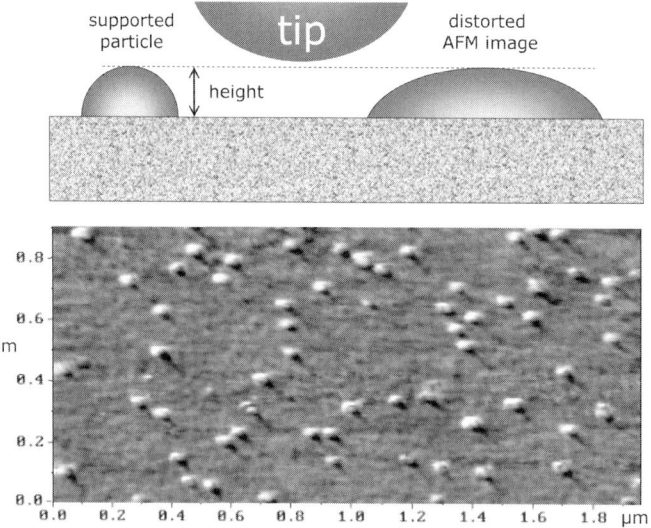

Fig. 7.18 AFM image of a planar model catalyst consisting of copper particles deposited by spin-coat impregnation on a flat SiO$_2$ on Si(100) substrate, after reduction in hydrogen. Because the image is a convolution of the shape of the particle and that of the tip, the particles appear larger than they are. The height, however, is correctly reproduced. (AFM image courtesy of V. Pushkarev and A. Borgna, Eindhoven).

on the top of a conventional pyramidal tip. An additional advantage of using sharp tips is that the forces between tip and surface are much smaller because they are limited to a much smaller interaction volume. As a consequence, they are better suited to the scanning of soft biological molecules and polymers. Most manufacturers also offer a number of cantilever–tip combinations with different stiffness. For example, when working with soft biological molecules, one selects a more flexible cantilever.

An even better solution is to operate the microscope in tapping mode (see Section 7.5.3). Here, one applies an oscillation to the cantilever, which allow it to vibrate between contact and non-contact mode. The lateral movement occurs when the tip is out of touch, the great advantage being that the tip cannot scratch and damage the surface during scanning. Tapping mode is also very useful in imaging supported catalysts too. If images as shown in Figure 7.18 are taken in straightforward contact–constant force mode, one often sees in a subsequent image of the same area that particles have been swept to the side in the previous measurement.

Contrast in tapping-mode images originates from three different sources. First, a constant height image reflects the topography, as described above. Second, recording the changes in amplitude of the oscillating cantilever produces an image of gradients in the interaction force between sample and tip – simply said, the derivative of the topography. Third, analyzing the phase response of the oscillating cantilever when it is scanned over the surface produces a phase-contrast image, with contributions from mechanical as well as topographical properties of the imaged area.

The scanning probe techniques readily achieve subnanometer resolution and can sometimes even visualize atoms, provided that the following requirements are satisfied:

- The microscope must be isolated from all external *mechanical vibrations*, because these cause fluctuations in the distance between the tip and the surface, as well as in the horizontal position of the tip with respect to the surface. Isolating the system from vibrations is usually achieved by placing it on a solid metal plate, which is suspended between elastic springs. If the table is made from a non-magnetic material such as aluminum, the horizontal vibrations of the suspended system are effectively damped by a static magnetic field, owing to eddy currents, which resist any change in position with respect to the field.

- The effects of *acoustic noise* should be prevented by enclosing the microscope in a metal container and placing the entire instrument in a quiet room, preferably without air conditioning.

- The *voltages* used to control the piezo scanners should be sufficiently stable to keep the scanning fluctuations well below the 0.01 nm level.

By taking these precautions, atomic-scale images can easily be obtained from the surfaces of, for example, mica, molybdenum sulfide, or graphite (see Fig. 7.15). Careful inspection of this image shows sharp white areas, corresponding to atoms with a carbon neighbor in the second layer of the graphite crystal (A-sites), where the repulsion between the surface and the tip is maximum. The black areas are due to sites where the repulsion is minimal, corresponding to the middle of the carbon six-ring. Atoms without a neighbor in the second layer do not appear as distinct features but rather fall into the middle of the gray scale. Again, one should realize that AFM images the forces between the tip and the surface and that the extremes, A-sites, and the middle of the ring, show up best.

It should be noted, however, that although AFM produces images of graphite indeed with the correct geometry of the lattice, this does not necessarily imply that the image is the result of an atom-by-atom scan of the tip across the surface. Graphite, and also MoS_2 or mica, possess layered structures. The atomic image most probably arises because the tip drags one layer over the next, which produces a modulation of the force with the same periodicity as the lattice exhibits. The effect can easily be recognized: if AFM images show perfect surfaces without defects, the chances are high that the picture arises from two extended layers sliding across each other.

Nonetheless, it is possible to obtain truly atom-by-atom measurements from suitable surfaces. Iwasawa and co-workers reported atom-resolved images of the $TiO_2(110)$ and CeO_2 surfaces by non-contact AFM, revealing the correct topography and also irregularities and defects such as single oxygen vacancies [45, 46]. Even more promising with respect to the potential of AFM is that individual adsorbates could be discerned [46]. Figure 7.19 shows an example of these authors' impressive studies on CeO_2 surfaces [46].

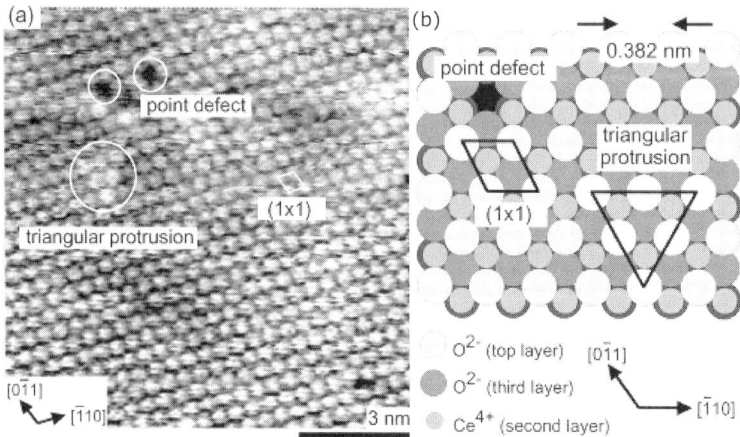

Fig. 7.19 (a) AFM image of a $CeO_2(111)$ surface and (b) a structure model of $CeO_2(111)$ terminated by an oxygen layer. The fact that oxygen defects are clearly detected forms convincing evidence that the image has true atomic resolution, unlike the AFM image of graphite in Figure 7.15. (From [46]).

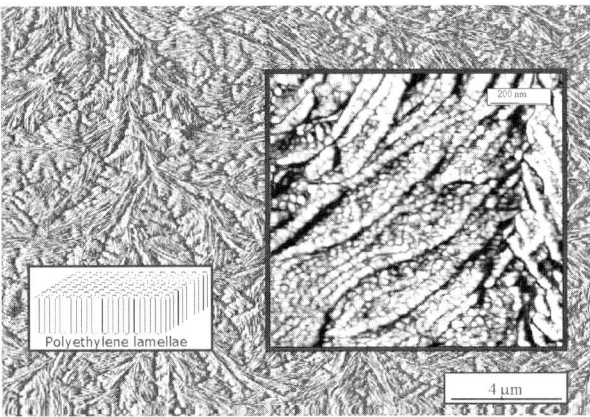

Fig. 7.20 AFM Image of polyethylene grown at 160 °C and subsequently crystallized during cooling on the surface of a planar CrOx/SiO$_2$ catalyst. The left inset indicates schematically how polyethylene molecules fold into lamellar structures. The AFM image shows how these lamellae have a tendency to order locally. The right inset is a measurement at higher magnification in phase contrast, and shows that lamellae contain a substructure, attributed to ordered and amorphous domains. (Adapted from [48]).

In principle, AFM can be used to image all surfaces, including those of a supported catalyst, provided that the powder particles can be immobilized (e.g., in glue). However, it remains difficult to identify any differences between a catalyst particle and the features of the support. Model systems of particles on a flat support offer better opportunities, as demonstrated in Figures 7.18 to 7.20.

AFM has produced spectacular images of biological samples such as cells, proteins and DNA, and also of polymers. Figure 7.20 shows the SFM image recorded in tapping mode at normal height contrast of polyethylene grown on a planar model of a polymerization catalyst, consisting of small amounts of chromium on a SiO$_2$/Si(100) surface [47]. The polymer was grown from ethylene at 160 °C – that is, above the melting temperature of polyethylene; hence, the image represents polymer crystallized from the melt. The white stripes are lamellae of polyethylene which, according to the enlarged part in the inset, contain further substructure [48]. We will return to the subject of Cr/SiO$_2$ polymerization catalysts in Chapter 9.

7.4.3
Scanning Tunneling Microscopy (STM)

STM is based on the tunneling of electrons between the surface and a very sharp tip [36, 37]. As explained in the Appendix, the cloud of electrons at the surface is not entirely confined to the surface atoms but rather extends into the vacuum (this effect causes the electric dipole layer at the surface that contributes to the work function). When an extremely fine tip (see Fig. 7.21) approaches the surface

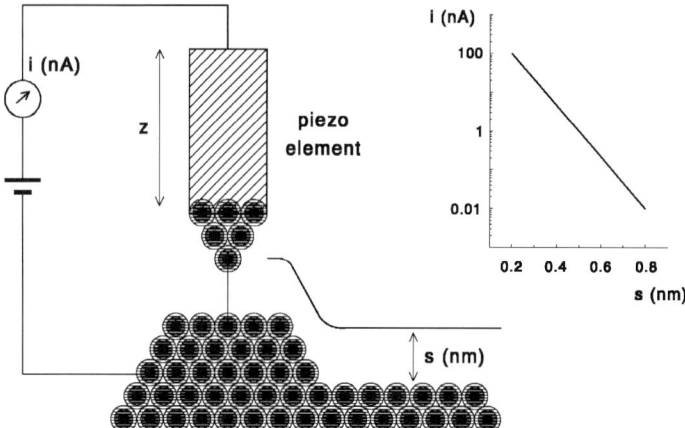

Fig. 7.21 Scanning tunneling microscopy is based on the tunneling of electrons between the surface and an atomically sharp tip positioned at a few Ångstrøms above the surface. The tunneling current depends sensitively on the distance s between the tip and the surface. An image of the surface is obtained by scanning the tip horizontally over the surface. A control system maintains the tunneling current, and therefore the distance between tip and surface, constant; the scan is a plot of the vertical position of the piezo-electrically driven tip versus its horizontal position.

to within a few Ångstrøms, the electron clouds of the two begin to overlap. A small positive potential on the tip is sufficient to cause a measurable tunneling current over the gap between the tip and the surface, the so-called "tunneling gap". The energy diagram in Figure 7.22 illustrates the principle behind STM on

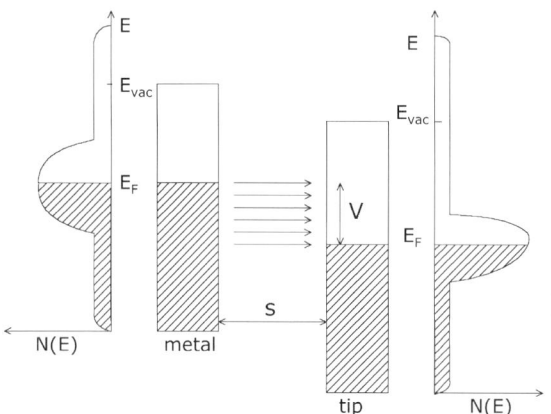

Fig. 7.22 The principle of tunneling between two metals with a potential difference V, separated by a gap s. Electrons tunnel horizontally in energy from occupied states of the metal to unoccupied states of the tip.

metals. The voltage V shifts the position of all electrons in the metal up with respect to those of the tip, and the Fermi levels are separated by an energy, $e \cdot V$. If the tunneling gap is sufficiently small, valence electrons from positions close to the Fermi level can tunnel to unoccupied states of the tip. The tunneling current is in the range of pico- to nanoamperes, and varies exponentially with the distance between tip and surface (see Fig. 7.21) according to the formula:

$$j = \alpha_1 \frac{\sqrt{\varphi_{av}}}{s} V e^{-\alpha_2 \sqrt{\varphi_{av}} s} \qquad (7\text{-}3)$$

where:
- j is the current density;
- s is the distance between tip and surface;
- V is the potential between tip and surface;
- φ_{av} is the average between the work functions of tip and sample;
- $\alpha_{1,2}$ are constants.

It is the steep variation of the tunneling current with distance that enables one to obtain an image of the atoms in the surface when the tip is rastered over the surface. The most common method of measuring STM images is to record the position of the tip from the surface necessary for keeping the tunneling current constant.

If the tunneling current is from the surface to the tip, the STM images the density of occupied states. However, if the potential is reversed the current flows in the other direction, and one images the unoccupied density of states, as can easily be understood from Figure 7.22. This figure also illustrates a necessary condition for STM: there must be levels within an energy $e \cdot V$ from the Fermi level on both sides of the tunneling gap, from and to which electrons can tunnel! In metals, such levels are practically always available, but when dealing with semiconductors or with adsorbed molecules, this criterion may be a limitation. A second condition is that the sample possesses conductivity; perfect electrical insulators cannot be measured with STM.

It is important to realize that STM does not necessarily image atoms, but rather a section of the density of states near the Fermi level. We have already encountered the example of graphite (see Fig. 7.15), where STM images only three of the six carbon atoms of the honeycomb lattice, in spite of the fact that all six atoms lie in the same plane. The reason for this is that the six-ring of graphite contains two types of atoms, three with a carbon atom directly underneath in the next lattice plane (shown as "A-site" in Fig. 7.15), and three above the empty center of the six-rings of the second layer (denoted "B-site"). Theoretical calculations with a tungsten cluster, W_{14}, as the tip, with a bias voltage of 0.5 V and at 0.52 nm distance above the graphite surface, predict a high tunnel current at the B-sites, a smaller current at A-sites, and the minimum current – as expected – in the center of the ring. The high tunneling current at B-sites is due to a local high density of occupied states just below the Fermi level, derived from p_z-orbitals perpendicular to the surface which are virtually non-bonding at B-sites, but have considerable overlap with p_z orbitals from neighbors below at A-sites [49]. This example, which is known by all STM users because graphite is a common test sample for demonstrating that an STM achieves atomic resolution, illustrates that one must be careful in translating STM pictures directly into ball models of the surface.

It is possible to obtain chemical information from STM when it is used in the spectroscopic mode, for example by measuring at a fixed distance the tunneling current I as a function of the voltage over the gap (I/V spectroscopy). This method of measurement is termed "scanning tunneling spectroscopy" (STS) [37].

7.4.4
Applications of STM in Catalytic Surface Science

In comparison to most other methods in surface science, STM offers two important advantages: (1) it provides local information on the atomic scale; and (2) it does so *in situ* [50]. As STM operates best on flat surfaces, applications of the technique in catalysis relate to models for catalysts, with the emphasis on metal single crystals. Several reviews have provided excellent overviews of the possibilities [51–54], and many studies of particles on model supports have been reported, such as graphite-supported Pt [55] and Pd [56] model catalysts. In the latter case, Humbert et al. [56] were able to recognize surface facets with (111) structure on palladium particles of 1.5 nm diameter, on an STM image taken in air. The use of ultra-thin oxide films, such as Al_2O_3 on a NiAl alloy, has enabled STM studies of oxide-supported metal particles to be performed, as reviewed by Freund [57].

Although STM is capable of imaging surfaces at atomic resolution, it does not automatically reveal the chemical identity of these atoms, except in very special cases. If STM can distinguish between atoms of different types, it has to be through a difference in tunneling current or a difference in the way in which the current changes with distance to the surface. Varga and co-workers have been able to distinguish between the metal atoms in binary alloys of Pt and Ni [58], Pt and Co [59], and Pt and Rh [60]. Figure 7.23 shows the STM image of a PtRh(100) surface. Although the bulk material contains equal amounts of each element, the surface consists of 69% platinum (dark) and 31% rhodium (bright), in agreement with the expected surface segregation of platinum on clean Pt–Rh alloys in ultra-high vacuum [61]. The black spots in Figure 7.23 are due to carbon impurities. It can be seen that platinum and rhodium have a tendency to cluster in small groups of the same elements.

Sulfide catalysts play an important role in the hydrotreating of crude oil (this is discussed in detail in Chapter 9), and STM has provided indispensable insight into the structure of the catalytically active phase. Sulfides such as MoS_2 and WS_2 form the basis of the catalyst, although in order to exhibit higher activity these sulfides are promoted by Co or Ni. Figure 7.24 shows the STM images of MoS_2 and Co-promoted MoS_2, as reported by Besenbacher and co-workers [62–65]. These authors prepared their sulfides as model systems on a gold substrate,

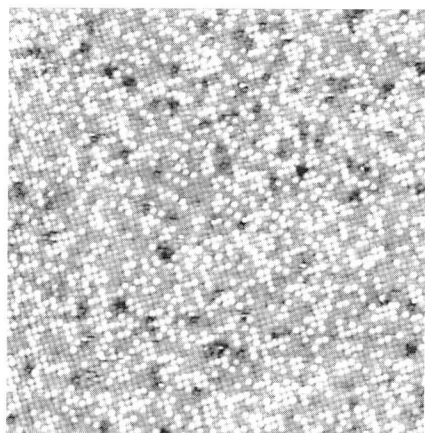

Fig. 7.23 STM image of a PtRh(100) surface, showing individual platinum (*dark*) and rhodium (*bright*) atoms. The black spots are due to carbon impurities. (From [60]).

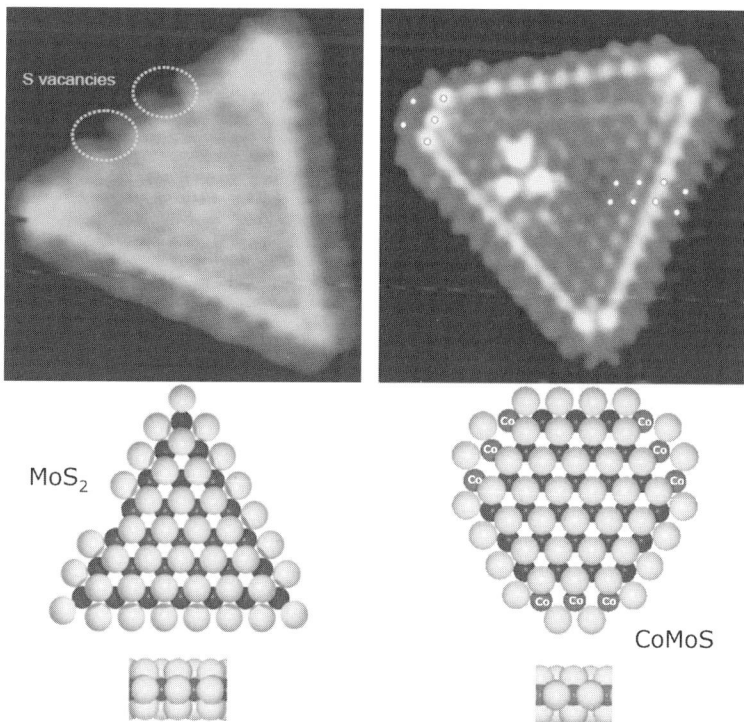

Fig. 7.24 STM images of MoS$_2$ and CoMoS particles on a gold substrate along with structure models. (Adapted from [62, 63]).

in order to have sufficient conductivity for being able to measure a tunneling current. The left image shows a triangular cluster consisting of a single layer of MoS_2; the corrugation is caused by the sulfur atoms. Interestingly, the S-atoms at the edge are out of registry with their bulk positions, as is clearly seen in the structure model below the STM image. S-vacancies, visible at the edge, are believed to represent sites where a sulfur-containing molecule such as thiophene may bind to the particle and become desulfurized. Note also the region of high intensity on the basal plane close to the edge, associated with a high local density of one-dimensional metal-like states. These brim states display catalytic hydrogenation activity, which is remarkable, because the basal plane of MoS_2 is generally assumed to be chemically inactive [65]. The cobalt-promoted MoS_2 cluster adopts an almost hexagonal shape with separate cobalt and molybdenum edges, both terminated by sulfur. The cobalt edge, however, is terminated by single sulfur atoms, giving the cobalt a tetrahedral coordination. The brim sites display even higher intensity than in the image of the MoS_2 cluster. Figure 7.24 represents the first atomically resolved image of the Co–Mo–S phase, which we discuss further in Chapter 9.

The next example shows how STM enables one to visualize the ordering and movement of adsorbate atoms on surfaces. Figure 7.25a shows a constant-height image of oxygen atoms on a Ru(0001) surface at 300 K, measured by Wintterlin et al. [66]. As the coverage of O-atoms is very low (<1%), the image exhibits mostly single atoms, although a few dimers also exist. In fact, this image is one from a video series taken at 15 frames per second, in which hopping of the atoms between adsorption sites can be monitored in real time. The authors derived a

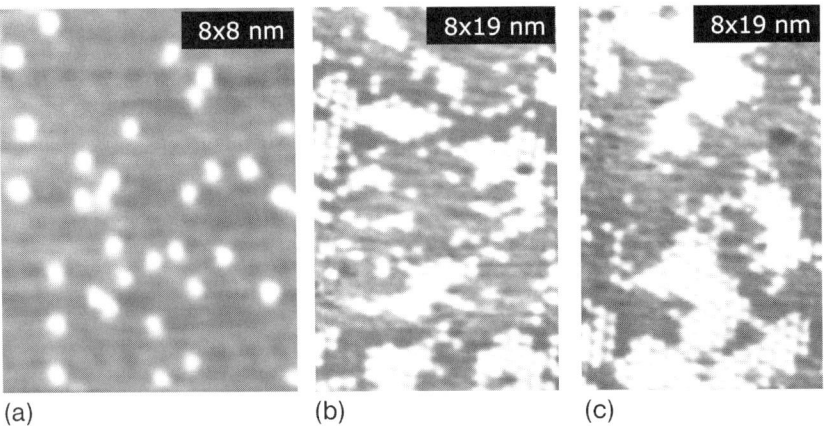

Fig. 7.25 STM images in constant height mode of (a) individual oxygen atoms on Ru(0001) at low coverage. Images (b) and (c) show how oxygen atoms order into islands at higher coverages (0.09 and 0.12 monolayer, respectively). The images are part of a video sequence which shows that the O-atoms exhibit considerable mobility at room temperature. (Courtesy J. Wintterlin [66]).

hopping rate of 14 3 s^{-1} for the motion of O-atoms on Ru(0001) at 300 K, corresponding to an activation barrier of 70 kJ mol^{-1} at a pre-exponential factor of 10^{13} s^{-1} [66]. At higher coverages of 9 and 12%, respectively, the atoms are seen to aggregate into islands of (2 × 2) structure (Fig. 7.25b and c) which, due to the significant mobility of the O-atoms, change shape continuously. Such information on the local ordering is of great interest for understanding many phenomena. In kinetic modeling, for example, one usually assumes that adsorbed species are distributed randomly across the surface. If however, the reacting species order in a fashion as shown in Figure 7.25, only the atoms at the perimeter of the island and those moving between islands, will be available for reaction. Hence, determination of the spatial distribution of adsorbed species across a surface is of the utmost importance for a correct kinetic description of catalytic reactions, and STM provides a means of measuring this distribution directly [66, 67]. STM holds great promise for studies of catalytic surface reactions, even under *in-situ* conditions [68].

STM relies on the tunneling current between a tip and the surface, and therefore it is mainly limited to conducting substrates, mainly metals and semiconductors. Nonetheless, possibilities do exist to study oxides. TiO_2, for example, is an insulator, but after heating in vacuum or bombardment by ions, oxygen vacancies arise which provide sufficient conductivity to the material to enable the application of STM. Thus, excellent images of the TiO_2 (110) surface and its (1 × 2) reconstruction have been reported [69]. The adsorption and decomposition of molecules such as pyridine on TiO_2 (110) have also been successfully visualized [70].

TiO_2 surfaces have been used as a model support for metal particles, which have successfully been imaged using STM [71–73]. In a very interesting study on gold particles, Goodman and co-workers used STS to show that particles below a certain size exhibit different electronic structure, which correlates with the catalytic activity of the particles for the CO oxidation reaction [72]. Also of note were some STM studies of metal–support interactions on systems made by depositing thin submonolayer films of oxide onto a metal surface. This approach has provided insight into the chemistry at metal–support interfaces [74].

7.5
Other Imaging Techniques

In the microscopic techniques discussed above, the challenge was to visualize the atomic detail. However, in catalysis one also encounters phenomena that occur on the scale of micrometers or millimeters which require imaging. In particular, these include the ordering of adsorbates in large islands and the development of spatiotemporal patterns in oscillating reactions [9]. This spectacular phenomenon has stimulated the exploration of imaging techniques that provide information on patterns on the micrometer to millimeter scale.

7.5.1
Low-Energy Electron Microscopy and Photoemission Electron Microscopy

Among the techniques discussed in this book, several lend themselves to imaging. For example, if one combines low-energy electron diffraction (LEED) with electron optics, one may obtain an image of the surface along with the LEED pattern. Suppose that a surface is partially covered by large islands of ordered adsorbates – say oxygen in a (2×2) structure – then low-energy electron microscopy (LEEM), as developed by Bauer [75], provides an image of the topography, in which the contrast is furnished by the different diffraction conditions of the islands and the bare surface.

Photoemission may also be taken as the basis for microscopy. Suppose a surface is covered by separated islands of oxygen atoms and carbon monoxide (as shown schematically in Fig. 7.26). The local work function above the O-atoms will be significantly larger than that above the CO molecules, which in turn will be higher than that of the bare surface. Hence, photoemission intensity from the O-islands is weaker than that from the CO, while the metal yields the most intense photoemission.

Several ways exist to image these regions of different work function. SEM and FEM have been discussed earlier in this chapter. As an alternative, scanning photoemission microscopy is carried out by scanning a focused UV beam (beam diameter 0.5 µm) over the surface and recording the photoemission intensity point by point. This is of course a slow procedure, but much faster imaging in real time becomes available if the electrons are collected from the entire surface in parallel, as is carried out in photoemission electron microscopy (PEEM). The lateral resolution of this technique is presently around 200 nm, but by using

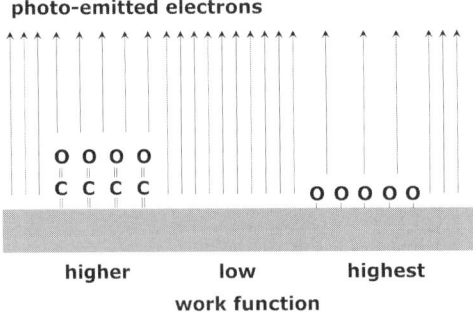

Fig. 7.26 Ordering of adsorbates on a surface into islands gives rise to regions of different work function, which can be imaged due to the associated differences in photoelectron intensity. The principle forms the basis of photoemission electron microscopy (PEEM). The same principle underlies the imaging of single molecules in the field electron microscope (FEM) (see also Fig 7.12).

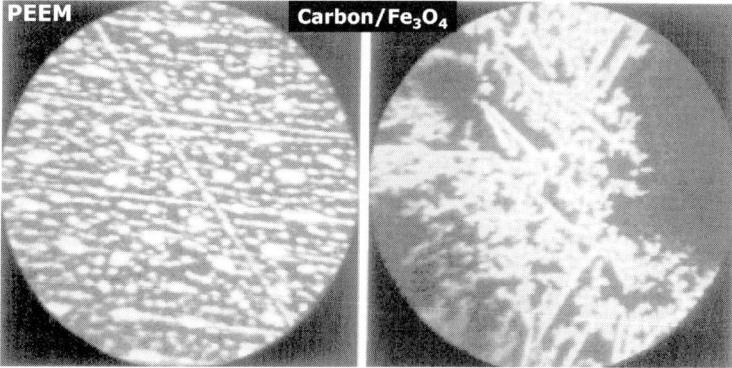

Fig. 7.27 Photoelectron emission microscopy (PEFM) images of two Fe$_3$O$_4$ surfaces that were used as model catalyst in the dehydrogenation of ethylbenzene to styrene at 870 K, showing carbonaceous deposits (*bright*). These graphitic deposits grow in dots and streaks on a surface of low defect density, but form dendritic structures on surfaces rich in point and step defects. (From [78]).

intense UV beams at synchrotrons as the source, the perspective is opened of resolving details on the nanometer scale. The interested reader is referred to extensive reviews by Rotermund [76] and by Günther et al. [77] for descriptions of photoemission microscopy and its applications. Figure 7.27 shows the PEEM image of carbonaceous deposits (bright) on an Fe$_3$O$_4$ surface (dark) used as a model catalyst for the reaction of ethylbenzene to styrene [78]. The two surfaces

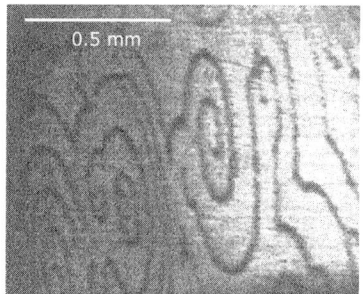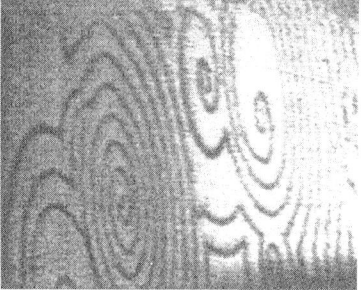

Fig. 7.28 Ellipsometry for surface imaging (EMSI) images from a Pt(110) surface during CO oxidation at 480 K and total pressure in the 10^{-2} mbar range. The image represents an area of about 1 mm^2. Electron microscopic techniques such as photoelectron emission microscopy would be able to image at higher resolution, but cannot be applied at these relatively high pressures. (From [80]).

differ in their densities of point and step defects, with the dendritic pattern of carbonaceous deposit growing on the defect-rich surface.

Optical methods are eminently suitable for imaging, the most attractive property of photons being that they do not require a vacuum. Hence, imaging under reaction conditions becomes possible. Among the successfully applied methods are infrared imaging (IRI), reflection anisotropy microscopy (RAM) [76], and ellipsomicroscopy for surface imaging (EMSI) [79].

Figure 7.28 shows a beautiful snapshot, taken by Ertl and Rotermund [80], of spiraling patterns caused by ordering of CO molecules and O-atoms, and reaction at the boundaries in the oxidation of CO on a platinum single-crystal surface. In these measurements one sets the optics such that a uniform surface produces a dark image; local deviations of the adsorbate composition give rise to differences in the ellipsometric parameters, which appear bright in the image. The EMSI picture of Figure 7.28 was recorded during reaction at 480 K and at relatively high partial pressures (10^{-3} mbar CO, 10^{-2} mbar O_2), at which electron microscopy techniques such as PEEM cannot be applied [80]).

References

1 S. Amelinckx, D. van Dyck, J. van Landuyt, and G. van Tendeloo, *Handbook of Microscopy*. VCH, Weinheim, 1997.
2 D.B. Williams and C.B. Carter, *Transmission Electron Microscopy, a textbook for Materials Science*. Kluwer Academic/Plenum Publishers, New York, 1996.
3 M. José-Yacamán and M. Avalos-Borja, *Catal. Rev. – Sci. Eng.* **34** (1992) 55.
4 A.K. Datye and D.J. Smith, *Catal. Rev. – Sci. Eng.* **34** (1992) 129.
5 J.M. Thomas and P.L. Gai, *Adv. Catal.* **48** (2004) 171.
6 P.L. Hansen, S. Helveg, and A.K. Datye, *Adv. Catal.* **50** (2006) 77.
7 J. Liu, *J. Electron Microsc.* **54** (2005) 251.
8 A.M. Saib, A. Borgna, J. van de Loosdrecht, P.J. van Berge, J.W. Geus, and J.W. Niemantsverdriet, *J. Catal.* **239** (2006) 326.
9 R. Imbihl and G. Ertl, *Chem. Rev.* **95** (1995) 697.
10 J.V. Sanders, in: *Catalysis, Science and Technology*, J.R. Anderson and M. Boudart (Eds.), Vol. 7. Springer, Berlin, 1985, p. 51.
11 J.V. Sanders and K.C. Pratt, *J. Catal.* **67** (1981) 331.
12 M.M.J. Treacy and A. Howie, *J. Catal.* **63** (1980) 265.
13 S. Bernal, J.J. Calvino, M.A. Cauqui, J.M. Gatica, C. López Cartes, J.A. Pérez Omil, and J.M. Pintado, *Catal. Today* **77** (2003) 385.
14 S. Giorgio, C.R. Henry, C. Chapon, G. Nihoul, and J.M. Penisson, *Ultramicroscopy* **38** (1991) 1.
15 J. Goldstein, D. Newbury, D. Joy, C. Lyman, P. Echlin, E. Fifshin, L. Sawyer, and J. Michael, *Scanning Electron Microscopy and X-ray Microanalysis*, 3rd edition. Springer, New York, 2003.
16 M.P. McDaniel, *Adv. Catal.* **33** (1985) 47.
17 R. van Hardeveld and F. Hartog, *Surface Sci.* **15** (1969) 189.
18 A. Carlsson, A. Puig-Molina, and T.V.W. Janssens, *J. Phys. Chem. B* **110** (2006) 5286.
19 R.T.K. Baker, *Catal. Rev. – Sci. Eng.* **19** (1979) 161.
20 P.L. Gai, *Topics Catal.* **8** (1999) 97–113.
21 L.C. Feldman and J.W. Mayer, *Fundamentals of Surface and Thin Film Analysis*. North-Holland, Amsterdam, 1986.
22 S.J.B. Reed, *Electron Microprobe Analysis*. Cambridge University Press, Cambridge 1993.
23 D. Wang, A. Villa, F. Porta, D. Su, and L. Prati, *Chem. Commun.* (2006), 1956.

24 E.W. Müller, *Z. Phys.* **37** (1936) 838; *Z. Phys.* **131** (1951) 136.
25 E.W. Müller and T.T. Tsong, *Field Ion Microscopy*. Elsevier, New York, 1969.
26 R. Gomer, *Field Emission and Field Ion Microscopy*. Harvard University Press, 1961.
27 N. Kruse, *Ultramicroscopy* **89** (2001) 51.
28 G. Ehrlich and F.G. Hudda, *J. Chem. Phys.* **44** (1966) 1039.
29 G. Ehrlich and C.F. Kirk, *J. Chem. Phys.* **48** (1968) 1465.
30 D.W. Bassett and M.J. Parseley, *Nature (London)* **221** (1969) 1046.
31 M.F.H. van Tol, F.A. Hondsmerk, J.W. Bakker, and B.E. Nieuwenhuys, *Surface Sci.* **266** (1992) 529.
32 M.F.H. van Tol, A. Gielbert, and B.E. Nieuwenhuys, *Catal. Lett.* **16** (1992) 297.
33 T.T. Tsong, *Atom-Probe and Field Ion Microscopy: Field-Ion Emission and Surfaces and Interfaces at Atomic Resolution*. Cambridge University Press, Cambridge, 1990.
34 T.T. Tsong, *Surface Sci. Rep.* **8** (1988) 127.
35 N. Kruse, G. Abend, and J.H. Block, *J. Chem. Phys.* **88** (1988) 1307.
36 R. Wiesendanger, *Scanning Probe Microscopy and Spectroscopy*. Cambridge University Press, Cambridge, 1994.
37 E. Meyer, H.J. Hug, and R. Bennewitz, *Scanning Probe Microscopy, The Lab on a Tip*. Springer-Verlag, Berlin, 2004.
38 G. Binnig and H. Rohrer, *Helv. Phys. Acta* **55** (1982) 726.
39 G. Binnig, C.F. Quate, and C. Gerber, *Phys. Rev. Lett.* **56** (1986) 930.
40 J. Israelachvili, *Intermolecular and Surface Forces*. Academic Press, London, 1991.
41 D. Sarid, *Scanning Force Microscopy with Applications to Electric, Magnetic and Atomic Forces*. Oxford University Press, New York, 1991.
42 S. Morita, R. Wiesendanger, and E. Meyer (Eds.), *Noncontact Atomic Force Microscopy*. Springer-Verlag, Berlin, 2002.
43 D. Sarid and V. Elings, *J. Vac. Sci. Technol.* **B9** (1991) 431.
44 G. Meyer and N. Amer, *Appl. Phys. Lett.* **53** (1988) 1045.
45 K. Fukui, H. Onishi, and Y. Iwasawa, *Phys. Rev. Lett.* **79** (1997) 4202.
46 K. Fukui, Y. Namai, and Y. Iwasawa, *Appl. Surface Sci.* **188** (2002) 252.
47 P.C. Thüne, J. Loos, P.J. Lemstra, and J.W. Niemantsverdriet, *J. Catal.* **183** (1999) 1.
48 J. Loos, P.C. Thüne, J.W. Niemantsverdriet, and P.J. Lemstra, *Macromolecules* **32** (1999) 8910.
49 N. Isshiki, K. Kobayashi, and M. Tsukuda, *Surface Sci.* **238** (1990) L439.
50 G.A. Somorjai, *CaTTech* **3** (1999) 84.
51 J.V. Lauritsen and F. Besenbacher, *Adv. Catal.* **50** (2006) 97.
52 J. Wintterlin, *Adv. Catal.* **45** (2000) 131.
53 D.W. Goodman, *J. Catal.* **216** (2003) 213.
54 G. Ertl, *J. Mol. Catal. A* **182–183** (2003) 5.
55 K.L. Yeung and E.E. Wolf, *J. Catal.* **135** (1992) 13; *Catal. Lett.* **12** (1992) 213.
56 A. Humbert, M. Dayez, S. Grandjeaud, P. Ricci, C. Chapon, and C.R. Henry, *J. Vac. Sci. Technol.* **B9** (1991) 804.
57 H.-J. Freund, *Angew. Chemie Int. Ed. Engl.* **36** (1997) 452.
58 M. Schmid, H. Stadler, and P. Varga, *Phys. Rev. Lett.* **70** (1993) 1441.
59 Y. Gauthier, P. Dolle, R. Baudoing-Savois, W. Hebenstreit, E. Platzgummer, M. Schmid, and P. Varga, *Surface Sci.* **396** (1998) 137.
60 P.T. Wouda, B.E. Nieuwenhuys, M. Schmid, and P. Varga, *Surface Sci.* **359** (1996) 17.
61 A.D. van Langeveld and J.W. Niemantsverdriet, *J. Vac. Sci. Technol.* **A5** (1987) 558.
62 S. Helveg, J.V. Lauritsen, E. Laegsgaard, I. Stensgaard, J.K. Nørskov, B.S. Clausen, H. Topsøe, and F. Besenbacher, *Phys. Rev. Lett.* **84** (2000) 951.
63 J.V. Lauritsen, S. Helveg, E. Laegsgaard, I. Stensgaard, B.S. Clausen, H. Topsøe, and F. Besenbacher, *J. Catal.* **197** (2001) 1.
64 M.V. Bollinger, J.V. Lauritsen, K.W. Jacobsen, J.K. Nørskov, S. Helveg, and F. Besenbacher, *Phys. Rev. Lett.* **87** (2001) 196803.
65 J.V. Lauritsen, M. Nyberg, J.K. Nørskov, B.S. Clausen, H. Topsøe, E. Lægsgaard, and F. Besenbacher, *J. Catal.* **224** (2004) 94.
66 J. Wintterlin, J. Trost, S. Renisch, R. Schuster, T. Zambelli, and G. Ertl, *Surface Sci.* **394** (1997) 159.
67 J. Trost, T. Zambelli, J. Wintterlin, and G. Ertl, *Phys. Rev. B* **54** (1996) 17850.

68 B.L.M. Hendriksen, S.C. Bobaru, and J.W.M. Frenken, *Topics Catal.* **36** (2005) 43.
69 R.A. Bennett, P. Stone, N.J. Price, and M. Bowker, *Phys. Rev. Lett.* **82** (1999) 3831.
70 S. Suzuki, Y. Yamaguchi, H. Onishi, K. Fukui, T. Sasaki, and Y. Iwasawa, *Catal. Lett.* **50** (1998) 117.
71 C. Xu, X. Lai, D.W. Goodman, and G.W. Zajac, *Phys. Rev. B* **56** (1997) 13464.
72 M. Valden, X. Lai, and D.W. Goodman, *Science* **281** (1998) 1647.
73 D.E. Starr, S.K. Shaikhutdinov, and H.-J. Freund, *Topics Catal.* **36** (2005) 33.
74 J. Schoiswohl, S. Surnev, and F.P. Netzer, *Topics Catal.* **36** (2005) 91.
75 E. Bauer, *Rep. Progr. Physics* **57** (1994) 895.
76 H.H. Rotermund, *Surface Sci. Rep.* **29** (1997) 265; *Surface Sci.* **386** (1997) 10.
77 S. Günther, B. Kaulich, L. Gregoratti, and M. Kiskinova, *Progr. Surf. Sci.* **70** (2002) 187.
78 W. Weiss, D. Zscherpel, and R. Schlögl, *Catal. Lett.* **52** (1998) 215.
79 H.H. Rotermund, G. Haas, R.U. Franz, R.M. Tromp, and G. Ertl, *Science* **270** (1995) 608.
80 G. Ertl and H.H. Rotermund, *Curr. Opin. Solid Sate Mater. Sci.* **1** (1996) 617.

8
Vibrational Spectroscopy

Keywords

Infrared spectroscopy
Transmission infrared spectroscopy
Diffuse reflectance infrared spectroscopy (DRIFTS)
Attenuated total reflection (ATR)
Infrared emission spectroscopy (IRES)
Reflection absorption infrared spectroscopy (RAIRS)
Sum-frequency generation (SFG)
Raman spectroscopy
Electron energy loss spectroscopy (EELS)

8.1
Introduction

Infrared spectroscopy is the first modern spectroscopic technique to have found general acceptance in catalysis. The most common application of the technique in catalysis is to identify adsorbed species and to study the way in which these species are chemisorbed onto the surface of the catalyst. In addition, the procedure is useful for identifying phases that are present in the precursor stages of the catalyst, during its preparation. On occasion, the infrared spectra of adsorbed probe molecules such as CO and NO can provide valuable information with regards to the adsorption sites that are present on a catalyst.

Vibrations in molecules or in solid lattices are excited by the absorption of photons (infrared spectroscopy), or by the scattering of photons (Raman spectroscopy), electrons (electron energy loss spectroscopy; EELS), or neutrons (inelastic neutron scattering). In case the vibration is excited by the interaction of the bond with a wave field – as with photons and electrons – the excitation is subject to strict selection rules. Collisions, on the other hand, excite all vibrational modes.

Infrared spectroscopy is the most common form of vibrational spectroscopy. Although infrared radiation falls into three categories (see Table 8.1), it is the mid-infrared region that is of most interest to us.

Spectroscopy in Catalysis: An Introduction, Third Edition
J. W. Niemantsverdriet
Copyright © 2007 WILEY-VCH Verlag GmbH & Co. KGaA, Weinheim
ISBN: 978-3-527-31651-9

Table 8.1 Classification of infrared radiation.

Region	Wavelength [µm]	Energy [meV][a]	Wavenumber [cm^{-1}]	Detection of
Infrared	1000–1	1.2–1240	10–10 000	–
Far	1000–50	1.2–25	10–200	Lattice vibrations
Mid	50–2.5	25–496	200–4000	Molecular vibrations
Near	2.5–1	496–1240	4000–10 000	Overtones

[a] 1 meV = 8.0655 cm^{-1}.

It is important first to mention a brief history of infrared spectroscopy. The first report of the systematic use of infrared radiation was by Coblentz in 1905, who investigated water in minerals [1]. Subsequent investigations of hydroxyl groups and adsorbed molecules on oxides were made by Terenin and co-workers in Russia during the 1940s, as described by Kiselev and Lygin [2]. These investigations were conducted in the near infrared, which offers the advantage that glass is transparent for radiation in this region, and that the construction of cells is straightforward. Commercial infrared instruments also became available during the 1940s, until which time Raman spectroscopy had dominated the study of vibrations in molecules. The first studies in the mid-infrared region in catalysis were performed with the pioneering investigations of Eischens and Pliskin, as described in their review of 1958 [3], which remains a recommended publication to this day. Since then, infrared spectroscopy has developed into one of the most widely applied spectroscopic techniques in catalysis.

8.2
Theory of Molecular Vibrations

Molecules possess discrete levels of rotational and vibrational energy. Transitions between vibrational levels occur by absorption of photons with frequency v in the mid-infrared range (see Table 8.1). The C–O stretch vibration, for example, is at 2143 cm^{-1}. For small deviations of the constituent atoms from their equilibrium positions, the potential energy $V(r)$ can be approximated by that of the harmonic oscillator:

$$V(r) = \frac{1}{2}k(r - r_{eq})^2 \tag{8-1}$$

where:
$V(r)$ is the interatomic potential;
r is the distance between the vibrating atoms;
r_{eq} is the equilibrium distance between the atoms;
k is the force constant of the vibrating bond.

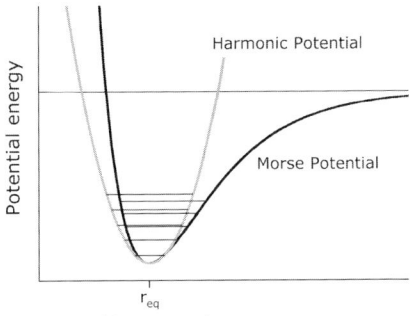

Fig. 8.1 The harmonic potential and the Morse potential, together with vibrational energy levels. The harmonic potential is an acceptable approximation for molecular separations close to the equilibrium distance and vibrations up to the first excited level, but fails for higher excitations. The Morse potential is more realistic. Note that the separation between the vibrational levels decreases with increasing quantum number, implying, for example, that the second overtone occurs at a frequency slightly less than twice that of the fundamental vibration.

The corresponding vibrational energy levels are equidistant (see Fig. 8.1):

$$E_n = \left(n + \frac{1}{2}\right) h\nu \tag{8-2}$$

$$\nu = \frac{1}{2\pi}\sqrt{\frac{k}{\mu}} \tag{8-3}$$

$$\frac{1}{\mu} = \frac{1}{m_1} + \frac{1}{m_2} \quad \text{or} \quad \mu = \frac{m_1 m_2}{m_1 + m_2} \tag{8-4}$$

where:
E_n is the energy of the *n*-th vibrational level;
n is an integer;
h is Planck's constant;
ν is the frequency of the vibration;
k is the force constant of the bond;
μ is the reduced mass;
m_i is the mass of the vibrating atoms.

Thus, vibrational frequencies increase with increasing bond strength and with decreasing mass of the vibrating atoms.

Allowed transitions in the harmonic approximation are those for which the vibrational quantum number changes by one unit. Overtones – that is, the absorption of light at a whole number times the fundamental frequency – would not be possible. A general selection rule for the absorption of a photon is that the dipole moment of the molecule must change during the vibration. This distinguishes infrared from Raman spectroscopy, where the selection rule requires that the molecular polarizability change during the vibration.

The harmonic approximation is only valid for small deviations of the atoms from their equilibrium positions. The most obvious shortcoming of the harmonic potential is that its walls are infinitely high, and thus the bond between two atoms cannot break. A physically more realistic potential is the Morse potential (Fig. 8.1):

$$V(r) = D(1 - e^{-a(r-r_{eq})})^2 - D \tag{8-5}$$

where:
$V(r)$ is the interatomic potential;
r is the distance between the vibrating atoms;
r_{eq} is the equilibrium distance between the atoms;
D is the dissociation energy of the vibrating bond;
a is a parameter which controls the steepness of the potential well.

In this potential the energy levels are no longer equally spaced and overtones (i.e., vibrational transitions with $\Delta n > 1$) become allowed. The overtone of gaseous CO at 4260 cm^{-1} (slightly less than $2 \times 2143 = 4286$ cm^{-1}) is an example. For small deviations of r from equilibrium, however, the Morse potential is satisfactorily approximated by a parabola and for the interpretation of IR spectra the harmonic oscillator description is usually sufficient.

The simple harmonic oscillator picture of a vibrating molecule has important implications. First, by knowing the frequency one can immediately calculate the force constant of the bond. Note that the latter corresponds to the curvature of the interatomic potential and *not* to its depth, the bond energy. However, as the depth and the curvature of a potential usually change hand in hand, it is often permissible to take the infrared frequency as an indicator for the strength of the bond. Second, isotopic substitution can be useful in the assignment of frequencies to bonds in adsorbed species, because frequency shifts due to isotopic substitution (of for example D for H in adsorbed ethylene, or OD for OH in methanol) can be predicted directly. An example is provided in Table 8.2.

Table 8.2 C–O stretch frequency of different isotopic combinations.

Molecule	ν [cm^{-1}]
$^{12}C^{16}O$	2143
$^{13}C^{16}O$	2096
$^{12}C^{18}O$	2091
$^{13}C^{18}O$	2042

The number of different vibrations that a molecule possesses follows from the following considerations. A molecule consisting of N atoms has 3N degrees of freedom. Three of these are translational degrees of freedom of the molecule, and three are rotations of the molecule along the three principal axes of inertia. Linear molecules have only two rotational degrees of freedom, as no energy change is involved in the rotation along the main axis. Thus, the number of fundamental vibrations is 3N − 6 for a non-linear and 3N − 5 for a linear molecule. In addition, there are overtones and combinations of fundamental vibrations. Fortunately, however, not all vibrations are visible.

There are four types of vibration (as illustrated by Fig. 8.2), each with a characteristic symbol:

- Stretch vibrations (symbol v), changing the length of a bond.
- Bending vibrations in one plane (symbol δ), changing bond angles but leaving bond lengths unaltered (in larger molecules further divided into rock, twist and wag vibrations).
- Bending vibrations out of plane (symbol γ), in which one atom oscillates through a plane defined by at least three neighboring atoms.
- Torsion vibrations (symbol τ), changing the angle between two planes through atoms.

Generally, the frequencies of these vibrations decrease in the order $v > \delta > \gamma > \tau$. In addition, vibrations are divided into symmetric and asymmetric vibrations (v_s and v_{as}). For more details, the reader is referred to textbooks on infrared spectroscopy [4–7].

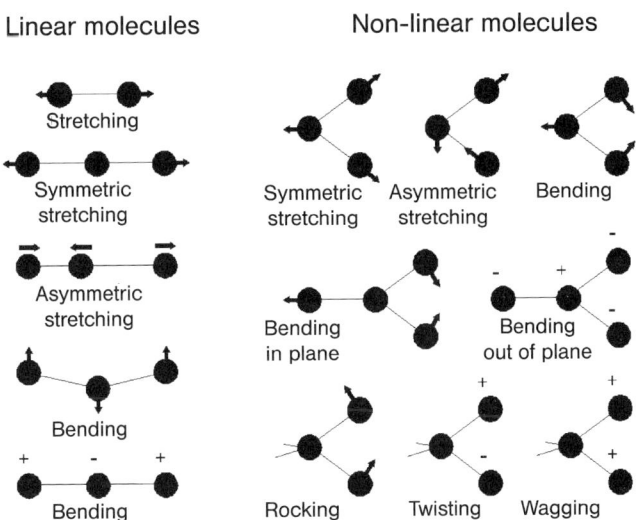

Fig. 8.2 Fundamental vibrations of several molecules.

Not all vibrations can be observed: the absorption of an infrared photon occurs only if a dipole moment changes during the vibration. It is not necessary that the molecule possesses a permanent dipole – it is sufficient if a dipole moment changes during the vibration. The intensity of the infrared band is proportional to the change in dipole moment. Although this statement is of little practical value, as the magnitude of dipole moments during the vibration are not known, it explains why species with polar bonds, such as CO, NO and OH, exhibit strong infrared bands, whereas covalent bonds such as C–C or N=N absorb infrared light only weakly, and molecules such as H_2, N_2 are not infrared-active at all.

The group frequency concept states that functional groups in molecules may be treated as independent oscillators, irrespective of the larger structure to which they belong. For example, the C=C double bond of the –CH=CH_2 group varies no more than from 1651 cm^{-1} in propylene, where it is bound to a methyl group, to 1632 cm^{-1} when it is bound to the much heavier CH_2Br group in CH_2Br–CH=CH_2 [5]. As a consequence, infrared frequencies are characteristic for certain bonds in molecules, and can also be used to identify species on surfaces. The following classification of characteristic stretching frequencies is a good starting point for interpreting vibrational spectra.

The infra red region between 4000 and 200 cm^{-1} can be divided roughly into five regions:

- The X–H stretch region (4000–2500 cm^{-1}), where strong contributions from O–H, N–H, C–H, and S–H stretch vibrations are observed.
- The triple bond region (2500–2000 cm^{-1}), where contributions from gas-phase CO (2143 cm^{-1}) and linearly adsorbed CO (2000–2200 cm^{-1}) are seen.
- The double bond region (2000–1500 cm^{-1}), where in catalytic studies bridge-bonded CO, as well as carbonyl groups in adsorbed molecules (around 1700 cm^{-1}), absorb.
- The fingerprint region (1500–500 cm^{-1}), where all single bonds between carbon and elements such as nitrogen, oxygen, sulfur, and halogens absorb.
- The M–X or metal-adsorbate region (around 200–450 cm^{-1}), where the metal–carbon, metal–oxygen and metal–nitrogen stretch frequencies in the spectra of adsorbed species are observed.

Correlation charts should be consulted for more precise assignments [4–6].

Although we are mainly interested in adsorbed molecules, spectra often contain contributions from gas-phase species, and therefore some knowledge of gas-phase spectra is essential. Molecules in the gas phase have rotational freedom, and as a consequence the vibrational transitions are accompanied by rotational transitions. For a rigid rotor that vibrates as a harmonic oscillator, the expression for the available energy levels is:

$$E_{n,j} = \left(n + \frac{1}{2}\right)h\nu + \frac{h^2}{8\pi^2 I}j(j+1); \quad I = \mu r^2 \tag{8-6}$$

where:
$E_{n,j}$ is the energy of a level with quantum numbers n and j;
n is the vibrational quantum number;
j is the rotational quantum number;
h is Planck's constant;
v is the frequency;
I is the moment of inertia;
μ is the reduced mass;
r is the distance between the vibrating atoms.

Here, a third selection rule applies: For linear molecules, transitions corresponding to vibrations along the main axis are allowed if $\Delta j = \pm 1$. The $\Delta j = 0$ transition is only allowed for vibrations perpendicular to the main axis. Note that because of this selection rule the purely vibrational transition (called Q branch) appears in the gas-phase spectrum of CO_2, but is absent in that of CO. In both cases, two branches of rotational side bands appear (called P and R branches) (see Fig. 8.3 for gas-phase CO).

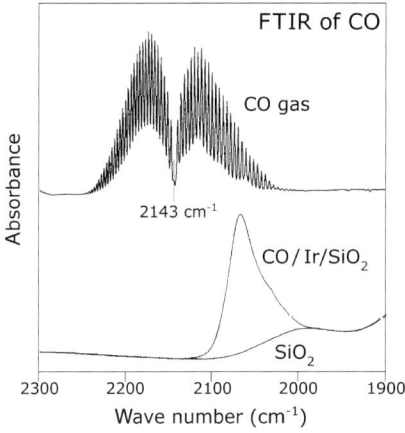

Fig. 8.3 The infrared spectrum of gas-phase CO shows rotational fine structure, which is absent in the spectrum of CO adsorbed onto an Ir/SiO$_2$ catalyst. (Courtesy of L.M.P. van Gruijthuijsen, Eindhoven).

Upon adsorption, the molecule loses its rotational freedom, and now only the vibrational transition is observed, albeit at a different frequency (see Fig. 8.3). In the case of CO, four factors contribute to this shift:

- Mechanical coupling of the C–O molecule to the heavy substrate increases the C–O frequency, about 30 cm^{-1} for CO adsorbed on platinum [8].
- The interaction between the C–O dipole and its image in the conducting, polarizable metal weakens the C–O frequency by 25–50 cm^{-1} [9].

- The formation of a chemisorption bond between C–O and the substrate alters the distribution of electrons over the molecular orbitals and weakens the C–O bond, as explained in the Appendix.
- In case the adsorbed CO molecule is surrounded by other CO molecules in the same bonding geometry, the dipoles couple, causing an additional upward shift in the frequency.

Thus, one must be careful to interpret the frequency difference between adsorbed and gas-phase C–O in terms of chemisorption bond strength only!

8.3
Infrared Spectroscopy

Currently, several forms of infrared spectroscopy are in general use, as illustrated in Figure 8.4. The most common form of the technique is transmission infrared spectroscopy, in which the sample consists typically of 10 to 100 mg of catalyst, pressed into a self-supporting disk of approximately 1 cm^2 and a few tenths of a millimeter thickness. Transmission infrared spectroscopy can be applied if the bulk of the catalyst absorbs weakly. This is usually the case with typical oxide supports for wavenumbers above about 1000 cm^{-1}, whereas carbon-supported catalysts cannot be measured in transmission mode. Another condition is that the support particles are smaller than the wavelength of the infrared radiation, otherwise scattering losses become important.

One major advantage of infrared spectroscopy is that the technique can be used to study catalysts *in situ*, and several cells for such investigations have been de-

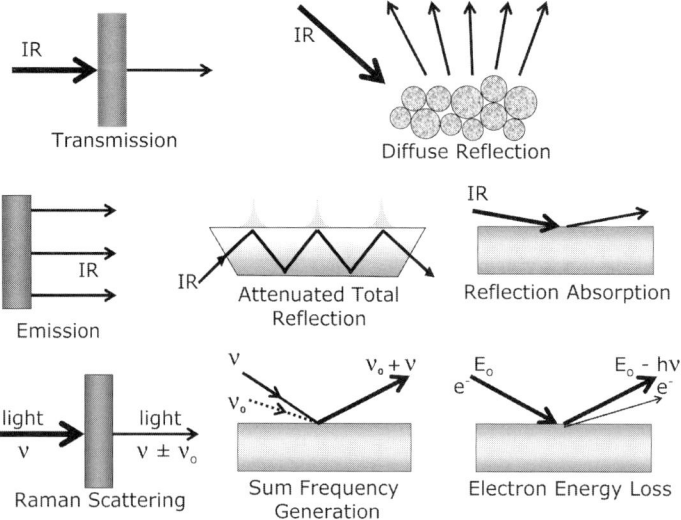

Fig. 8.4 Eight different ways to perform vibrational spectroscopy (for details, see text).

scribed in the literature [4]. The critical point here is the construction of infrared transparent windows that can withstand high temperatures and pressures.

In the diffuse reflectance mode, samples can be measured as loose powders, with the advantages that not only the tedious preparation of wafers is unnecessary but also that diffusion limitations associated with tightly pressed samples are avoided. Diffuse reflectance is also the indicated technique for strongly scattering or absorbing particles. (The often-used acronyms DRIFT or DRIFTS represent diffuse reflectance infrared Fourier transform spectroscopy.) The diffusely scattered radiation is collected by an ellipsoidal mirror and focused on the detector. The infrared absorption spectrum is described the Kubelka–Munk function:

$$\frac{K}{S} = \frac{(1 - R_\infty)^2}{2R_\infty} \tag{8-7}$$

where:
K is the absorption coefficient, a function of the frequency v;
S is the scattering coefficient;
R_∞ is the reflectivity of a sample of infinite thickness, measured as a function of v.

If the scattering coefficient does not depend on the infrared frequency, the Kubelka–Munk function transforms the measured spectrum $R_\infty(v)$ into the absorption spectrum $K(v)$. *In-situ* cells for DRIFT studies of catalysts have been described [10], and are commercially available.

Attenuated total reflection (ATR) is sometimes used to measure the infrared spectra of catalysts inside a reactor. The infrared light is coupled into an ATR crystal, which can be either a flat plate (e.g., the wall of a reactor) or a cylindrical rod (surrounded by catalyst particles). The evanescent wave that protrudes outside the crystal when the infrared beam reflects on the inside of its surface is used for the measurement. A review of ATR in catalysis has been published by Bürgi and Baiker [11], and a catalytic cell to apply the method *in situ* inside a catalyst bed reported by Moser and co-workers [12]. An example of ATR is discussed later in this chapter.

Measurements of supported catalysts in diffuse reflection and transmission mode are, in practice, limited to frequencies above those where the support absorbs (below about 1250 cm^{-1}). Infrared emission spectroscopy (IRES) offers an alternative in this case. When a material is heated to about 100 °C or higher, it emits a spectrum of infrared radiation in which all the characteristic vibrations appear as clearly recognizable peaks. Although measuring in this mode offers the attractive advantage that low frequencies such as those of metal–oxygen or sulfur–sulfur bonds are easily accessible, the technique has hardly been explored for the purpose of catalyst characterization. An *in-situ* cell for IRES measurements and some experiments on Mo–O–S clusters of interest for hydrodesulfurization catalysts have been described by Weber et al. [13].

For measuring the infrared absorption spectra of gases adsorbed on the surfaces of metal single crystals or polycrystalline foils, one uses reflection absorption infrared spectroscopy (RAIRS), sometimes also referred to as infrared reflection absorption spectroscopy (IRAS). In RAIRS, the infrared beam enters at the grazing angle – that is, almost parallel to the surface (see Fig. 8.4). During reflection, the p-component of the infrared light (the component perpendicular to the surface) excites those vibrations of the chemisorbed molecule for which the component of the dipole moment perpendicular to the surface changes. This rather strict metal surface selection rule is typical for RAIRS. Although absorption bands in RAIRS have intensities that are some two orders of magnitude weaker than in transmission studies on supported catalysts, RAIRS spectra can be measured accurately with standard spectrometers. For reviews, the reader is referred to Hoffmann [14] and Chabal [15].

8.3.1
Equipment

The first generation of infrared spectrometers was of the energy dispersive type. Here a monochromator (initially a prism, but after the mid-1960s a grating) selects the wavelength of interest from the continuum emitted by the infrared source, and the transmission corresponding to that particularly frequency by the sample can be measured. Nowadays, energy-dispersive instruments have largely been abandoned in favor of Fourier transform infrared (FTIR) spectrometers, which operate on the principle of the Michelson interferometer. These instruments have the major advantage that the entire spectrum is obtained for each scan that the interferometer makes, with the result that the total collection time needed to measure a spectrum is much less. The treatment of the Fourier transform technique is beyond the scope of this book, but the interested the reader is referred elsewhere [6, 16].

The optical components can be made from NaCl (which is transparent from 650 to 4000 cm^{-1}), from KBr, with a low energy cut-off of 400 cm^{-1}, or from CsI, with an even more favorable cut-off of 200 cm^{-1}. The source is usually a temperature-stabilized ceramic filament operating around 1500 K. The detector may be a slowly reacting thermocouple in energy dispersive instruments, but in FTIR it must be a rapidly responding device. The standard for routine applications is the deuterium triglycine sulfate (DTGS) detector, while the liquid nitrogen-cooled mercury cadmium telluride (MCT) detector is used for more demanding applications, as for example in RAIRS.

8.3.2
Applications of Infrared Spectroscopy

Carbon monoxide on metals forms the best-studied adsorption system in vibrational spectroscopy. The strong dipole associated with the C–O bond makes this molecule a particularly easy one to study. Moreover, the C–O stretch frequency is very informative about the direct environment of the molecule. The metal–carbon bond, however, falling at frequencies between 200 and 450 cm^{-1}, is more difficult to measure with infrared spectroscopy. First, its detection requires special optical parts made from CsI, but even with suitable equipment the peak may be invisible, due to absorption by the catalyst support. In reflection experiments on single crystal surfaces the metal–carbon peak is difficult to obtain due to the low sensitivity of RAIRS at low frequencies [14, 15]. EELS, on the other hand, has no difficulty in detecting the metal–carbon bond, as will be seen later on.

The C–O stretch frequency is often an excellent indicator for the way in which CO binds to the substrate. Linearly adsorbed CO absorbs at frequencies between 2000 and 2130 cm^{-1}, twofold or bridge-bonded CO between 1880 and 2000 cm^{-1}, threefold CO between 1800 and 1880 cm^{-1}, and finally fourfold-bonded CO at wavenumbers below 1800 cm^{-1} [17]. The precise absorption frequency depends on the substrate metal, its surface structure and, importantly, the CO coverage. The latter dependence is due to mutual interactions between the dipoles of the CO molecule, with the effect that the CO stretch frequency increases with increasing coverage. The presence of dipole coupling effects can be investigated by using mixtures of ^{12}CO and ^{13}CO, because there is no resonance between the dipoles of the isotopes. For reviews, the reader is referred to Hollins and Pritchard [18, 19].

8.3.3
Transmission Infrared Spectroscopy

The sensitivity of the C–O stretching frequency for the bonding configuration is illustrated with a perhaps somewhat dated (but still very instructive) study on the adsorption sites of alloy surfaces. Soma-Noto and Sachtler [20] reported an infrared investigation of CO adsorbed on silica-supported Pd–Ag alloys; some of their spectra are shown in Figure 8.5. On pure palladium, CO adsorbs mainly in a twofold position, as evidenced by the intense peak around 1980 cm^{-1}, although some CO appears also to be present in threefold and linear geometries. This is a common feature in adsorption studies on supported catalysts, where particles exhibit a variety of surface planes and defects. The addition of Ag to Pd leads to a pronounced increase in the fraction of linearly adsorbed CO, which is the dominant species in Pd–Ag alloy particles with ≥30% Ag content. It should be noted that the absorption frequencies themselves do not shift significantly. The results are easily explained by the so-called "ensemble effect", without the need to invoke any major electronic effect of Ag on Pd. The incorporation of Ag into the Pd lattice rapidly decreases the fraction of Pd ensembles with atoms in the correct geometry for twofold adsorption of CO, leaving linear adsorption on single Pd atoms as the only alternative.

Experiments with CO adsorption on Ni–Cu/SiO$_2$ [21] and Pd–Au/SiO$_2$ catalysts [22] produced very similar results. In the latter case, two CO absorption bands were also seen, one at around 1900 cm^{-1} (this was assigned to bridge-bonded CO) and another at 2070 cm^{-1}, which was characteristic of linear CO on

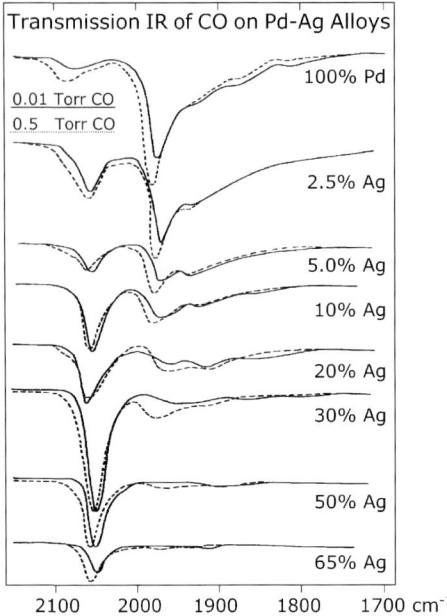

Fig. 8.5 Transmission infrared spectra of CO adsorbed at room temperature on a series of silica-supported Pd–Ag alloys. CO adsorption on Ag is negligible. The dashed spectra are for higher CO exposures (0.5 Torr), the full spectra for low exposures (0.01 Torr). On pure Pd, CO adsorbs predominantly on multiple sites (frequencies below 2000 cm^{-1}). As Pd becomes diluted with Ag, linearly bonded CO becomes dominant, indicating that ensembles of two or more Pd atoms to accommodate multiply bonded CO are no longer present. (From [20]).

single Pd atoms. Figure 8.6 shows the intensities of the two bands as a function of Pd content in the Pd–Au alloys. The almost quadratic increase of the twofold CO intensity reflects the probability of finding two adjacent Pd atoms at the surface of the alloy particles. The intensity of linearly adsorbed CO is more or less proportional to the Pd concentration, as would be expected for adsorption on single Pd atoms.

Accurate analysis of the band positions revealed another interesting point: whereas the frequency of linear CO on pure Pd increases substantially from 2070 to 2095 cm^{-1} with increasing CO coverage due to dipole–dipole interactions between adjacent molecules, the frequencies of linear CO on the alloys increased much less with coverage. Kugler and Boudart [22] took this as evidence for a small ligand effect of Au on Pd. However, Toolenaar et al. [23], when observing similar phenomena in the infrared spectra of CO on Pt–Cu alloys, proved that the lower frequencies of linear CO on the alloy were caused by diminished dipole–dipole coupling between adsorbed CO molecules, which necessarily form less-dense overlayers on alloys than on pure Pt. Thus, the ensemble effect (which is illustrated schematically in Fig. 8.6) provides an entirely geometric explanation

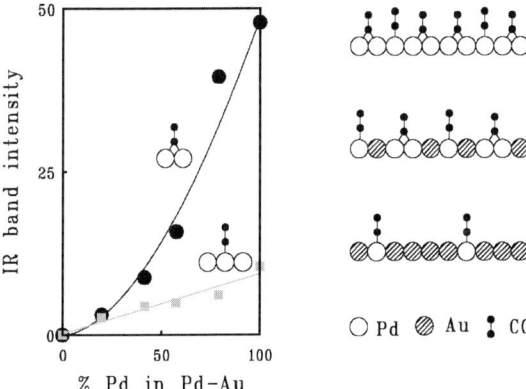

Fig. 8.6 Left: Intensities of the CO bands of linear and bridged CO on Pd–Au alloys, as a function of Pd content. The almost parabolic fit to the intensities for CO in twofold positions reflects the probability of finding two adjacent Pd atoms in the Pd–Au surface. (Data from [22]). Right: Schematic illustration of how alloying destroys ensembles of Pd where CO adsorbs in the twofold position.

for the variation of both intensities and peak positions in infrared spectra of CO on alloy surfaces.

The use of infrared spectroscopy of adsorbed molecules to probe oxide surfaces has been reviewed extensively by Davydov [24]. Such approaches are also effective for sulfide catalysts. The infrared signal of NO has been used successfully to identify sites on the surface of a hydrodesulfurization catalyst, as the following example shows [25].

Sulfided Mo and Co–Mo catalysts, which are used in hydrotreating reactions, contain Mo as MoS_2. This compound has a layer structure consisting of sandwiches, each of a Mo layer between two S layers. The chemical activity of MoS_2 is associated with the edges of the sandwich where Mo can be exposed to the gas phase; the basal plane of the S^{2-} anions is largely unreactive. The infrared spectrum of NO on a sulfided Mo/Al_2O_3 catalyst (Fig. 8.7, curve a) shows two peaks at frequencies which agree with those observed in organometallic clusters of Mo and NO groups [26]. NO on sulfided Co/Al_2O_3 gives rise to two infrared peaks (Fig. 8.7, curve b), but at different frequencies from those observed on MoS_2. These results suggest that NO can be used as a probe to titrate the number of Co and Mo sites in the $Co-Mo/Al_2O_3$ catalyst. Curve c in Figure 8.7 confirms that this concept is correct. Moreover, a comparison of the intensities of the NO/Mo infrared signals on Mo/Al_2O_3 and $Co-Mo/Al_2O_3$ reveals that the presence of cobalt decreases the number of molybdenum atoms that are accessible to NO. This means that cobalt most probably decorates the edges of the MoS_2 sandwiches, because the edges constitute the adsorption sites for NO. The structure of the Co–Mo catalyst is further discussed in Chapter 9.

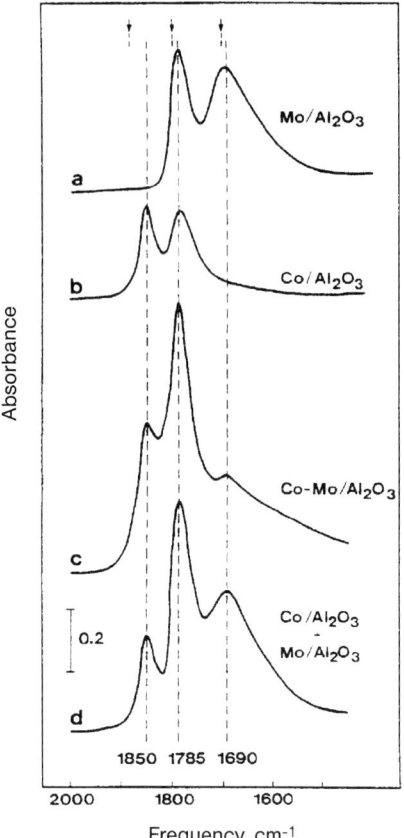

Fig. 8.7 Infrared spectra of NO probe molecules on sulfided Mo, Co, and Co–Mo hydrodesulfurization catalysts. The peak assignments are supported by the infrared spectra of organometallic model compounds. These spectra allow for a quantitative titration of Co and Mo sites in the Co–Mo catalyst. (From [25]).

Infrared spectroscopy is an important tool in catalyst preparation for studying the decomposition of infrared active catalyst precursors as a result of drying, calcination, or reduction procedures. In particular, if catalysts are prepared from organometallic precursors, infrared spectroscopy is the indicated technique for investigation [27]. Here, we discuss an example from the impressive investigations of Gates and co-workers [28, 29].

8.3.4
Diffuse Reflectance Infrared Fourier Transform Spectroscopy (DRIFTS)

Figure 8.8 shows a series of infrared spectra obtained in DRIFTS mode of an organometallic gold complex, dimethyl gold acetyl acetonate, $Au(CH_3)_2(C_5H_7O_2)$

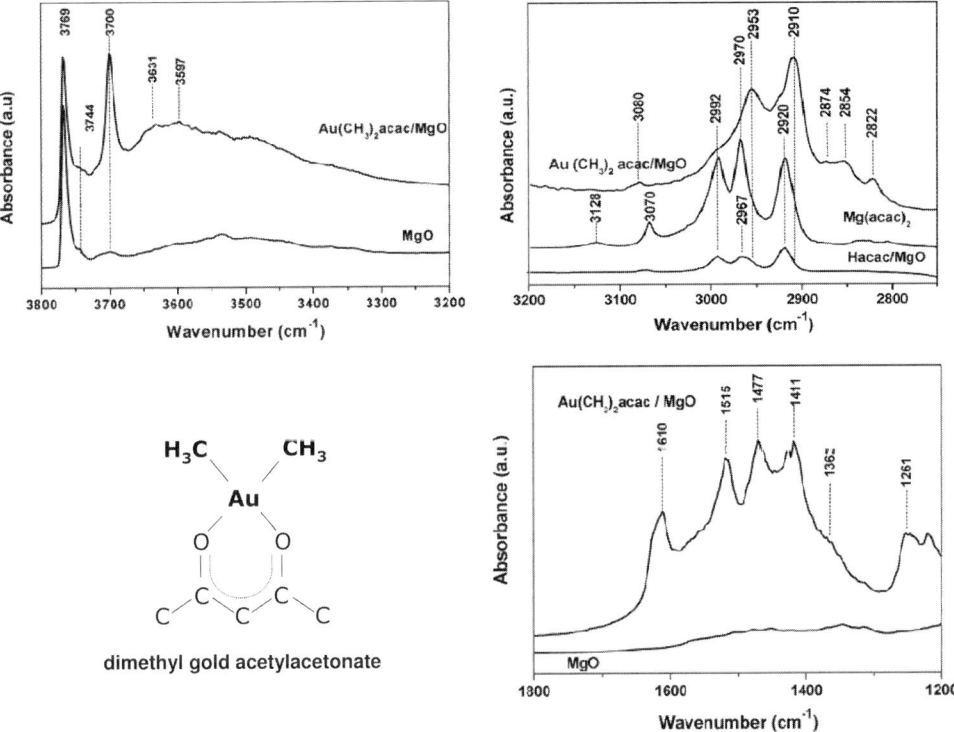

Fig. 8.8 Diffuse reflectance infrared spectra of an MgO-supported mononuclear gold compound, dimethyl gold acetyl acetonate, along with spectra of several reference compounds in the O–H and C–H stretch and the fingerprint region. Detailed assignments are shown in Table 8.3. (Adapted from [29]).

or Au(CH$_3$)$_2$(acac), on an MgO support, prepared by impregnation in a heptane solvent [29]. The data in Figure 8.8 also illustrate how the acetyl acetonate (acac) serves as a bidentate ligand for the gold. Spectra of the partially dehydroxylated MgO support are also included. The spectra have been divided into the hydroxyl region between 3800 and 3200 cm^{-1}, the C–H stretch region between 3200 and 2700 cm^{-1}, and the fingerprint region at lower wavenumbers. The C–H region is the most characteristic for the ligands, and in order to facilitate the interpretation several reference compounds have been measured (Fig. 8.8). The O–H spectrum of an MgO support normally contains several peaks for isolated and multiply coordinated OH groups, but the heavily dehydroxylated MgO support used for the impregnation of the gold complex is dominated by the isolated OH groups coordinated to a magnesium cation of the support. The infrared spectra clearly indicate that the interaction with the gold complex leads to more OH groups, notably the peaks characteristic of multiply coordinated hydroxyls, and these are most likely due to reaction of the H-atoms in the ligand with O-atoms of the support

Table 8.3 Assignment of the most important bands in the diffuse reflectance infrared (DRIFT) spectra of Figure 8.8 [29].

$Au(CH_3)_2(acac)$ on MgO	MgO dehydrox.	Hacac on MgO	Mg(acac)	Assignment
O–H region				
3769 vs	3765 vs			OH on Mg; single coordination
3744	3745			OH on Mg; single coordination
3736	3737 sh			OH; fourfold coordination
3700 vs	3700 vw			OH; threefold coordination
C–H region				
3080		3080 sh	3070	C_3–H in acac
2992		2993	2993	CH_3 in acac
2967		2965	2970	CH_3 in acac
2953				Asymmetric C–H in CH_3 on Au
2928		2920	2925	CH_3 in acac
2910				Symmetric C–H in CH_3 on Au
2854			2866	Combination band
2822			2836	Combination band

vs: Very strong; sh: Shoulder; vw: Very weak.

(see also Table 8.3). The broad, structureless infrared absorption between 3675 and 3200 cm^{-1} is generally assigned to multiple OH groups that interact through H-bonding [24, 30].

The interpretation of the rather complex C–H spectra (Fig. 8.8) is facilitated by the reference spectra. The three strongest bands in the lower two spectra are the methyl vibrations of the acetyl acetone and the Mg(acac) reference complex. These bands are visible as shoulders in the DRIFT spectrum of the supported gold complex which, however, is dominated by the symmetric–asymmetric duo of the methyl groups bound to the gold atom. The bands in the fingerprint region can also approximately be understood on the basis of reference compounds (for details, see [29]). Assignments of peaks in the C–H region are listed in Table 8.3, which nicely illustrates the validity of the group frequency concept, described earlier in this chapter.

Guzman et al. [29] also reported extended X-ray absorption fine structure (EXAFS) and X-ray absorption near edge spectroscopy (XANES) data which, together with the DRIFTS results, provided a highly complete account of the interaction of the gold acetyl acetonate complex with the support. In brief: $Au^{III}(CH_3)_2(acac)$ in heptane solution reacts at room temperature with the MgO surface to form $Au^{III}(CH_3)_2(MgO)_2$ with a structure that actually resembles the bonding of the dimethyl gold moiety to the acetyl acetonate complex in Figure 8.8. Acetylacetonate and acetylacetone (Hacac) species remain on the support, and interact with existing and newly formed OH groups via H-bonding interactions. DRIFT, when coupled to a mass spectrometer, was used to follow the decomposition of the supported species. At higher temperatures, the acac residues and hydroxyls decomposed to CO_2, acetate ligands, and water, while the dimethyl gold species reacted with OH species to CH_4 and gold (III) oxide aggregated into clusters. These were reduced to gold metal clusters at increasing temperatures. This mechanism of interaction and decomposition most probably also applies to acetyl acetonate complexes on other supports [31], although few systems have been reported in the same detail as the dimethyl gold acetyl acetonate complex on MgO as described above.

The above example also highlights the role of hydroxyl groups, which occur on the surface of all oxidic supports and are extremely important in catalyst preparation, as they provide sites where catalyst precursors may anchor to the support. These OH groups may posses positive, zero, or negative charge, and are referred to as acidic, neutral or basic, respectively. In solution, these groups either exchange with metal ion complexes (OH^- with a negative ion complex, H^+ with a cation) or they provide sites where ions of opposite charge adsorb by electrostatic interaction. The different hydroxyls can be distinguished by their infrared spectra [32]. Mestl and Knözinger [33] presented an extensive overview of hydroxyls on various oxide surfaces, along with vibrational frequencies.

8.3.5
Attenuated Total Reflection

When the sample of interest is a planar model catalyst with particles on a flat substrate, the possibilities for performing infrared spectroscopy may become limited, especially if the support is a thin oxide film on a silicon wafer. For such cases, attenuated total reflection (ATR) is an option. Figure 8.9 shows an example reported by Leewis et al. [34], where the ATR crystal, which itself is used as the substrate for the support, consists of a silicon crystal with a thin SiO_2 top layer. Rhodium is deposited on the system by spincoat impregnation, after which the sample is reduced in hydrogen to convert the rhodium to the metallic state. The particles have dimensions in the range of 1 to 3 nm, as measured by AFM. Figure 8.9 also shows the ATR-IR spectra of this model system after exposure to CO at different pressures. The signal is small (absorption on the order of $10^{-2}\%$ only), but the peaks are easily measurable. The spectra show that the evanescent wave probes not only the CO adsorbed onto the rhodium particles, but also the CO in the gas phase (compare with Fig. 8.3). As the intensity of the infrared absorption is roughly proportional to the number of internal reflections of the infrared beam inside the ATR crystal, the method is highly sensitive; in fact, the authors estimate that the minimum detectable coverage of CO corresponds to 0.005 molecules per nm^2, or the equivalent of 5×10^{-4} monolayers of the support. Such sensitivity is clearly needed, because the available metallic surface area in these model systems corresponds at most to the equivalent of a few percent of a monolayer on the support.

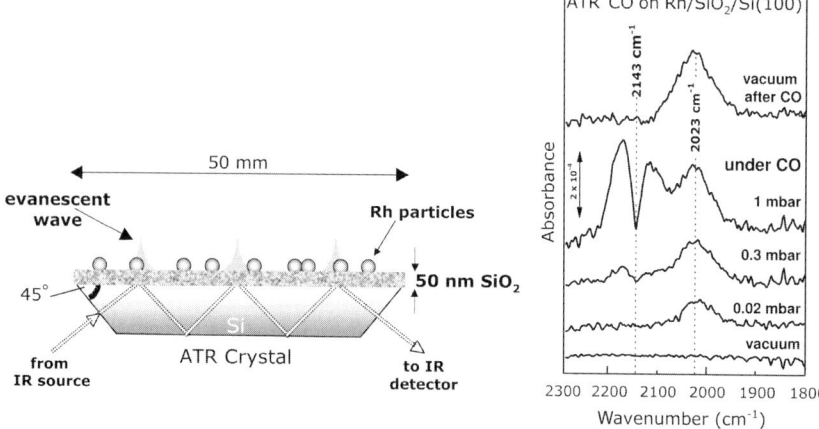

Fig. 8.9 Set-up for infrared measurements in attenuated total reflection mode on a planar model catalyst. The silicon ATR crystal has a top layer of SiO$_2$ which serves as the supported for the catalyst particles. In reality, the infrared beam undergoes 25 reflections with the surface in the crystal; the evanescent waves protruding outside the crystal are used to probe gases on and around the particles. The spectra in the right-hand part of the figure show a band around 2023 cm^{-1} characteristic of linearly adsorbed CO on rhodium, as well as the two rotational bands of gas-phase CO around 2143 cm^{-1}. The latter disappears upon evacuation. (Adapted from [34]).

8.3.6
Reflection Absorption Infrared Spectroscopy (RAIRS)

Sensitivity becomes less of an issue when the entire surface is available for adsorbed species, as is the case for single crystal studies. When the substrate is a metal, one can apply RAIRS, as described above. Here, this technique is illustrated with an example described by Raval and co-workers [35].

The surface composition and availability of certain adsorption sites are not the only factors that determine how CO binds to the surface; rather, interactions between CO and co-adsorbed molecules also play an important part. The RAIRS study conducted by Raval et al. [35] showed how NO forces CO to leave its favored binding site on palladium (see Fig. 8.10). When only CO is present, it occupies the twofold bridge site, as the infrared frequency of about 1930 cm^{-1} indicates. However, if NO is co-adsorbed, then CO leaves the twofold site and ultimately appears in a linear mode with a frequency of approximately 2070 cm^{-1}. Raval and colleagues [35] attributed the move of adsorbed CO to the top sites to the electrostatic repulsion between negatively charged NO and CO, which decreases the back-donation of electrons from the substrate into the $2\pi^*$ orbitals of CO. In this interpretation, NO has the opposite effect that a potassium promoter would have (see Chapter 9 and the Appendix).

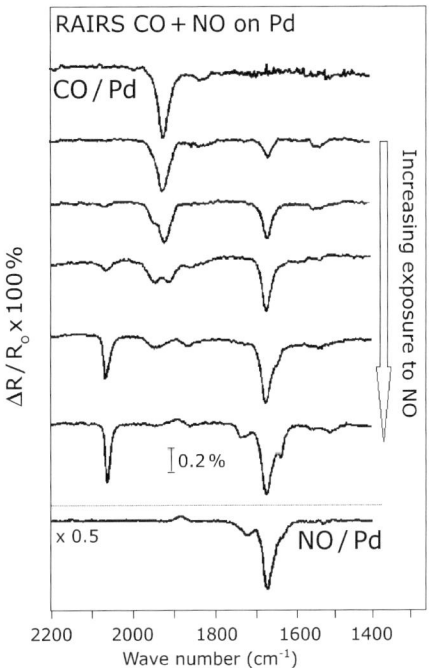

Fig. 8.10 Reflection absorption infrared spectroscopy (RAIRS) spectra show that lateral interactions force CO to leave the twofold adsorption sites on palladium (IR frequency ca. 1920 cm^{-1}) when NO is co-adsorbed, and push it to the on top site (adsorption frequencies above 2000 cm^{-1}). Adsorbed NO gives rise to the absorption peaks below 1800 cm^{-1}. (Adapted from [35]).

It should be noted that, in all the examples discussed so far, infrared spectroscopy produces information relating to the catalyst in an indirect manner, via hydroxyl groups on the support, or via the adsorption of probe molecules such as CO and NO. The reason why it is often difficult to measure the metal–oxide or metal–sulfide vibrations of the catalytically active phase in transmission infrared spectroscopy is that the frequencies are well below 1000 cm^{-1}, where measurements are difficult due to absorption by the support. In this respect, infrared emission and Raman spectroscopy (see below) offer better opportunities.

8.4
Sum-Frequency Generation

Optical sum-frequency generation (SFG) is a specialty in vibrational spectroscopy that probes adsorbates at interfaces both in real-time and *in situ*. The complicated set-up of SFG, together with the considerable expertise it requires in laser optics, make it unlikely that the technique will be widely adopted. However, this method has vast potential for studying catalytic surfaces under working conditions.

SFG [36–38] is the conversion of two photons with frequencies ω_1 and ω_2 to a new photon of frequency $\omega = \omega_1 + \omega_2$. One frequency is tunable in the infrared region, and is used to excite vibrational excitations in the adsorbed molecules, while the second frequency is fixed in the green part of the optical spectrum. The resulting photon has a frequency in the blue region of the spectrum, and can easily be detected using a photomultiplier. In fact, both incoming beams of photons – infrared and green light – originate from one light source, namely a Nd:YAG laser, which emits photons at a wavelength of 1064 nm. The beam is split, and one part is frequency-doubled using a KDP (KH_2PO_4) dye which generates second-harmonics at 532 nm in the green. The second part of the initial laser beam falls on a so-called "optical parametric amplification crystal" (e.g., $LiNbO_3$, a non-linear, uni-directional material), which is tunable in the infrared by changing its direction. The two beams are incident on the surface and overlap at a common spot of a few hundred micrometers in diameter.

The strong point of SFG is that the process is forbidden in centrosymmetric media (i.e., media with an inversion center). Therefore, it occurs only at interfaces, where the sum-frequency response forms within a region of typically 1 nm thickness. Hence, neither the bulk of the catalyst nor the molecules in the gas phase contribute to the SFG spectrum.

The SFG process involves the simultaneous absorption of infrared and the scattering of visible light, and consequently the selection rules include those of both infrared and Raman spectroscopy. In essence, the event is described by a third-rank tensor, the non-linear susceptibility, $\chi^{(2)}$, which links the two vectors of the incoming photons to the outgoing photons. Mathematically, the tensor is a $3 \times 3 \times 3$ matrix with 27 elements, 18 of which in general are non-zero. For surfaces that are rotationally isotropic around the surface normal, there are only three elements, however. The susceptibility has a resonant part, owing to infrared transitions in the adsorbate, and a non-resonant part which, when working with molecules adsorbed onto metals, is mainly due to the substrate and results in a background to the vibrational peak.

The intensity of SFG is low, but readily detectable, for a photomultiplier. It should be noted that although the sensitivity increases quadratically with laser intensity, the possibility of damage requires that the energy per unit area remains below a certain threshold value.

Figure 8.11 shows the application of SFG on adsorbed hydrocarbons [39]. Ethylene was adsorbed onto the (111) surface of platinum at 240 K, and subsequently heated to different temperatures. The spectra monitor the conversion of "di-σ"-bonded ethylene to ethylidyne ($\equiv C-CH_3$), via an intermediate characterized by a frequency of 2957 cm^{-1} attributed to the asymmetric C–H stretch of a CH_3 group in the ethylidene ($=CH-CH_3$) fragment. Somorjai and co-workers have demonstrated the usefulness of the SFG technique for *in-situ* studies of ethylene hydrogenation and CO oxidation at atmospheric pressure [40].

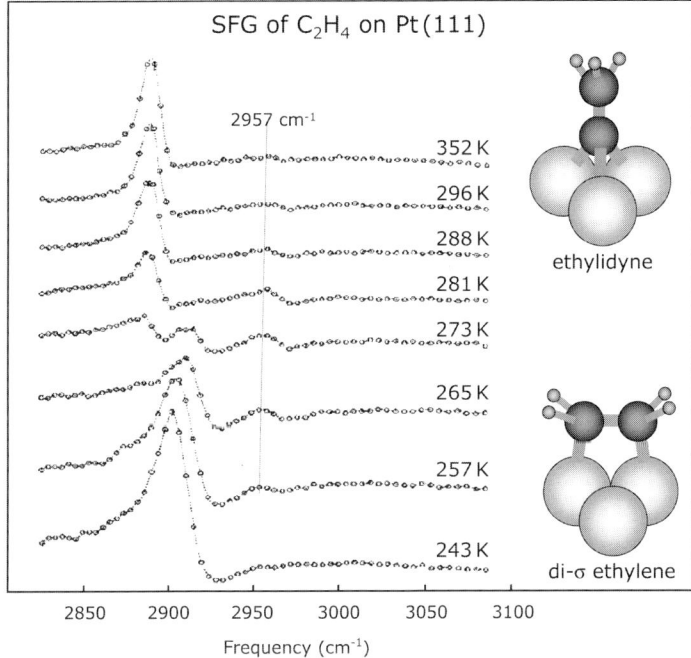

Fig. 8.11 Sum-frequency generation spectra of ethylene adsorbed on Pt(111) at 200 K after heating to the temperature indicated. The spectra indicate the conversion of di-σ-bonded ethylene to ethylidyne via an intermediate attributed to ethylidene. (Adapted from [39]).

SFG is not limited to single crystal studies. Indeed, Rupprechter et al. [41] have shown the technique to be very effective when studying the adsorption of CO onto palladium model catalysts, which consisted of Pd particles on a thin film Al_2O_3 support. The study was particularly interesting because the authors compared CO adsorption on model catalysts with that on Pd(111) single crystals over a wide range of CO pressures, from ultra-high vacuum to 1 bar. They found no evidence for high-pressure species which differed from those observed under ultra-high vacuum conditions; nor did they find any evidence of major rearrangements of the surface. Another interesting conclusion is that the vibrational frequencies of CO on the particles differed from those on the single crystal, even in the case of well-facetted particles exposing (111) planes; this indicated that the single crystal could hardly be considered a realistic model for the vibrational properties of molecules adsorbed onto particles [41].

Another recent report by the same group [42] addressed the shape and intensity of SFG peaks in the spectra of gases on supported particles. As mentioned above, the spectra are affected by non-resonant absorption in the metal, although a detailed discussion of this finding is beyond the scope of this text. Finally, a promising new development is that the SFG technique has also been shown to be effective on high-surface-area powders [43]. This may lead to an important method for *in-situ* studies of supported catalysts, without the need to resort to planar models.

8.5
Raman Spectroscopy

Whilst in infrared spectroscopy a molecule absorbs photons with the same frequency as its vibrations, Raman spectroscopy – in contrast – is based on the *inelastic scattering* of photons, which lose energy by exciting vibrations in the sample. The scattering process is illustrated schematically in Figure 8.12. In this process, monochromatic light of frequency v_o falls onto a sample, where the majority of the photons undergoes Rayleigh scattering (i.e., scattering without energy exchange). In a quantum mechanical picture it is as if the molecule is excited to an unstable state with energy hv_o above the ground state, from which it decays back to the ground state. No energy is exchanged between the molecule and the photon. However, when the excited molecule decays to the first vibrational level with frequency v_{vib}, it effectively takes an amount of energy equal to hv_{vib} away from the photon. Hence, the scattered light exhibits intensity at the frequency $v_o - v_{vib}$; this Raman peak is called the "Stokes band".

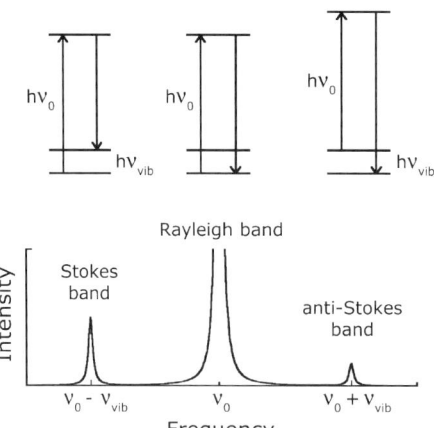

Fig. 8.12 The Raman effect. Monochromatic light of frequency v_o is scattered by a sample, either without losing energy (Rayleigh band) or inelastically, in which a vibration is excited (Stokes band), or a vibrationally excited mode in the sample is de-excited (anti-Stokes band). The spectrum is that of the light scattered by the sample. The energy level diagrams illustrate that the scattering process occurs via highly unstable states of high energy.

Yet, the reverse process may also take place. If the collision with a photon brings a vibrationally excited molecule to the unstable state of energy $h\nu_o + h\nu_{vib}$, it may decay to the ground state, transferring a net amount of energy $h\nu_{vib}$ to the photon, which leaves the sample with a higher frequency equal to $\nu_o + \nu_{vib}$. This peak, which is termed the "anti-Stokes band", has much lower intensity than the Stokes band, because the fraction of vibrationally excited molecules is usually small.

Inelastic scattering of light due to the excitation of vibrations was predicted as early as 1923 [44], and confirmed experimentally a few years later by Raman [45]. Because at that time the Raman effect was much easier to measure than infrared absorption, Raman spectroscopy dominated the field of molecular structure determination until commercial infrared spectrometers became available during the 1940s [10].

As in infrared spectroscopy, not all vibrations are observable. A vibration is Raman active if it changes the polarizability of the molecule; this requires in general that the molecule changes its shape. For example, the vibration of a hypothetical spherical molecule between the extremes of a disk-shaped and a cigar-shaped ellipsoid would be Raman active. We recall that the selection rule for infrared spectroscopy is that a dipole moment must change during the vibration. As a consequence, the stretch vibrations of for example H_2 (4160.2 cm^{-1}), N_2 (2330.7 cm^{-1}) and O_2 (1554.7 cm^{-1}) are observed in Raman spectroscopy but not in infrared spectroscopy. Thus, the two techniques complement each other, in particular for highly symmetrical molecules.

One strong point of Raman spectroscopy for research in catalysis is that the technique is highly suitable for *in-situ* studies. The spectra of adsorbed species interfere weakly with signals from the gas phase, enabling studies to be performed under reaction conditions. A second advantage is that typical supports such as silica and alumina are weak Raman scatterers, with the consequence that adsorbed species can be measured at wavenumbers as low as 50 cm^{-1}. These benefits render Raman spectroscopy a powerful tool for studying catalytically active phases on a support.

One disadvantage of the technique is the small cross-sections for Raman scattering, as most of the scattered intensity passes into the Rayleigh band which, typically, is about three orders of magnitude stronger than the Stokes bands. Of course using an intense laser increases all intensities, but this has the disadvantage that the sample may be heated up during measurement and that surface species may either decompose or desorb. Finally, fluorescence of the sample, giving rise to spectral backgrounds, may seriously limit the detectability of weak signals [10, 46].

One interesting development in the Raman spectroscopy of catalysts has been the use of an ultraviolet laser to excite the sample. This has two major advantages: (1) the scattering cross-section, which varies with the fourth power of the frequency, is substantially increased; and (2) the Raman peaks shift out of the visible region of the spectrum where fluorescence occurs. The interested reader is referred to the reports of Li and Stair for the applications of UV Raman spectroscopy on catalysts [47].

8.5.1
Applications of Raman Spectroscopy

Raman spectroscopy has been used successfully to investigate oxidic catalysts. According to Wachs, the annual number of Raman spectroscopy-related publications rose to about 80 to 100 by the late 1990s, with typically two-thirds of reports devoted to oxides [48]. Raman spectroscopy provides insight into the structure of oxides, their crystallinity, the coordination of metal oxide sites, and even the spatial distribution of phases through a sample, when the technique is used in microprobe mode. As the frequencies of metal–oxygen vibrations in a lattice are typically found to range between a few hundred and 1000 cm^{-1}, and thus are difficult to investigate with infrared spectroscopy, the Raman approach is clearly the best for this purpose.

Molybdenum, being a constituent of many hydrotreating and partial oxidation catalysts, is among those elements most studied by Raman spectroscopy. Molybdena catalysts are usually prepared by impregnating the support with a solution of ammonium heptamolybdate $(NH_4)_6Mo_7O_{24} \cdot 4H_2O$, in water. A few characteristic vibrations are listed in Table 8.4. The first three of these are the symmetric and antisymmetric stretch and the bending vibrations of the terminal M=O groups, while the others are stretch and bend vibrations of the internal Mo–O–Mo bonds. The pH determines which type of Mo complex is present in the solution [49]: $Mo_7O_{24}^{6-}$ at almost neutral pH (4.8–6.8), even larger $Mo_8O_{26}^{4-}$ complexes at low pH (≤ 2.2) and isolated MoO_4^{2-} units at high pH (>8). As the data in Table 8.4 show, all species are readily distinguished by their Raman spectra.

Kim et al. [50] impregnated a series of supports with $(NH_4)_6Mo_7O_{24} \cdot 4H_2O$ in H_2O, and used Raman spectroscopy to investigate the molybdenum structures on the support. In order to avoid confusion, it should be mentioned that it is customary to refer to oxidic molybdena catalysts as MoO_3/SiO_2 or MoO_3/Al_2O_3, without implying that crystalline MoO_3 is actually present. Figure 8.13 shows the Raman spectra of alumina- and silica-supported MoO_3 catalysts. The freshly impregnated MoO_3/Al_2O_3 catalysts all contain $Mo_7O_{24}^{6-}$ species (compare Table 8.4), whereas the samples prepared at pH values of 6 and 8.5 also contain some MoO_4^{2-}, as indicated by the characteristic Raman peak at 320 cm^{-1}. The fresh MoO_3/SiO_2 catalyst prepared at pH 8.5 exhibits the complete Raman spectrum of the $(NH_4)_6Mo_7O_{24} \cdot 4H_2O$ salt, while the spectrum of the sample prepared at pH 6 contains the peaks of both $Mo_7O_{24}^{6-}$ and the $(NH_4)_6Mo_7O_{24} \cdot 4H_2O$ salt. At low pH, the peaks of $Mo_7O_{24}^{6-}$ and $Mo_8O_{26}^{4-}$ arise. The weak signal around 1047 cm^{-1} is due to nitrate groups from HNO_3, which were added to lower the pH. Thus, remarkable differences exist between the molybdenum complexes that form on alumina and silica, and Raman spectroscopy successfully reveals all of these.

Table 8.4 Raman frequencies (in cm^{-1}) of molybdate species [49, 50].

Vibration	Species			
	$(NH_4)_6Mo_7O_{24} \cdot 4H_2O$	$Mo_7O_{24}^{6-}$	$Mo_8O_{26}^{4-}$	MoO_4^{2-}
v_s (Mo=O)	931	943	965	897
v_{as} (Mo=O)	879	903	925	837
δ (Mo=O)	354	362	370	317
v (Mo–O–Mo)	–	564	860	–
δ (Mo–O–Mo)	217	219	230	–

Even more remarkable differences arise after calcination at 775 K. The sharp Raman spectra of the calcined MoO_3/SiO_2 catalysts are those of crystalline MoO_3. No crystalline MoO_3 is present in the calcined MoO_3/Al_2O_3 catalysts, however. Here, the Raman spectra show similarities to those of the polymolybdates, which indicates that these species have a sufficiently strong interaction with the alumina support to withstand calcination at 775 K.

The fact that the negative molybdenum complexes in the solution interact much better with alumina than with silica is not difficult to understand. Alumina has an isoelectric point (point of zero charge) at pH $\approx$ 6–7, implying that alumina is positively charged in acidic solutions. This produces an excellent electrostatic interaction between the alumina surface and the negative $Mo_7O_{24}^{6-}$ species in a solution of pH 3.5. Silica, on the other hand, has a much lower isoelectric point (pH = 1.8), implying that the surface is either neutral or negatively charged. Hence, silica has little opportunity for interaction with the negative molybdenum complexes in solution, and as a result these species are loosely deposited on the surface when the catalyst is dried. Calcination leads to the formation of crystalline MoO_3.

This picture, although conceptually useful, is too simple. Kim et al. [50] showed that it is not the overall pH of the solution that dictates which species deposit on the surface, but rather the local pH at the support interface. The latter depends on the isoelectric point of the support, the coverage of molybdenum species, and the number of NH_4^+ or H^+ counter ions of the negative complexes. The interested reader is referred to Ref. [50] for details.

Before the molybdenum catalysts can be used in hydrotreating reactions, they must be sulfided, and Raman spectroscopy sensitively reveals this transition. The characteristic Mo–S frequencies of MoS_2 are at 389 and 411 cm^{-1} [51], much lower than the Mo–O bands. This occurs because, first, sulfur is twice as heavy as oxygen and, second, the Mo–S bond is weaker than the Mo–O bond [see Eq. (8-3)]. The Raman spectra of a sulfided MoO_3/Al_2O_3 also show a band at 529 cm^{-1}, due to the $(S-S)^{2-}$ disulfide species, which is probably located at the edges of the

242 | 8 Vibrational Spectroscopy

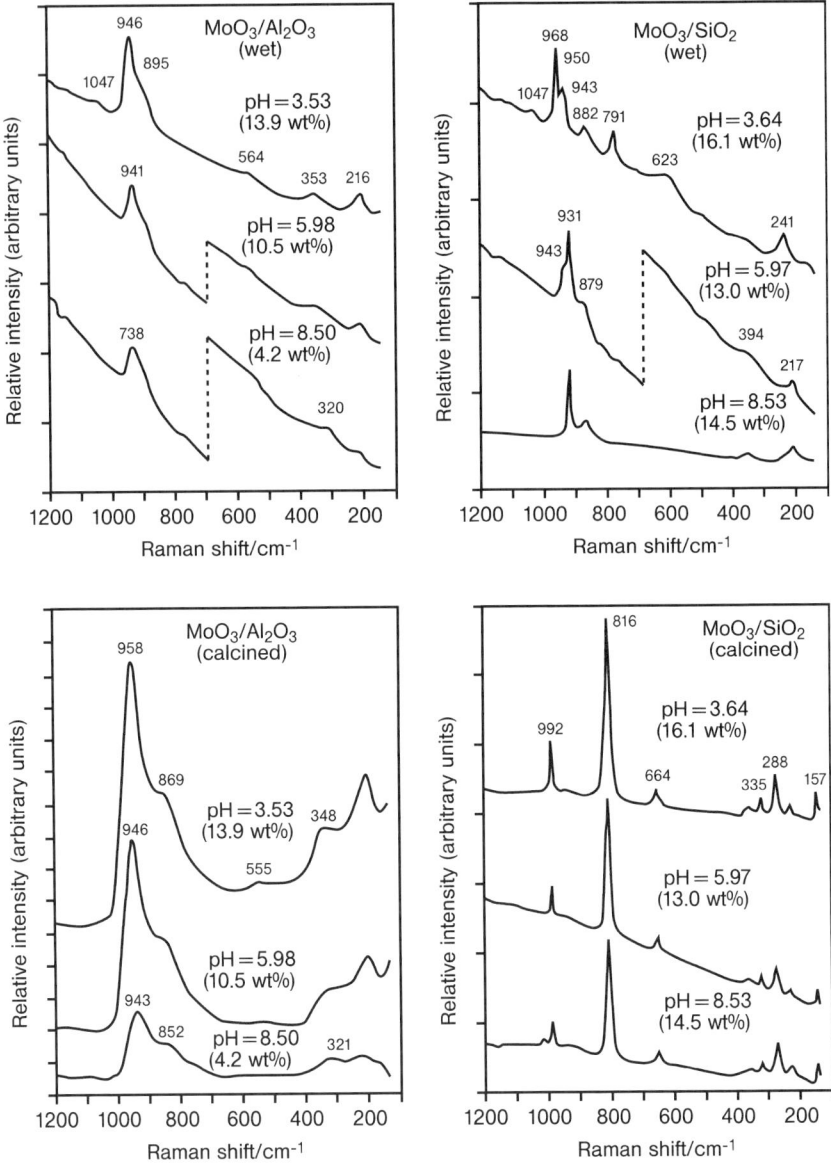

Fig. 8.13 Raman spectra of alumina- and silica-supported molybdena catalysts after impregnation of the supports with solutions of ammonium heptamolybdate, $(NH_4)_6Mo_7O_{24} \cdot 4\,H_2O$ of different pH values, and after calcination in air at 775 K. (See Table 8.4 for a list of characteristic Raman frequencies of molybdate species). The sharp peaks in the spectra of the calcined MoO_3/SiO_2 catalyst are those of crystalline MoO_3. (From [50]).

MoS$_2$ phase and thought to play a role in the splitting of hydrogen [51]. All of these bands would be difficult to detect in routine infrared spectroscopy.

These examples, together with many others discussed in a review by Mestl and Srinivasan [52], indicate that Raman spectroscopy is indispensable in investigating the preparation of molybdenum catalysts. Several other applications of Raman spectroscopy in catalysis have been discussed by Stencel [53], and reviewed by Bañares [54, 55].

8.6
Electron Energy Loss Spectroscopy (EELS)

As noted in Section 8.1, vibrations in molecules can be excited by interaction with waves and with particles. In electron energy loss spectroscopy (EELS; also termed HREELS for high-resolution EELS), a beam of monochromatic, low-energy electrons falls on the surface, where it excites lattice vibrations of the substrate, molecular vibrations of adsorbed species, and even electronic transitions. An energy spectrum of the scattered electrons reveals how much energy the electrons have lost to vibrations, according to the formula:

$$E = E_o - h\nu \qquad (8\text{-}8)$$

where:
- E is the energy of the scattered electron;
- E_o is the energy of the incident electrons;
- h is Planck's constant;
- ν is the frequency of the excited vibration.

The use of electrons requires that experiments are conducted in ultra-high vacuum, and preferably on the flat surfaces of single crystals or foils.

The first EELS experiments were reported by Propst and Piper in 1967, and concerned the adsorption of H$_2$, N$_2$, CO, H$_2$O on the (100) surface of tungsten [56]. Later, during the early 1970s, Ibach studied the energy losses of electrons to phonons in ZnO surfaces [57], and continued to develop the technique for studying adsorbates on metal surfaces [58, 59]. During the 1980s, EELS grew steadily into an important tool in surface science.

Where infrared and Raman spectroscopy are limited to vibrations in which a dipole moment or the molecular polarizability changes, EELS detects all vibrations. Two excitation mechanisms play a role in EELS, namely dipole scattering and impact scattering.

In *dipole scattering* we are dealing with the wave character of the electron. When it comes close to the surface, the electron sets up an electric field with its image charge in the metal. This oscillating field is perpendicular to the surface, and excites only those vibrations in which a dipole moment changes in a direction normal to the surface. The outgoing electron wave has lost an amount of energy equal to $h\nu$ [see Eq. (8-8)], and travels mainly into the specular direction, just as infrared radiation in RAIRS would do. Although the selection rules for RAIRS and dipole scattering in EELS are the same, the excitation mechanism of the latter is different in that there is no resonance between the frequencies of the electron wave and the excited vibration.

Impact scattering involves a short-range interaction between the electron and the molecule (simply said, a collision), which scatters the electrons over a wide range of angles. The useful feature of impact scattering is that *all* vibrations may be excited, and not only the dipole active ones. As in Raman spectroscopy, the electron may also take an amount of energy $h\nu$ away from excited molecules and leave the surface with an energy equal to $E_o + h\nu$.

The two scattering modes are illustrated in Figure 8.14 for a hypothetical adsorption system consisting of an atom on a metal. The stretch vibration of the atom perpendicular to the surface is accompanied by a change in dipole moment, while the bending mode parallel to the surface has no such change. As explained above, the EELS spectrum of electrons scattered in the specular direction detects only the dipole active vibration. The isotropically scattered electrons, however, undergo impact scattering and excite both vibrational modes. Note that the comparison of EELS spectra recorded in specular and off-specular directions yields information about the orientation of an adsorbed molecule. A slightly more complex situation (as shown in Fig. 8.14) has been observed experimentally in the adsorption of atomic hydrogen on a tungsten (100) surface [60].

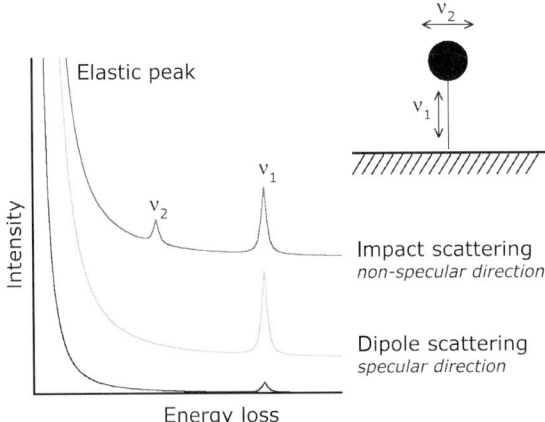

Fig. 8.14 Excitation mechanisms in electron energy loss spectroscopy (EELS) for a simple adsorbate system. Dipole scattering excites only the vibration perpendicular to the surface (v_1) in which a dipole moment normal to the surface changes; the electron wave is reflected by the surface into the specular direction. Impact scattering excites also the bending mode v_2 in which the atom moves parallel to the surface; electrons are scattered over a wide range of angles. The EELS spectra show the highly intense elastic peak and the relatively weak loss peaks. Off-specular loss peaks are in general one to two orders of magnitude weaker than specular loss peaks.

8.6 Electron Energy Loss Spectroscopy (EELS)

One strong point of EELS is that it detects losses over a very broad energy range, which comprises the entire infrared regime of Table 8.1, and extends even to electronic transitions at several electron volts. For example, EELS has no problems in detecting vibrations in the range between 50 and 800 cm^{-1} where, for example, metal–carbon and metal–oxygen stretch vibrations, as well as many skeletal vibrations of larger molecules, occur. This range is hardly accessible to a study in RAIRS.

In order to be effective, EELS spectrometers must satisfy a number of stringent requirements. First, the primary electrons should be monochromatic, with as little spread in energy as possible, preferably around 1 meV or better (1 meV = 8 cm^{-1}). Second, the energy of the scattered electrons should be measured with an accuracy of 1 meV, or better. Third, the low-energy electrons must effectively be shielded from magnetic fields. The resolution of EELS has steadily been improved, from typically 50 to 100 cm^{-1} around 1975 to better than 20 cm^{-1} for the currently available spectrometers. When the latter value comes close to the line width of a molecular vibration, the technique is usually referred to as high-resolution EELS (HREELS).

EELS is not suited for investigating catalysts but, similar to RAIRS, is a typical surface science technique. The two methods can be compared on a number of points:

- Environment: EELS can be applied only in a vacuum, while RAIRS operates both in a vacuum and at high pressures.
- Mechanism: EELS, when applied off-specular, detects all vibrations, whereas RAIRS detects only those vibrations in which a dipole moment changes perpendicular to the surface.
- Spectral range: EELS scans the entire vibrational region, whereas RAIRS is limited in practice to a range between about 800 and 4000 cm^{-1}.
- Resolution: today, the energy resolution of EELS is about 15 cm^{-1} at best, although many instruments operate at lower resolution. The resolution of RAIRS is not limited by the spectrometer (as low as 0.5 cm^{-1}) but rather by the linewidth of the vibration under study.

EELS has been used to study the kinetics of relatively slow surface reactions, such as the hydrogen–deuterium exchange in benzene adsorbed onto platinum [61]. In this case, the reaction occurs on a time scale of about 1 hour, and a series of EELS spectra (the recording of which took 5 minutes each) reflects the extent of reaction with sufficient accuracy. Most surface reactions proceed at significantly higher rates, however. In order to make EELS a faster technique, Ho and co-workers [62] incorporated a multichannel detector in their EELS analyzer and reduced the acquisition time for a spectrum to milliseconds. In this way, they could monitor the dissociation of CO on an iron (111) surface during temperature-programmed heating at a not-too-rapid heating rate of 0.05 K s^{-1}.

Figure 8.15 illustrates the HREELS and LEED measurements of CO adsorbed onto Rh(111) [63]. The bottom spectrum is that of the empty surface and shows only the elastic peak, which has a full width at half-maximum of about 2 meV, or

246 | 8 Vibrational Spectroscopy

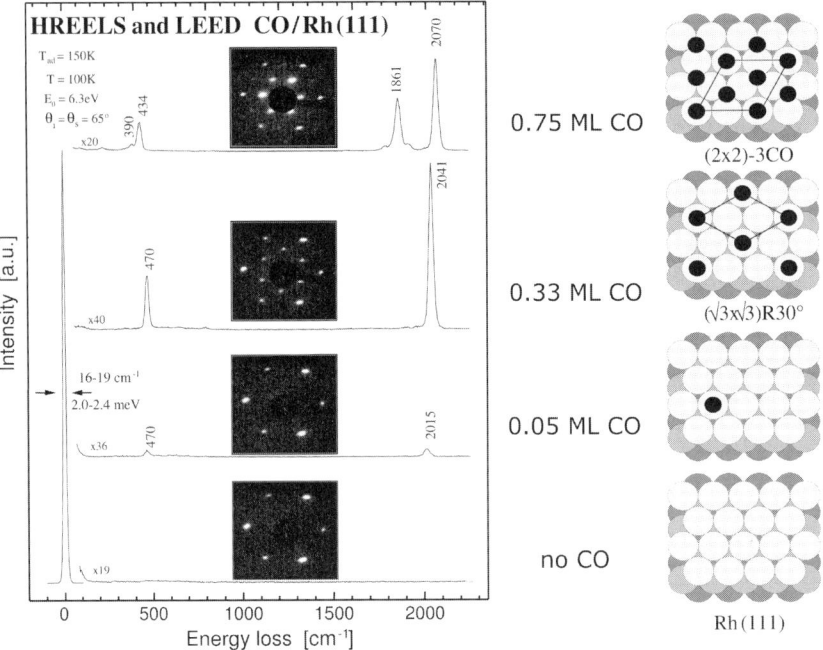

Fig. 8.15 High-resolution electron energy loss spectroscopy (HREELS) and low-energy electron diffraction of CO adsorbed on a Rh(111) surface, along with structure models. The HREELS spectra show the C–O and metal–CO stretch vibrations of linear and threefold CO on rhodium. (From [63]).

16 cm^{-1}. In fact, such a spectrum provides a stringent test for the cleanliness of the surface, as any impurity (e.g., of adsorbed carbon or oxygen) would immediately result in a peak in the 300 to 600 cm^{-1} region of the HREELS spectrum. The LEED image shows the expected pattern of a hexagonal surface, with three of the six spots having stronger intensity than the other three due to the a-b-c stacking of the fcc lattice (see the Appendix) resulting in two types of threefold hollow sites on the fcc(111) surface.

The adsorption of 0.05 monolayers (ML) of CO onto this surface gives rise to a peak at 2015 cm^{-1}, corresponding to the internal C–O stretch frequency of the molecule in the on-top adsorption site, and one at 470 cm^{-1}, being due to metal–molecule bond. The latter is not easily observable in infrared spectroscopy. Increasing the CO coverage to 0.33 ML enhances the intensity of the HREELS peaks. In addition, the C–O stretch frequency shifts upward due to dipole–dipole coupling [18, 19]. The LEED pattern corresponds to an ordered $(\sqrt{3} \times \sqrt{3})R30°$ overlayer in Wood's notation (see the Appendix), in accordance with the coverage of 0.33 ML.

The top spectrum in Figure 8.15 is that of a saturated CO overlayer on Rh(111), corresponding to a coverage of 0.75 ML. The LEED pattern is that of a (2 × 2) periodicity, implying that the unit cell contains three CO molecules. The HREELS spectrum indicates that CO is now present in two adsorption states, linear (2070 cm^{-1}) and threefold (1861 cm^{-1}). In spite of the fact that one unit cell contains one linear and two threefold CO molecules, the HREELS intensity of the linear CO peak is larger than that of the threefold CO. Indeed, infrared and EELS intensities are often not proportional to surface coverages. In addition, the transfer of intensity from peaks at lower to peaks at higher frequencies can occur.

The interpretation of the HREELS spectrum and the structure belonging to the (2 × 2)-3CO LEED pattern has been the subject of some debate in the literature [64–67]. The CO stretch peak at the lower frequency had previously been assigned to a bridge-bonded CO [64], with obvious consequences for the way that CO fills the (2 × 2) unit cell. A recent structural analysis from the same laboratory on the basis of tensor LEED has confirmed the structures of both the ($\sqrt{3} \times \sqrt{3}$)R30° and the (2 × 2)-3CO, as given in Figure 8.14 [65]. The assignments have also been supported by high-resolution XPS measurements [66] and by computational chemistry [67].

The intensity of the HREELS peaks, including that of the elastic peak, depends heavily on the adsorbate and the degree of ordering on the surface. Therefore, it is common practice to scale all spectra with respect to the elastic peak, as has been done for the spectra in Figure 8.15.

Although HREELS is very powerful and informative, the measurement of spectra of good quality is by no means a matter of routine. Time-consuming tuning of the electronics and aligning of the sample are needed to obtain meaningful results, and in this sense its infrared counterpart, RAIRS, is more convenient. The one slight disadvantage of the latter procedure – that the low-frequency vibrations of metal–molecule bonds cannot be observed – is usually taken for granted.

8.7
Concluding Remarks

In summarizing, infrared spectroscopy measures – in principle – the force constants of chemical bonds. It is a powerful tool for identifying adsorbed species and their bonding mode; moreover, it is an *in-situ* technique and is applicable in either transmission or diffuse reflection mode on real catalysts, and in reflection–absorption mode on single crystal surfaces. Sum-frequency generation is a promising specialty which focuses exclusively on interfaces, and offers great opportunities to study adsorbates on catalytic surfaces under high-pressure conditions. Raman spectroscopy has proven very useful in studying the structure of oxidic and sulfidic catalysts. Electron energy loss spectroscopy can only be applied on flat surfaces in ultra-high vacuum, but has the advantages that it can detect vibrations over a wider range of frequencies, and that its selection rules are less strict than for infrared spectroscopy. The latter technique, however, is easier to apply in practice.

References

1 W.W. Coblentz, *J. Franklin Inst.* **172** (1911) 309.
2 A.V. Kiselev and V.I. Lygin, in: *Infrared Spectroscopy of Adsorbed Species*, L.H. Little (Ed.). Academic Press, New York, 1966.
3 R.P. Eischens and W.A. Pliskin, *Adv. Catal.* **10** (1958) 1.
4 L.H. Little, *Infrared Spectroscopy of Adsorbed Species*. Academic Press, New York, 1966.
5 A. Fadini and F.-M. Schnepel, *Vibrational Spectroscopy*. Ellis-Horwood Ltd, Chichester, 1989.
6 W.O. George and P.S. McIntyre, *Infrared Spectroscopy*. Wiley, Chichester, 1987.
7 B.M. Weckhuysen (Ed.). *In Situ Characterization of Catalytic Materials*. American Scientific Publishers, 2004.
8 P.A. Lund, D.E. Tevault, and R.R. Smardzewski, *J. Phys. Chem.* **88** (1984) 1731.
9 J. Heidberg and H. Weiss, in: *Thin Metal Films and Gas Chemisorption*, P. Wissmann (Ed.). Elsevier, Amsterdam, 1987, p. 196.
10 W.N. Delgass, G.L. Haller, R. Kellerman, and J.H. Lunsford, *Spectroscopy in Heterogeneous Catalysis*. Academic Press, New York, 1979.
11 T. Bürgi and A. Baiker, *Adv. Catal.* **50** (2006) 227.
12 W.R. Moser, B.J. Marshik-Guerts, and S.J. Okrasinski, *J. Mol. Catal. A* **143** (1999) 57.
13 T. Weber, J.C. Muijsers, and J.W. Niemantsverdriet, *J. Phys. Chem.* **99** (1995) 9194.
14 F.M. Hoffman, *Surface Sci. Rep.* **3** (1983) 107.
15 Y.J. Chabal, *Surface Sci. Rep.* **8** (1988) 211.
16 P.R. Griffiths, *Chemical Infrared Fourier Transform Spectroscopy*. Wiley, Chichester, 1975.
17 N. Sheppard and T.T. Nguyen, in: *Advances in Infrared and Raman Spectroscopy*, Vol. 5, R.J.H. Clark and R. Hester (Eds.). Heyden, London, 1978, p. 67.
18 P. Hollins, *Surface Sci. Rep.* **16** (1992) 51.
19 P. Hollins and J. Pritchard, *Progr. Surface Sci.* **19** (1985) 275.
20 Y. Soma-Noto and W.M.H. Sachtler, *J. Catal.* **32** (1974) 315.
21 Y. Soma-Noto and W.M.H. Sachtler, *J. Catal.* **34** (1974) 162.
22 E.L. Kugler and M. Boudart, *J. Catal.* **59** (1979) 201.
23 F.J.C.M. Toolenaar, D. Reinalda, and V. Ponec, *J. Catal.* **64** (1980) 110.
24 A.A. Davydov, *Molecular Spectroscopy of Oxide Catalyst Surfaces*. Wiley, Chichester, 2003.
25 N.Y. Topsøe and H. Topsøe, *J. Catal.* **84** (1983) 386.
26 N.Y. Topsøe and H. Topsøe, *J. Electron. Spectrosc. Relat. Phenom.* **39** (1986) 11.
27 B.C. Gates, L. Guczi, and H. Knözinger (Eds.), *Metal Clusters in Catalysis*. Elsevier, Amsterdam, 1987.
28 J.C. Fierro-Gonzalez, S. Kuba, Y. Hao, and B.C. Gates, *J. Phys. Chem. B* **110** (2006) 13326.
29 J. Guzman, B.G. Anderson, C.P. Vinod, K. Ramesh, J.W. Niemantsverdriet, and B.C. Gates, *Langmuir* **21** (2005) 3675.
30 O. Diwald, M. Sterrer, and H. Knözinger, *Phys. Chem. Chem. Phys.* **4** (2002) 2811.
31 M. Okumura, S. Tsubota, and M. Haruta, *J. Mol. Catal. A* **199** (2003) 73.
32 H. Knözinger and P. Ratnasamy, *Catal. Rev. – Sci. Eng.* **17** (1978) 31.
33 G. Mestl and H. Knözinger, in: *Handbook of Heterogeneous Catalysis*, Vol. II, G. Ertl, H. Knözinger and J. Weitkamp (Eds.). Wiley-VCH, Weinheim, 1997, p. 539.
34 C.M. Leewis, W.M.M. Kessels, M.C.M. van de Sanden, and J.W. Niemantsverdriet, *J. Vac. Sci. Technol. A* **24** (2006) 296.
35 R. Raval, G. Blyholder, S. Haq, and D.A. King, *J. Phys. Cond. Mater.* **1** (1989) SB165.

36 Y.R. Shen, *The Principles of Nonlinear Optics*. Wiley, New York, 1984.
37 Y.R. Shen, *Nature* **337** (1989) 519.
38 R.M. Corn, *Anal. Chem.* **63** (1991) 285 A.
39 P.S. Cremer, C. Stanners, J.W. Niemantsverdriet, Y.R. Shen, and G.A. Somorjai, *Surface Sci.* **328** (1995) 111.
40 G.A. Somorjai, *CaTTech* **3** (1999) 84.
41 G. Rupprechter, H. Unterhalt, M. Morkel, P. Galletto, L. Hu, and H.-J. Freund, *Surface Sci.* **502–503** (2002) 109.
42 M. Morkel, H. Unterhalt, T. Klüner, G. Rupprechter, and H.-J. Freund, *Surface Sci.* **586** (2005) 146.
43 M.V. Yeganeh, S.M. Dougal, and B.G. Silbernagel, *Langmuir* **22** (2006) 637.
44 A. Smekal, *Naturwissenschaften* **11** (1923) 873.
45 C.V. Raman, *Indian J. Phys.* **2** (1928) 387.
46 S.K. Freeman, *Applications of Laser Raman Spectroscopy*. Wiley, New York, 1974.
47 C. Li and P.C. Stair, in: *Proceedings, 11th International Congress of Catalysis*, J.W. Hightower, W.N. Delgass, E. Iglesia, and A.T. Bell (Eds.). Elsevier, Amsterdam, 1996, p. 881.
48 I.E. Wachs, *Topics Catal.* **8** (1999) 57.
49 L. Wang and W.K. Hall, *J. Catal.* **77** (1982) 232; *J. Catal.* **66** (1980) 251.
50 D.S. Kim, K. Segawa, T. Soeya, and I.E. Wachs, *J. Catal.* **136** (1992) 539.
51 J. Polz, H. Zeilinger, B. Müller, and H. Knözinger, *J. Catal.* **120** (1989) 22.
52 G. Mestl and T.K.K. Srinivasan, *Catal. Rev. – Sci. Eng.* **40** (1998) 451.
53 J.M. Stencel, *Raman Spectroscopy for Catalysis*. Van Nostrand Reinhold, New York, 1989.
54 M.O. Guerrero-Pérez and M. Bañares, *Catal. Today* **113** (2006) 48.
55 M. Bañares, in: *In Situ Characterization of Catalytic Materials*, B.M. Weckhuysen (Ed.). American Scientific Publishers, 2004.
56 F.M. Propst and T.C. Piper, *J. Vac. Sci. Technol.* **4** (1967) 53.
57 H. Ibach and D.L. Mills, *Electron Energy Loss Spectroscopy and Surface Vibrations*. Academic Press, New York, 1982.
58 H. Ibach, *Electron Energy Loss Spectrometers: The Technology of High Performance*. Springer-Verlag, Berlin, 1991.
59 G. Ertl and J. Küppers, *Low Energy Electrons and Surface Chemistry*. VCH, Weinheim, 1985.
60 W. Ho, R.F. Willis, and E.W. Plummer, *Phys. Rev. Lett.* **40** (1978) 146.
61 M. Surman, S.R. Bare, P. Hoffman, and D.A. King, *Surface Sci.* **126** (1983) 349.
62 W. Ho, *J. Phys. Chem.* **91** (1987) 766.
63 R. Linke, D. Curulla, M.J.P. Hopstaken, and J.W. Niemantsverdriet, *J. Chem. Phys.* **115** (2001) 8209.
64 L.H. Dubois and G.A. Somorjai, *Surface Sci.* **91** (1980) 514.
65 M. Gierer, A. Barbieri, M.A. van Hove, and G.A. Somorjai, *Surface Sci.* **391** (1997) 176.
66 A. Beutler, E. Lundgren, R. Nyholm, J.N. Andersen, B. Setlik, and D. Heskett, *Surface Sci.* **371** (1997) 381.
67 D. Curulla, R. Linke, A. Clotet, J.M. Ricart, and J.W. Niemantsverdriet, *Phys. Chem. Chem. Phys.* **4** (2002) 5372.

9
Case Studies in Catalyst Characterization

Keywords

Supported rhodium catalysts
Alkali promoters on metal surfaces
Cobalt–molybdenum sulfide hydrodesulfurization catalysts
Chromium oxide polymerization catalysts

9.1
Introduction

In this chapter we present four case studies to illustrate catalyst characterization from a problem-oriented approach. The intention is to show what combinations of techniques can achieve, without any pretence to provide an exhaustive overview. The selected studies are all aimed at determining the composition and structure of a catalyst or a catalytic surface in the smallest possible detail, and to provide insight at the atomic/molecular level.

We begin with the structure of a noble metal catalyst, where the emphasis is placed on the preparation of rhodium on aluminum oxide and the nature of the metal support interaction. Next, we focus on a promoted surface in a review of potassium on noble metals. This section illustrates how single crystal techniques have been applied to investigate to what extent promoters perturb the surface of a catalyst. The third study deals with the sulfidic cobalt–molybdenum catalysts used in hydrotreating reactions. Here, we are concerned with the composition and structure of the catalytically active surface, and how it evolves as a result of the preparation. In the final study we discuss the structure of chromium oxide catalysts in the polymerization of ethylene, along with the polymer product that builds up on the surface of the catalyst.

9.2
Supported Rhodium Catalysts

The structure of supported rhodium catalysts has been the subject of intensive research during the past decade. Rhodium is the component of the automotive

Spectroscopy in Catalysis: An Introduction, Third Edition
J. W. Niemantsverdriet
Copyright © 2007 WILEY-VCH Verlag GmbH & Co. KGaA, Weinheim
ISBN: 978-3-527-31651-9

exhaust catalyst (the three-way catalyst) responsible for the reduction of NO by CO [1]. In addition, it exhibits a number of fundamentally interesting phenomena, such as "strong metal support interaction" after high-temperature treatment in hydrogen [2], and particle disintegration under CO [3]. In this section we illustrate how techniques such as X-ray photoelectron spectroscopy (XPS), secondary ion mass spectrometry (SIMS), extended X-ray absorption fine structure (EXAFS), transmission electron microscopy (TEM) and infrared spectroscopy have led to a fairly detailed understanding of supported rhodium catalysts.

9.2.1
Preparation of Alumina-Supported Rhodium Model Catalysts

Rhodium on alumina catalysts are usually made by impregnating the Al_2O_3 support with a solution of rhodium trichloride. In water, $RhCl_3 \cdot xH_2O$ forms a range of ionic and neutral complexes, in which the Rh^{3+} ion is surrounded by six ligands, which are Cl^-, OH^- and H_2O. Negatively charged complexes of the form $[RhCl_x(H_2O)_y(OH)_z]^{n-}$ adsorb as the result of an electrostatic interaction on the positively charged sites of the alumina surface, whereas the neutral complexes may attach to the support through a reaction with surface OH groups. The first point to determine is what type of complex exists on the support after impregnation. The studies of Borg et al. [4] have shown that XPS and SIMS can yield the desired information, provided that one uses a suitable model system.

Catalysts on high-surface-area supports offer limited possibilities for surface analysis by techniques that make use of charged particles, such as Auger electron spectroscopy (AES), ultraviolet photoelectron spectroscopy (UPS), XPS, and SIMS. The main reason is that typical oxide supports are electrical insulators, which charge up during measurement. This leads to shifted and broadened peaks in the electron spectroscopies and to decreased secondary ion emission in SIMS. Charging can successfully be avoided by using model supports consisting of flat silicon or aluminum crystals with a thin oxide layer of a few nanometers thickness on top [5]. Several authors have used such systems to prepare model catalysts either by evaporating rhodium onto the support [6] or by depositing organometallic complexes such as $Rh_6(CO)_{16}$, which are subsequently decomposed [7, 8]. From the point of view of catalyst preparation, however, it is much more attractive to create model catalysts on flat, conducting supports by using the same wet chemical techniques that are used to make technical catalysts [5, 9, 10]. In this way, one can study the effect of essential steps such as drying, calcination and reduction by means of surface spectroscopies, in particular with those that are not easily applicable to powder catalysts.

SIMS spectra of the $RhCl_3$ salt in Figure 9.1 show clear molecular peaks characteristic of rhodium coordinated by chlorine. In particular, the $RhCl_2^-$ signal is very intense. As explained in Chapter 4, there is little doubt that molecular cluster ions from compounds other than alloys are the result of a direct emission process. Hence, Figure 9.1 implies that if a sample contains rhodium atoms with more than one chlorine ligand, then SIMS is capable of detecting this combination with high sensitivity.

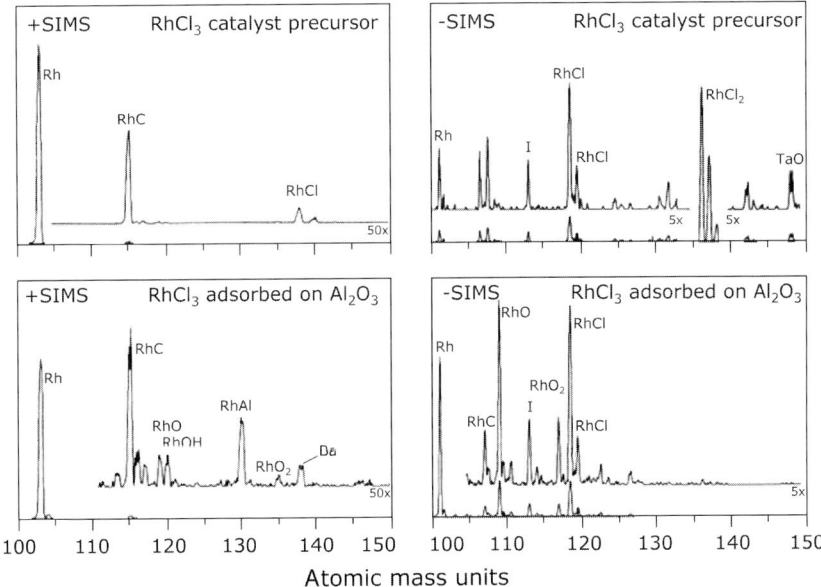

Fig. 9.1 Positive (left) and negative (right) static SIMS spectra of RhCl$_3$·xH$_2$O on tantalum (top) and of a model catalyst prepared by adsorbing Rh complexes derived from RhCl$_3$·xH$_2$O in water on an Al$_2$O$_3$/Al model support. (From [4]).

The SIMS spectra of the freshly prepared Rh/Al$_2$O$_3$/Al catalyst of Figure 9.1 contain several interesting clusters that provide an indication of the Rh ligands present after adsorption. With respect to the number of Cl ligands of Rh, the negative SIMS spectrum shows that the RhCl$_2^-$ ions, which dominated the RhCl$_3$ spectrum, have disappeared completely. XPS spectra indicate that rhodium and chlorine are present as Rh^{3+} and Cl$^-$, respectively, and that the atomic Cl/Rh ratio is about 1.4. As the SIMS spectra confirm that Rh–Cl contact exists, whereas RhCl$_2$ and RhCl$_3$ species (for which SIMS is particularly sensitive through the RhCl$_2^-$ signal) do not appear to be present, XPS and SIMS together indicate that the majority of the Cl-containing Rh complexes on the catalyst contains one Cl$^-$ ion only, and is thus most likely of the form [RhCl(OH)$_3$(H$_2$O)$_2$]$^-$. It should be noted that this conclusion cannot be based on the XPS spectra alone, as alumina is known to bind Cl$^-$ ions strongly. The SIMS spectra are of great value here because they confirm that at least a part of the Cl$^-$ is coordinated to rhodium.

The fact that adsorbed rhodium complexes contain one Cl$^-$ ion at most is remarkable, if one realizes that the actual concentration of Rh-complexes with one Cl$^-$ in moderately acid solutions cannot be high. According to Fenoglio et al. [11], exchange reactions between Cl$^-$ ions and hydroxyls of the support of the type

$$\text{Al–OH}_2^+ \cdots [\text{RhCl}_3(\text{OH})(\text{H}_2\text{O})_2]^- + \text{Al–OH}$$
$$\rightarrow \text{Al–OH}_2^+ \cdots [\text{RhCl}_2(\text{OH})_2(\text{H}_2\text{O})_2]^- + \text{Al–Cl} \quad (9\text{-}1)$$

may occur several times, because OH⁻ forms a stronger bond with rhodium than Cl⁻ does. Borg et al. [4] propose that the remaining Cl⁻ ion is at the ligand position pointing away from the support surface, where it cannot participate in the surface-mediated exchange reaction [Eq. (9-1)].

The pH of the impregnating solution is an important parameter in the preparation of alumina-supported catalysts. In order to adsorb negatively charged rhodium complexes, the pH of the solution must be well below the isoelectric point of alumina (pH $\approx$ 7) in order to have a positively charged surface. The pH, on the other hand, should not be too low either, because Al_2O_3 dissolves below pH $\approx$ 4. Dissolved alumina species react with rhodium complexes in solution, and precipitate in an uncontrolled manner during drying. The occurrence of "acid attack" (as the process is called) is easily recognized in the XPS spectra of the thin-film Al_2O_3/Al model supports by a decreased Al^{3+}/Al intensity ratio [4]. Acid attack must be avoided when preparing alumina-supported catalysts because it results in ill-defined systems [11].

9.2.2
Reduction of Supported Rhodium Catalysts

The reduction of catalysts in hydrogen can be followed using many techniques, including temperature-programmed reduction (TPR), XPS, SIMS, EXAFS, X-ray absorption near edge spectroscopy (XANES), and electron microscopy. TPR patterns (e.g., see Fig. 2.4) indicate that supported rhodium catalysts reduce at relatively low temperatures, often below 200 °C. The XPS data in Figure 9.2, taken with a monochromated X-ray source, confirm this: reduction at 200 °C or higher gives rise to a sharp rhodium 3d XPS doublet, with a Rh $3d_{5/2}$ binding energy of 307.4 eV. This value is in good agreement with binding energies measured for well-dispersed rhodium catalysts, but is 0.4 eV higher than the value of the bulk metal (indicated by a vertical, dotted line in Fig. 9.2). This is a well-known small-particle effect, described in detail by Mason [12], which nowadays is explained as an initial state effect: particles in the 1 to 2-nm range have not yet attained the normal bulk band structure. In addition, a final state effect contributes, which is related to the reduced capacity of the valence electrons in a small particle to screen the core hole formed in the photoemission process [13].

Huizinga et al. [14] reported Rh $3d_{5/2}$ binding energies of alumina-supported rhodium catalysts as a function of dispersion determined with hydrogen chemisorption (H/M). The H/M scale was later calibrated in terms of particle size by Kip et al. [15] (see Fig. 6.15). Figure 9.3 combines their data into a calibration of binding energies versus particle size, applicable to Rh/Al_2O_3 catalysts. According to Figure 9.3, the $Rh/Al_2O_3/Al$ model catalyst of Figure 9.2 contains particles equivalent to half-spheres with a diameter of about 0.8 nm.

As chlorine may affect the catalytic behavior of rhodium, it is important to know how residual chlorine from the preparation is affected by the reduction process. Chlorine contents down to about 1% are easily determined by XPS, but in order to detect lower concentrations then SIMS is the best choice.

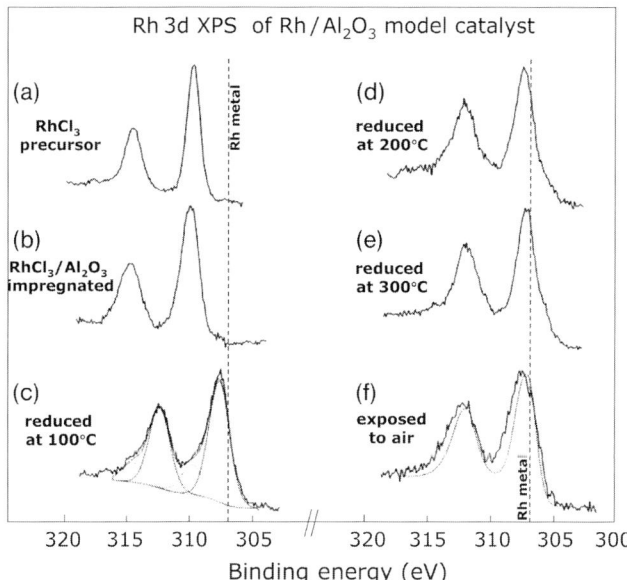

Fig. 9.2 Rh 3d XPS spectra of: (a) $RhCl_3 \cdot xH_2O$ on tantalum, and of a model catalyst prepared by adsorbing Rh complexes derived from $RhCl_3 \cdot xH_2O$ in water on an Al_2O_3/Al model support after (b) adsorption and drying, (c–e) reduction at the indicated temperatures, and (f) subsequent exposure to air at room temperature. The dashed line represents the Rh 3d spectrum of the reduced catalyst. (From [4]).

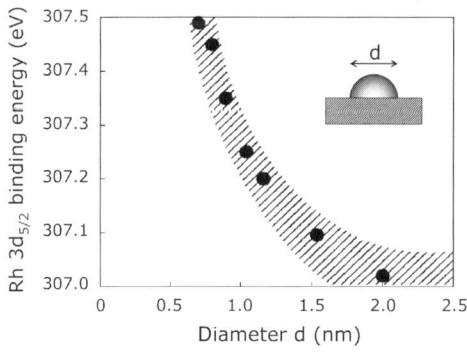

Fig. 9.3 Rh $3d_{5/2}$ binding energy versus particle size for half-spherical rhodium particles on an alumina support. (Data from [14, 15]; see also Fig. 6.15).

XPS indicates that the atomic ratio Cl/Rh of 1.4 in the fresh catalyst decreases to 0.6 after reduction at 100 °C, and to less than 0.1 after reduction at 200 °C. SIMS spectra in Figure 9.4 reveal whether the chlorine is in contact with rhodium, or not: the $RhCl^-$ signal at 138 and 140 amu is clearly present in the cata-

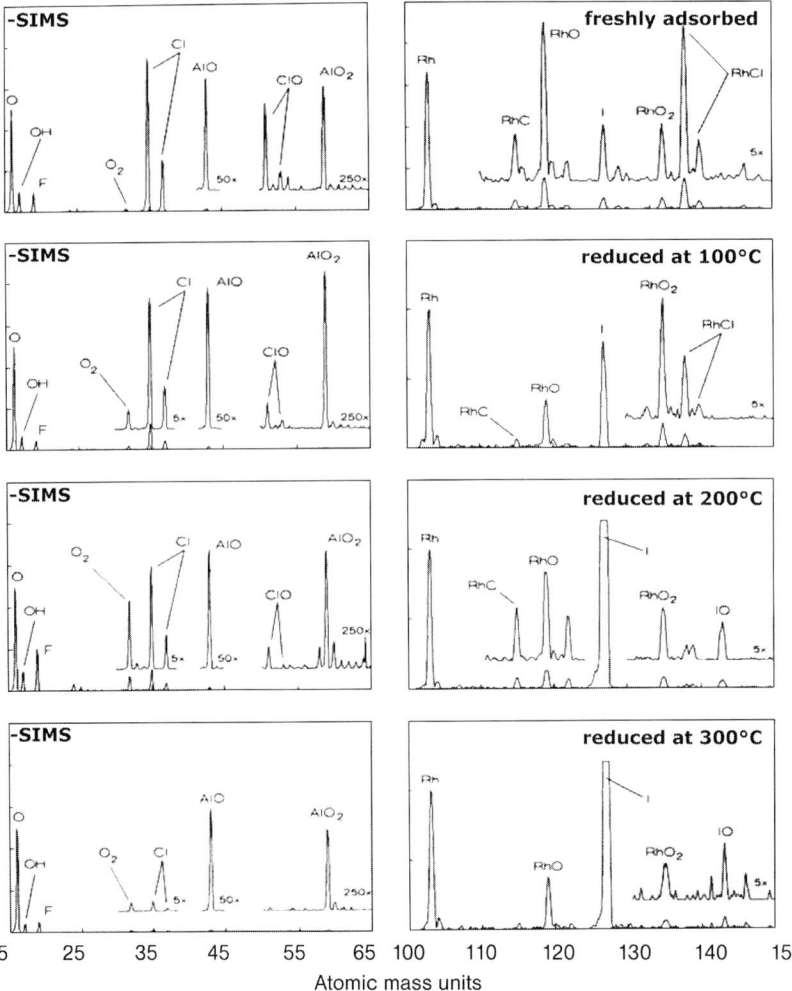

Fig. 9.4 Negative SIMS spectra in the mass ranges of 15–65 and 100–150 amu of a Rh/Al$_2$O$_3$/Al model catalyst prepared from RhCl$_3 \cdot$xH$_2$O, after reduction at different temperatures. Note the disappearance of the RhCl$^-$ signal after reduction at 200 °C, whereas significant signals from Cl$^-$ and ClO$^-$ are still observed. (From [4]).

lysts after reduction at 100 °C, but has disappeared after reduction at 200 °C, indicating that Rh–Cl species are absent. SIMS spectra in the elemental mass range, however, show that chlorine is still present on the catalyst. A significant ClO$^-$ signal indicates that this chlorine is in contact with the support. The ClO$^-$ intensity drops to zero after reduction at 300 °C. It is pertinent to note that porous alumina supports are known to retain significantly more chlorine, also after reduction at 300 °C [3, 16], due to diffusion limitations and readsorption of HCl in high-surface-area supports. Hence, we conclude that the reduction of Rh/ Al_2O_3 catalysts prepared from $RhCl_3$ proceeds in two steps:

- Step 1, in which Rh–Cl and Rh–O bonds are broken and Rh reduces to the metallic state at temperatures below 200 °C.
- Step 2, which involves the removal of residual chlorine from the alumina support, which requires higher temperatures.

The data in Figures 9.1, 9.2 and 9.4 nicely illustrate the complementarity of XPS and SIMS, and highlights the possibilities that thin-film oxide supports offer for surface investigations. Owing to the conducting properties of the support charging is virtually absent, and typical single crystal techniques such as monochromatic XPS and static SIMS can be applied to their full potential to answer questions on the preparation of supported catalysts.

9.2.3
Structure of Supported Rhodium Catalysts

The size and shape of small metal particles determines the geometry in which atoms are available to reacting gases. Adsorbed and dissolved reactant molecules, on their part, often modify the atomic arrangement at the surface of a particle. The intricate interplay between the structure of catalytically active sites at the surface of a particle and the conditions under which the catalyst operates is only now beginning to be explored. As an example, we discuss the structure of small rhodium crystallites in a Rh/Al_2O_3 catalyst under hydrogen, and the changes in structure that occur if rhodium particles are exposed to CO. EXAFS, XPS and infrared spectroscopy have been the key tools in these investigations.

A central question with respect to supported metal catalysts is that of the structure of the metal support interface. Various possibilities have been proposed, varying from interfaces consisting of a mixed metal aluminate or silicate layer [17], or the presence of metal ions which serve as anchors between particle and support [18], to the attractive interaction between ions of the support and the dipoles that these ions induce in the metal particle [19]. EXAFS highlights the atomic surroundings of an atom in the catalyst, and if the supported metal particles are sufficiently small, the oxygen atoms in the metal support interface provide a measurable contribution to the EXAFS spectrum.

Figure 9.5 shows EXAFS data reported by Koningsberger et al. [19] of a highly dispersed (H/M = 1.2) Rh/Al_2O_3 catalyst under hydrogen, after reduction at 200 and 400 °C. The magnitude of the Fourier transform, corrected for the Rh–Rh phase and backscattering amplitude according to Eq. (6-11), shows the presence of Rh neighbors at 0.269 nm and oxygen neighbors at lesser distances. Note that the oxygen distances in Figure 9.5 must be corrected for the Rh–O phase shift, and therefore appear at distances that are too small. Computed fits based on the distances and coordination numbers in Table 9.1 agree very well with the isolated EXAFS signal of the coordination shells between 0.08 and 0.32 nm from the rhodium atoms.

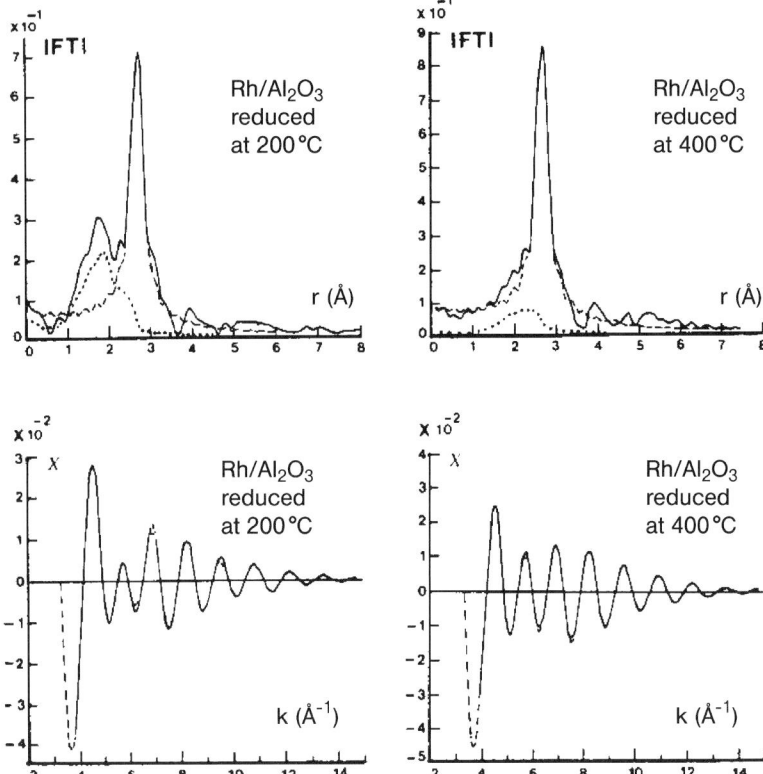

Fig. 9.5 EXAFS of Rh/Al$_2$O$_3$ catalysts after reduction at (left) 200 °C and (right) 400 °C. Top spectra: the magnitude of the Fourier transform of the measured EXAFS signal. Bottom spectra: back-transformed EXAFS corresponding to distances between 0.8 and 3.2 from Rh atoms. The lower Fourier transform contains a dominant contribution from Rh nearest neighbors at 0.27 nm and a minor contribution from oxygen neighbors in the metal support interface. After correction for the Rh–O phase shift, the oxygen ions are at a distance of 0.27 nm. (From [19]).

The data analysis indicates that the catalyst reduced at 200 °C still contains Rh–O contributions characteristic of Rh$_2$O$_3$, attributed to unreduced particles. The rate-determining step in the reduction of noble metal oxides is the nucleation – that is, the formation of a metal atom from *molecular* hydrogen and metal oxide – while further reduction by hydrogen *atoms* formed on already reduced metal atoms is a very rapid process. Therefore, it seems unlikely that noble metal catalysts under hydrogen contain particles that are only partially reduced. After reduction at 400 °C, a separate Rh$_2$O$_3$ phase is no longer observed, and the rhodium atoms have only two types of nearest scattering neighbors, namely rhodium and oxygen. Both are found at a distance of 0.269 nm.

Only rhodium atoms at the interface with the alumina support have oxygen neighbors; hence, the Rh–O coordination number, representing the average

9.2 Supported Rhodium Catalysts

Table 9.1 EXAFS parameters of a 2.4 wt% Rh/Al$_2$O$_3$ catalyst. (From [19]).

Treatment	Coordination	R [Å]	N	N$_{corr}$
Reduction 473 K	Rh^0–Rh^0	2.69	5.2	6.3
	Rh^{3+}–O^{2-}	2.05	1	5.9
	Rh^0–O$_s^{2-}$	2.68	1.3	2.6
Reduction 673 K	Rh^0–Rh^0	2.69	6.3	6.3
	Rh^0–O$_s^{2-}$	2.68	1.2	2.5

R = distance; N = number of neighbors; N_{corr} = corrected number of neighbors (see text).
Accuracy: N, ±10–20%; R, ±0.5–1%.

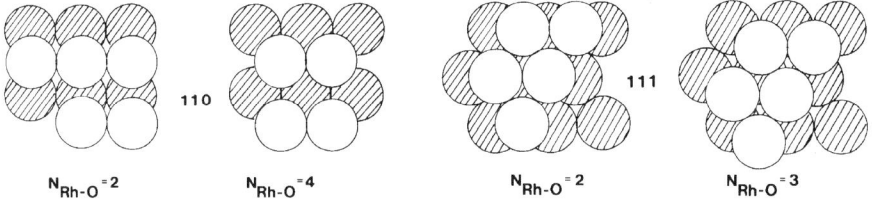

Fig. 9.6 Possible configurations of Rh atoms (open circles) in the metal support interface on γ-Al$_2$O$_3$ (hatched circles), corresponding to Rh–O$_s$ coordination numbers between 2 and 4. (From [20]).

number of O neighbors of *all* Rh atoms in the particle, needs to be corrected. In order to do so, one must assume a certain particle shape. According to Figure 6.15, a Rh–Rh coordination number of 6.3 corresponds to a half-spherical particle with a diameter of about 1.1 nm, while the corrected Rh–O coordination number of atoms in the interface becomes 2.5. A few possible configurations with Rh–O coordination numbers between 2 and 4 are shown in Figure 9.6, assuming epitaxy between corresponding lattice planes of γ-Al$_2$O$_3$ and fcc rhodium metal [20]. Of course, these are ideal situations. It is also possible that rhodium particles attach to defect sites on the support, such that the overall Rh–O coordination averages between 2 and 3.

The EXAFS results have implications for the metal support interaction. The data in Table 9.1 indicate that the main interaction between rhodium and alumina occurs between reduced metal atoms and two to three oxygen ions in the surface of the support at a distance of 0.27 nm. It appears logical, therefore, to attribute the metal support interaction to bonding between oxygen ions of the support and induced dipoles inside the rhodium particle [19]. Although such bonding is weak on a per atom basis, the cumulative bond for the whole particle may be significant.

During the late 1970s, scientists at Exxon discovered that metal particles supported on titania, alumina, ceria and a range of other oxides, lose their ability to chemisorb gases such as H_2 or CO, after reduction at temperatures of about 500 °C. Electron microscopy revealed that the decreased adsorption capacity was not caused by particle sintering as oxidation, followed by reduction at moderate temperatures, restored the adsorption properties of the metal in full. The suppression of adsorption after high-temperature reduction was attributed to a strong metal support interaction (SMSI) [2].

This SMSI in Pt/TiO_2 catalysts was initially explained by an electronic effect: charge transfer from the TiO_2 support to the Pt particles was thought to modify the electronic configuration of platinum towards that of gold, onto which H_2 and CO do not adsorb [21]. Unfortunately, this electronic picture does not explain why a metal such as rhodium exhibits SMSI behavior, as the addition of electrons to rhodium would make it similar to palladium, which activates both H_2 and CO successfully. Photoemission studies, undertaken to observe changes in the electronic structure of metal particles in the SMSI state, led to contradicting conclusions. Some investigators [22] suggested that charge was transferred from the support to the metal, while others [23] concluded that there was no charge transfer at all. The apparent contradiction stems largely from the fact discussed in relation with Figures 9.3 and 3.19, that binding energies and also the density of states of the valence electrons are a function of particle size, and likely also of morphology.

A more likely explanation for the SMSI phenomenon (which today is supported by substantial evidence) is that a layer of support material covers the metal particle in the SMSI state. During reduction at high temperature, TiO_2 reduces, possibly assisted by the noble metal, to a suboxide of composition Ti_4O_7 which, because of its partially unsaturated character, maximizes its contact with the metal by spreading over the particle [24]. As a consequence, all adsorption sites of the metal are blocked. The TEM image in Figure 9.7 of a Rh/TiO_2 catalyst after high-temperature reduction clearly shows the presence of amorphous surface layers on the metal particles. Such layers are absent after reduction at lower temperatures [25].

In addition, Auger and SIMS studies using thin layers of Rh or Pt evaporated onto an oxidized titanium model support provide clear evidence that TiO_x species migrate on top of the metals during heating at temperatures around 500 °C [23]. Interestingly, thermal desorption studies of CO and H_2 from these model catalysts indicated that the metal–adsorbate bond is affected by the TiO_x species on the metal. This effect, however, is not necessarily related to charge transfer between the metal and the oxide, but can also have electrostatic origins, such as the promoter effects described later in this chapter.

We conclude that strong metal support interaction is in fact an incorrect name for a phenomenon which is satisfactorily explained by the blocking of adsorption sites due to the covering of metal particles by mobile oxide species from the support. Additionally, these oxide species may act as promoters in catalytic reactions.

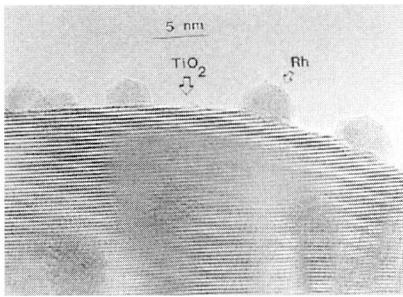
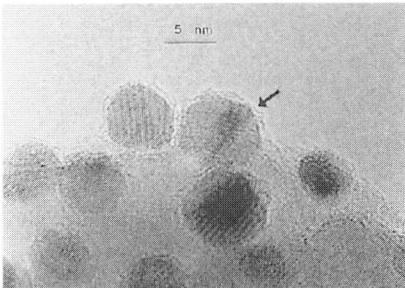

Fig. 9.7 Transmission electron microscopy of rhodium particles on a model titania support after reduction in H_2 at 200 °C (top) and the same catalyst in the SMSI state after reduction at 500 °C (bottom). An amorphous overlayer on the surface of the SMSI catalyst is clearly discerned (arrow). (From [25]).

9.2.4
Disintegration of Rhodium Particles Under CO

The state of a given small particle in a catalyst depends on the composition of the surrounding gas atmosphere, the temperature, and the pressure. A rather extreme illustration is provided by the disintegration of rhodium particles in CO.

Infrared spectra of CO chemisorption on rhodium usually show the well-known peaks for the linear and bridged CO stretching modes at about 2070 and 1900 cm^{-1}, respectively. However, when rhodium particles are well dispersed, two additional bands arise at approximately 2090 and 2030 cm^{-1}. These are characteristic of the symmetric and asymmetric stretching modes of the rhodium gem-dicarbonyl (gem is derived from gemini, meaning twins), in which two CO molecules are bound to a Rh ion with a formal oxidation state of 1+. Coordination chemistry provides an analogue in the complex $[Rh^+(CO)_2Cl]_2$, in which the two rhodium ions are connected by two Cl bridges. Figure 9.8 shows the occurrence of the gem-dicarbonyl species in a silica-supported rhodium catalyst during heating in CO, as reported by Knözinger [26]. The spectrum at 80 K contains the bands of linear CO at 2078 cm^{-1} and bridged CO at 1915 cm^{-1}, as well as peaks associated with physically adsorbed and hydrogen-bonded CO at higher wavenumbers. The latter species disappear, and the dicarbonyl bands develop when the catalyst is brought to room temperature under CO, implying that the reactions leading to dicarbonyl formation form an activated process. The important question is what the state of rhodium is in catalysts for which infrared spectroscopy indicates the presence of the dicarbonyl.

Van't Blik et al. [3] exposed a highly dispersed 0.57 wt% Rh/Al_2O_3 catalyst (H/M = 1.7) to CO at room temperature, and measured a CO uptake of 1.9 molecules of CO per Rh atom. Binding energies for the Rh $3d_{5/2}$ XPS peak increased

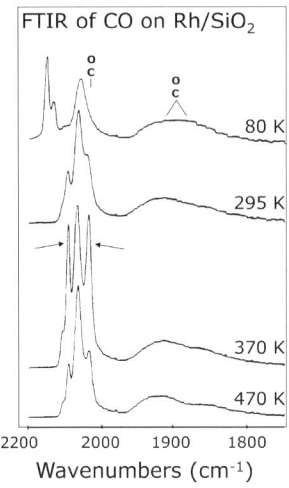

Fig. 9.8 Infrared spectra of CO adsorbed on a well-dispersed Rh/SiO$_2$ catalyst at 80 K, 295 K, 370 K, and 470 K. The spectra show the peaks of linear and bridged CO at 2078 and 1908 cm^{-1}, and the bands of the gem-dicarbonyl species at 2098 and 2030 cm^{-1}, respectively. The bands above 2100 cm^{-1} in the spectrum at 80 K are due to physisorbed and H-bonded CO. (From [26]).

from 307.5 eV for the reduced catalyst under H$_2$ to 308.7 eV for the catalyst under CO. The latter value equals that of the [Rh$^+$(CO)$_2$Cl]$_2$ complex, in which rhodium occurs as a Rh$^+$ ion. The infrared spectrum of the Rh/Al$_2$O$_3$ catalyst under CO showed exclusively the gem-dicarbonyl peaks at 2095 and 2023 cm^{-1}. All results point to the presence of rhodium in Rh$^+$(CO)$_2$ entities. The question must be asked, however, as to how a rhodium *particle* can accommodate so much CO?

Rhodium K-edge EXAFS spectra (see Fig. 9.9) indicate that the structure of the rhodium particles changes entirely after CO chemisorption. The rapid decrease of the EXAFS wiggles with increasing wave vectors is typical of coordination by light elements such as oxygen (see Figs. 6.11 and 6.12). The Fourier transform of the reduced catalyst shows the contribution of rhodium neighbors, and of oxygen ions in the metal support interface. The Fourier transform of the catalyst under CO is entirely different: the contribution from rhodium neighbors has disappeared, and instead contributions arise that are consistent with carbon, oxygen from CO, and a second oxygen from the support. The latter is at 0.21 nm, a distance characteristic of an ionic Rh–O pair, and has a coordination number equal to 3. The conclusion from the EXAFS is that the rhodium particle has disintegrated under CO to mononuclear Rh$^+$(CO)$_2$ complexes that coordinate to three oxygen ions of the support.

The driving force for the disruption of the particles is the strength of the Rh–CO bond, which with its energy of about 145 kJ mol^{-1} is stronger than the 121 kJ mol^{-1} of the Rh–Rh bond in metallic rhodium [26]. With respect to the mechanism of the disintegration, Basu et al. [27] presented infrared evidence that surface hydroxyl groups are involved:

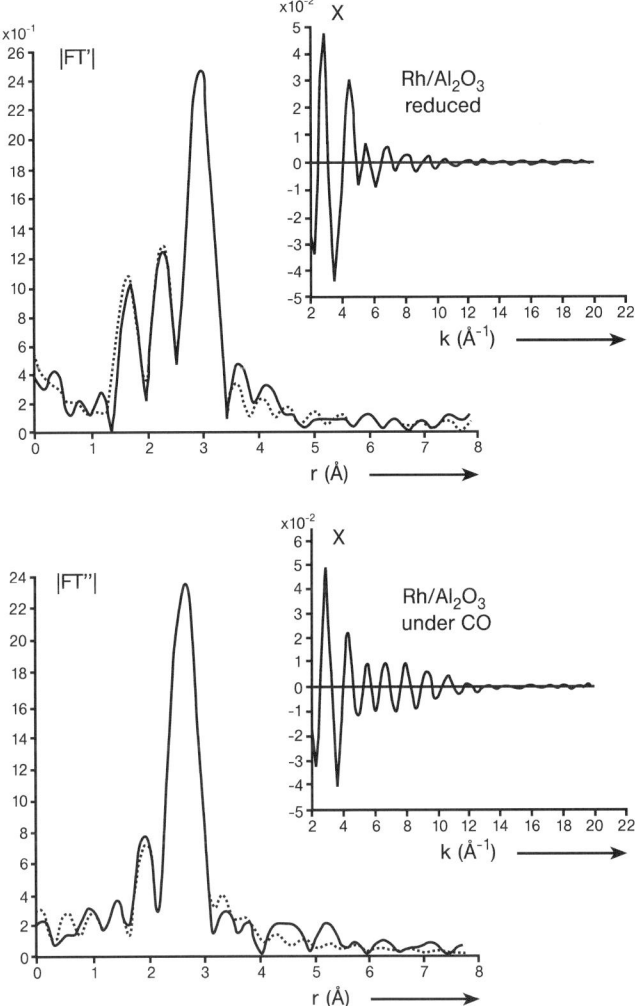

Fig. 9.9 EXAFS spectra and Fourier transforms of a highly dispersed Rh/Al$_2$O$_3$ catalyst after reduction (left) and after exposure to CO at room temperature (right). (Courtesy of H.F.J. van't Blik, Eindhoven).

$$\begin{aligned} &\text{Rh}_n + \text{CO}_g \rightarrow \text{Rh}_{n-1} + [\text{Rh}^\circ\text{CO}]_{\text{ads}} \\ &[\text{Rh}^\circ\text{CO}]_{\text{ads}} + \text{OH}_s^- \rightarrow [\text{O}^{2-}\text{-Rh}^+\text{-CO}]_s + \tfrac{1}{2}\text{H}_{2,g} \\ &[\text{O}^{2-}\text{-Rh}^+\text{-CO}]_s + \text{CO}_g \rightarrow [\text{O}^{2-}\text{-Rh}^+\text{-(CO)}_2]_s \end{aligned} \qquad (9\text{-}2)$$

where Rh$_n$ is a rhodium particle, and the subscripts s, ads, and g represent support, adsorbed, and gas phase, respectively. Thus, disruption of the Rh$_n$ particle is caused by the formation of a mononuclear, neutral rhodium carbonyl, which is subsequently oxidized by a hydroxyl group on the support. The first reaction is thought to proceed on coordinatively highly unsaturated rhodium atoms only, which would explain why the disintegration of rhodium has only been observed on highly dispersed particles. CO adsorption on the surface of larger particles is accompanied by adsorption energies that are significantly smaller than the Rh–CO bond energies of carbonyls, implying that the activation energy for particle disintegration becomes high.

9.2.5
Concluding Remarks

The studies discussed above deal with highly dispersed and therefore well-defined rhodium particles with which fundamental questions on particle shape, chemisorption, and metal support interactions can be addressed. Often, such catalysts are prepared from different starting materials, such as organometallic clusters [28] and carbonyls [29] than are used in industry. Noteworthy are also the studies on evaporated rhodium clusters on planar supports, as have been studied extensively in surface science approaches [5, 30–32].

Practical rhodium catalysts – for example, those used in the three-way catalyst for reduction of NO by CO – are prepared by wet impregnation and have in general larger particle sizes than the catalysts discussed here. In fact, large rhodium particles with diameters in excess of 10 nm are much more active for the NO + CO reaction than the particles we discussed here, due to the large ensembles of Rh surface atoms needed for this reaction [33]. Such particles have also extensively been characterized with spectroscopic techniques and electron microscopy; we mention in particular the studies of Wong and McCabe [34] and Burkhardt and Schmidt [35]. These investigations deal with the materials science of rhodium catalysts that are closer to those used in practice, and which is of great interest from an industrial point of view.

9.3
Alkali Promoters on Metal Surfaces

Potassium is a well-known promoter in the synthesis of ammonia and in Fischer–Tropsch syntheses, where it is thought to assist the dissociation of the reactants, N_2 and CO, respectively [36, 37]. Empirical knowledge concerning the promoting effect of many elements has been available since the development of the iron ammonia synthesis catalyst, for which a few thousand different catalyst formulations have been tested. Recent research into surface science and theoretical chemistry has led to a rather complete understanding of how the promoter functions [38, 39].

Promoters are generally divided in two classes:

- Structural promoters, which help to stabilize certain surface structures of the catalyst, or to prevent sintering. These promoters are not involved in the catalytic reaction itself and have no interaction with the reacting species.
- Chemical promoters which, in contrast, directly influence the reacting species on the surface of the catalyst.

Clearly, alkali promoters fall in the latter category.

In practice, alkali promoters are introduced in the catalyst preparation stage in the form of KOH or K_2CO_3. Because potassium is highly mobile, it is gradually lost and consequently catalysts must often be reactivated after having been used for some time. This can be done by impregnating the catalyst with a solution of a potassium salt, if necessary even inside the reactor. Owing to its mobility, the promoter spreads over the entire catalyst bed. For surface science studies, the potassium-promoted catalyst can be modeled by evaporating small amounts of potassium onto the surface of a single crystal. This type of experiment will be described in the following section.

The fact that evaporated potassium arrives at the surface as a neutral atom (whereas in real life it is applied as KOH) is not a real drawback, because atomically dispersed potassium is almost a K^+ ion. The reason for this is that alkali metals have a low ionization potential (see Table A.3), and consequently they tend to charge positively on many metal surfaces, as explained in the Appendix. A density-of-state calculation of a potassium atom adsorbed onto the model metal jellium (see Appendix) reveals that the 4s orbital of adsorbed K, occupied with one electron in the free atom, falls largely above the Fermi level of the metal, such that it is about 80% empty. Thus, adsorbed potassium is present as $K^{\delta+}$ with δ close to 1 [40]. Calculations with a more realistic substrate such as nickel show a similar result. The K 4s orbital shifts largely above the Fermi level of the substrate and potassium becomes positive [41]. The charges of $K^{\delta+}$ on several metals are listed in Table 9.2.

Table 9.2 Substrate work function, potassium dipole moment, maximum work function change and charge transfer for potassium adsorbed on transition metal surfaces.

	φ [eV][a]	μ_o [D][b]	$\Delta\varphi_{max}$ [eV][b]	δ^+[b]
Ni (100)	5.30	6.7	−3.4	0.58
Ru (001)	5.52	7.1	−3.9	0.67
Rh (111)[c]	5.60	7.8	−3.8	0.75
Pt (111)	6.40	9.4	−4.6	0.86

[a] From [42]. [b] From [38]. [c] From [43].

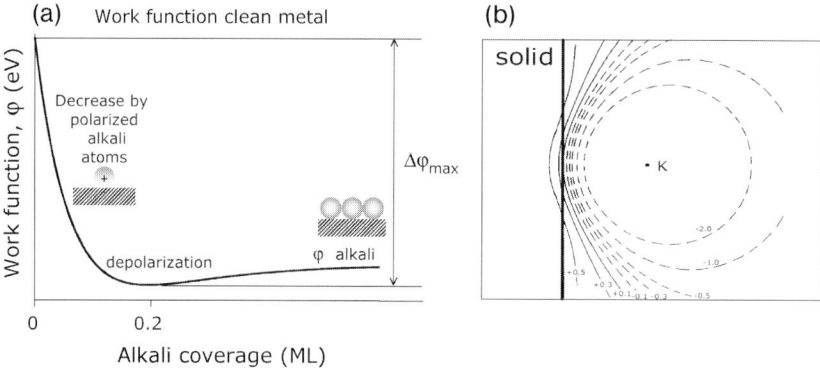

Fig. 9.10 (a) Work function of alkali-promoted metals as a function of alkali coverage; see also Table 9.2. (b) Electrostatic potential around a single alkali atom adsorbed on jellium. The effective local work function at each position is the sum of the substrate work function and the value of the electrostatic potential in the figure. (From [44]).

Experimental confirmation that alkali adsorbates form positive ions on metallic substrates was obtained from work function measurements. The adsorption of alkali metals produces a steep decrease in the macroscopic work function, until a minimum is found at potassium coverages of around 15–20% (expressed as potassium atoms per metal surface atom). At higher coverages, the work function increases somewhat to reach an ultimate value of the alkali metal (see Fig. 9.10a and Table 9.2). The interpretation of Figure 9.10a is that at low coverages the potassium forms a positive ion which is screened by electrons inside the metal and consequently forms a dipole with the positive end directed outwards. Figure 9.10b illustrates the electrostatic potential around such a potassium atom on jellium [44]. The dipole counteracts the dipole layer of the metal that constitutes the surface contribution to the work function of the substrate. At high coverages, potassium becomes metallic and hence the work function of potassium metal can be measured. At about the minimum is a regime where potassium ions come closer to each other than they would prefer, and this leads to mutual depolarization.

The slope of the work function curve near coverage zero enables one to calculate the dipole moment associated with a single adatom. The values are between 6 and 10 Debye, and are of a similar magnitude to the dipole moments of alkali halides such as KBr and NaCl. (Note that some authors quote the dipole moment as twice the value derived from the work function decrease, the argument for doing so being developed from the way in which the negative charge is distributed inside the substrate; the factor of two, however, is by no means well established.) A calculation of the electrostatic potential around the potassium atom on jellium by Lang et al. [44] illustrates the origin of the dipole: the dashed and solid lines in Figure 9.10b represent places where electron density is decreased and increased, respectively, as a result of potassium adsorption. Clearly, the positive potassium is

9.3 Alkali Promoters on Metal Surfaces

screened by electron density just below the surface of the substrate. The values in the plot should be added to the work function of the substrate in order to obtain the local work function on any point in space. This is the work function that an adsorbate would feel at that position.

The electrostatic potential in Figure 9.10b is that of a single potassium atom. It suggests that the promoter effect of the potassium atom is highly local and limited to the adjacent adsorption site only. But, what might happen if the surface coverage of the alkali were to become higher?

The positively charged alkali atoms repel each other, and as a result the atoms spread over the surface. The mutual repulsion leads to ordered overlayers at coverages above about 20%, as has been revealed by low-energy electron diffraction (LEED) studies [38, 39]. For instance, potassium orders in the p(2 × 2) structure on Ni(100) at coverages around 25%, and in the ($\sqrt{3} \times \sqrt{3}$)R30° on the hexagonal Pt(111) and Ru(001) surfaces at coverages around 33% (both structures are shown in Fig. A.3). In general, the alkali atoms prefer to bind in hollow sites, although they may shift to top positions when gases are co-adsorbed [46, 47]. For catalysis, however, we are interested in promoter coverages of a few percent only.

Janssens et al. [43, 45] utilized the photoemission of adsorbed noble gases to measure the electrostatic surface potential on the potassium-promoted (111) surface of rhodium, to estimate the range that is influenced by the promoter. As explained in Chapter 3, UPS of adsorbed Xe measures the local work function, or, equivalently, the electrostatic potential of adsorption sites. The idea of using Kr and Ar in addition to Xe was that, by using probe atoms of different size, one could vary the distance between potassium and the noble gas atom. Provided that the interpretation in terms of Eq. (3-13) is permitted (and this is a point that the authors checked [43]), one thus obtains information about the variation of the electrostatic potential around potassium promoter atoms.

Figure 9.11 shows the binding energy shifts as a function of the estimated distance between the adsorption site and the nearest potassium, along the line between two potassium atoms. The distance between the noble gas and the potassium has been estimated from van der Waals radii of the noble gases and the ionic radius of potassium. The general picture that emerges from Figure 9.11 is that the surface potential changes steeply in the immediate vicinity of the potassium atom, while it becomes more or less constant at distances greater than about 0.5 nm. However, this value is significantly lower than on clean Rh(111), and depends on the potassium coverage: the long-range effect of adsorbed potassium corresponds to a lowering of the local work function of about 0.4 eV for a potassium coverage of 2.7 atom%, and of about 1.0 eV for 5 atom% on the surface.

This long-range effect of the promoter is a result of the cumulative electrostatic effect of all potassium atoms on the surface. Figure 9.12 illustrates the result of a theoretical calculation of the electrostatic potential caused by an ordered network of dipoles, corresponding to a potassium coverage of 2.7% and an average distance between the promoter atoms of 1.6 nm. The value of the dipole strength needed to obtain agreement with the measured points equals 10 Debye, which is

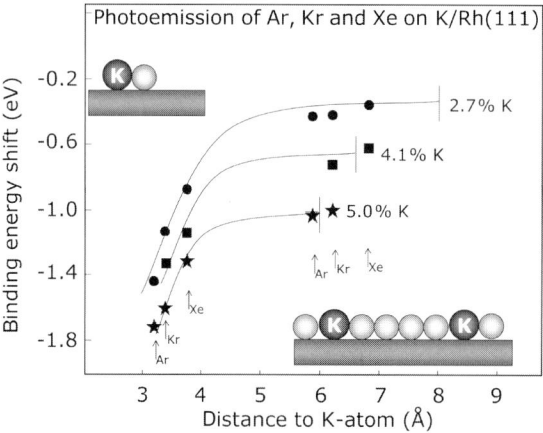

Fig. 9.11 Promoter-induced binding energy shifts of Ar, Kr and Xe photoemission peaks with respect to adsorption on the clean metal as a function of the distance of the adsorption site to the nearest potassium atom on a potassium-promoted Rh(111) surface. These curves reflect the variation of the surface potential (or local work function) around an adsorbed potassium atom. Note the strong and distance-dependent local work function at short distances and the constant local work function, which is lower than that of clean Rh(111) at larger distances from potassium. The lowering at larger distances depends on the potassium coverage. The averaged distances between the potassium atoms are 1.61, 1.32, and 1.20 nm for coverages of 2.7, 4.1, and 5.0%, respectively; the vertical lines mark the half-way distances. Lines are drawn as a guide to the eye. (Adapted from [43]).

quite reasonable in comparison to the values reported in Table 9.2. This calculation confirms that both the short- and long-range promoter effects of potassium have a purely electrostatic origin. Effects through the substrate, associated with charge donated by potassium to the substrate metal, do not have to be invoked.

The catalytic significance of Figure 9.12 is that it represents the differences in the effective work functions that a molecule experiences upon adsorption at different positions on the surface. As explained in the Appendix, a low work function of the substrate enhances the capability of the substrate to donate electrons into empty chemisorption orbitals of the adsorbate. If such an orbital is antibonding with respect to an intramolecular bond of the adsorbed molecule, the latter is weakened due to a higher electron occupation.

Weakening of the intramolecular CO bond by potassium has been observed in vibrational spectra of CO on Ni [48], Ru [49], Rh [50], and Pt [51]. Figure 9.13 shows a series of electron energy loss spectroscopy (EELS) spectra of CO adsorbed on a Pt(111) surface promoted with 9% potassium, as reported by Pirug and Bonzel [51]. In order to fully appreciate these results, we should know that CO on a clean Pt(111) surface yields two CO adsorption states characterized by CO stretch frequencies of 2120 and 1875 cm^{-1}, belonging to linear and bridge-bonded CO, respectively.

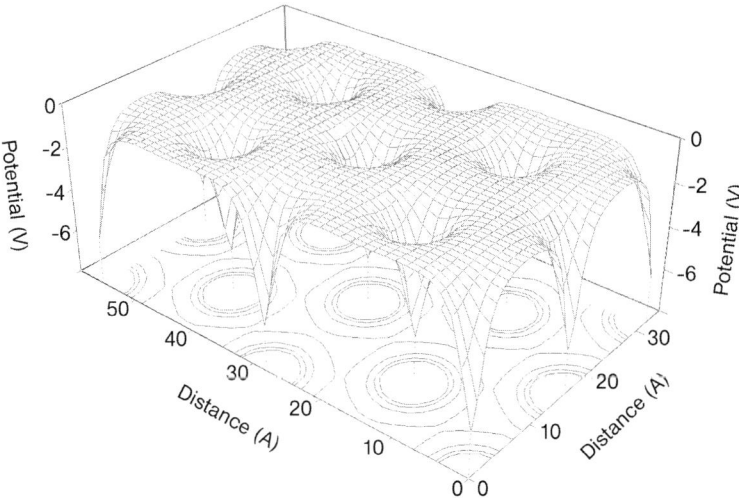

Fig. 9.12 Three-dimensional map of the calculated electrostatic potential at 0.25 nm above the symmetry plane in a hexagonally ordered network of dipoles with a dipole–dipole distance of 1.61 nm and a dipole moment of 10 Debye. The dipoles are positioned at the minima. Note that the potential is lowered at every position on the surface. Equipotential lines for −1.05, −0.84, −0.63, and −0.42 V are indicated in the bottom plane. The contours are circular at short distances from a potassium atom, indicating that at these sites the nearest potassium atom largely dominates the potential. The equipotential line for −0.42 V, however, has hexagonal symmetry due to the influence of the dipoles further away. (From [45]).

Figure 9.13 shows that potassium affects in particular the first adsorbed CO molecules to a great extent. At low CO coverages (Fig. 9.13, bottom spectrum) only one peak for the CO stretch vibration is seen at a remarkably low frequency of 1550 cm^{-1}, assigned to CO adjacent to a potassium atom. At higher coverages, other CO adsorption states develop until, at saturation, the multiply-bonded CO states coexist with linear CO, with a slightly lower frequency (2100 cm^{-1}) than on the unpromoted surface. The results of all vibrational studies of CO on promoted Group VIII metals indicate that a promoted surface contains several adsorption states and that the CO stretch frequencies are lower than on the unpromoted surface, while the magnitude of the downward shift increases with increasing potassium loading [39].

The decreased CO stretch frequencies are easily rationalized as the result of adsorption on sites with a lower work function, as explained in the Appendix. The effect of a lower work function is that all orbitals of CO shift downward with respect to the Fermi level of the substrate. This shift of the occupied CO levels to higher binding energy has been observed in UPS spectra (see Fig. 3.21), while the shift of the unoccupied part of the $2\pi^*$-derived chemisorption orbital

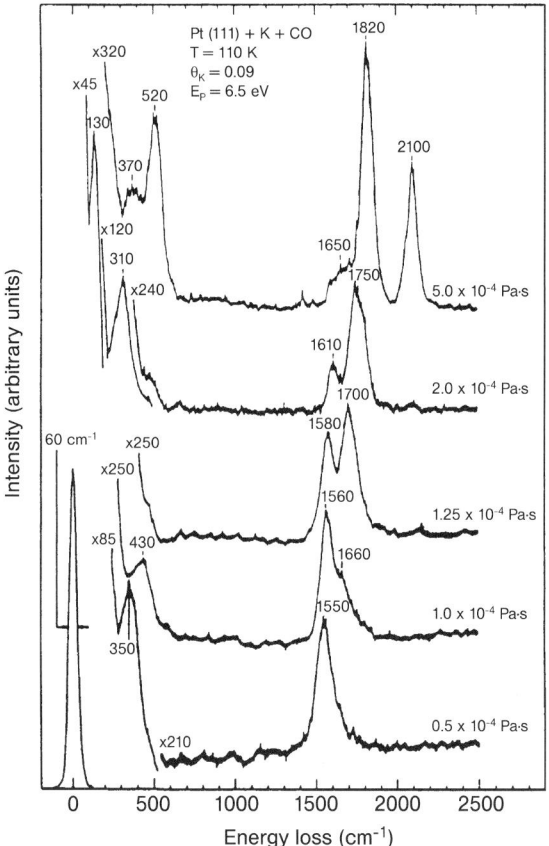

Fig. 9.13 EELS spectra of CO adsorbed on Pt(111) promoted with 9% potassium for increasing CO coverages. The influence of the promoter is most clearly seen in the spectra at low CO coverages. The CO stretch frequencies of CO adsorbed on clean Pt(111) are about 2120 and 1875 cm^{-1}. (From [51]).

has been observed in inverse photoemission [38, 39, 52]. The overall effect is that the bond between the metal and the CO becomes stronger, while at the same time the intramolecular CO bond is weakened.

The fact that potassium enhances the adsorption bond of CO with the substrate is revealed by thermal desorption, in a shift of the CO desorption peak to higher temperatures. Figure 9.14 shows thermal desorption spectra (TDS) of CO from potassium-promoted nickel, as reported by Whitman and Ho [53]. On clean Ni(110), CO desorbs in a first-order process with a peak maximum temperature of about 440 K. When potassium is present, CO desorbs out of two states,

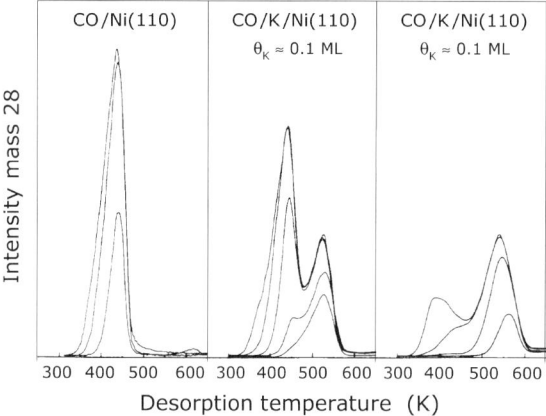

Fig. 9.14 Thermal desorption spectra of CO from clean (left) and potassium-promoted Ni(110) (center and right) measured at a heating rate of 13 K s^{-1}. The spectra exhibit two desorption states for CO on promoted surfaces, and indicate that CO binds stronger to sites adjacent to potassium. (From [53]).

with peak maxima at about 530 and 430 K. At low CO coverage, only the high-temperature peak is seen, while at saturation coverages of CO the intensity of the high-temperature peak increases with increasing potassium coverage. This again indicates that the promoted surface is heterogeneous and contains at least two adsorption sites – one adjacent to potassium, where CO binds stronger than on clean Ni(110), and one further away from the promoter, which resembles sites on clean Ni(110). CO fills these sites sequentially; as shown in Figure 9.14, the sites adjacent to potassium are populated first. The TDS results thus confirm that the dominant effect of potassium is short-ranged. The technique is not sensitive enough to reveal evidence for a promoting effect on longer distances. Indeed, it appears that vibrational spectroscopies such as EELS and reflection absorption infrared spectroscopy (RAIRS) are the only techniques that have been able to confirm the existence of a long-range effect of potassium on adsorbed CO [54].

In conclusion, UPS of adsorbed noble gases indicates that potassium adsorption on metal surfaces results in a dominant lowering of the surface potential (or local work function) on sites adjacent to a potassium atom, and a smaller (but still significant) lowering of the potential on sites located further away. The long-range effect is caused by a cumulative effect of all potassium atoms on the surface. The effect of potassium on a reacting molecule such as CO is twofold: (1) the adsorption bond between CO and the substrate becomes stronger, as follows from TDS; and (2) the intramolecular CO bond is weakened, as observed by vibrational spectroscopy.

9.4
Cobalt–Molybdenum Sulfide Hydrodesulfurization Catalysts

Catalysts based on molybdenum disulfide, MoS_2, and cobalt or nickel as promoters are used for the hydrodesulfurization (HDS) and hydrodenitrogenation (HDN) of heavy oil fractions [55, 56]. The catalyst, containing at least five elements (Mo, S, Co or Ni, as well as O and Al or Si of the support), is rather complex and represents a real challenge for the spectroscopist. Nevertheless, owing largely to research conducted during the past 20 years, the sulfided $Co-Mo/Al_2O_3$ system is one of the few industrial catalysts for which we know the structure in almost atomic detail [56, 57].

Let us take 1978 as the starting point, when Massoth [58] published an extensive review of what was known about the structure of HDS catalysts. Characterization was essentially based on techniques such as X-ray diffraction, electron microscopy, photoelectron spectroscopy, electron spin resonance and magnetic methods. Massoth was rather unhappy with the state of affairs in 1978. He was struck by the "... diversity and apparent contradictions of results and interpretations ... It almost seems as though everyone is working with a different catalyst."

Indeed, several models existed for the active $Co-Mo/Al_2O_3$ catalyst. There were, for example, the monolayer models [59], which assumed that the HDS activity was associated with highly dispersed molybdenum oxy-sulfides bound to the support. Other models appreciated the layer structure of MoS_2 and assumed that HDS activity was associated with Co species intercalated between or near the edges of the MoS_2 slabs [60, 61]. A completely different view was taken in the contact synergy model. Here, the active sites were thought to be at the interface of the well-known bulk sulfides MoS_2 and Co_9S_8 [62]. None of these structures was proven, however. In the following section we describe how a combination of methods such as Mössbauer spectroscopy, EXAFS, XPS, and infrared spectroscopy has led to a picture in which the active site of a sulfided $Co-Mo/Al_2O_3$ catalyst is known in almost atomic detail.

9.4.1
Sulfidation of Oxidic Catalysts

HDS catalysts are usually prepared by impregnating the support with aqueous solutions of ammonium heptamolybdate and cobalt or nickel nitrate. Some of the molybdenum phases that are formed on silica and alumina supports have been discussed in connection with the Raman spectra of Figure 8.13. The catalyst is brought in the sulfided state by treatment in a mixture of H_2S in H_2, or in the sulfur-containing feed. The process is conveniently studied with temperature-programmed sulfidation (TPS) (see Fig. 2.6). The TPS patterns indicate that supported molybdenum oxides begin to consume H_2S already at room temperature, by exchanging sulfur for oxygen. At about 225 °C, a remarkable uptake of H_2 takes place, accompanied by the formation of H_2S, indicating that the catalyst contains excess sulfur, which is removed through hydrogenation [63]. At higher

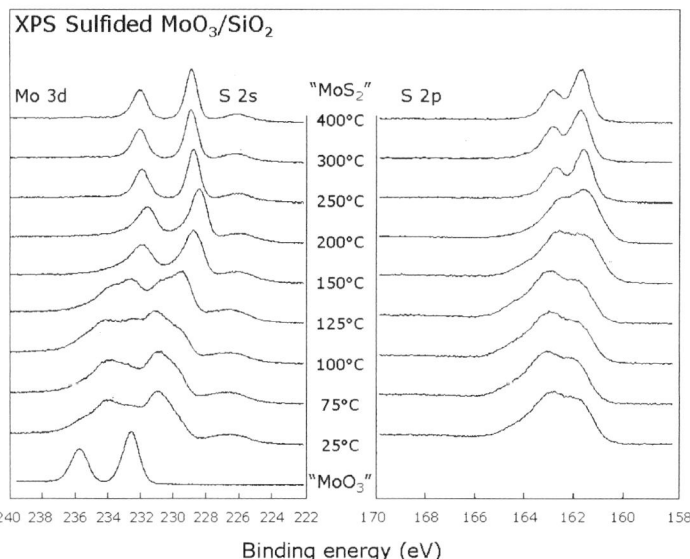

Fig. 9.15 XPS spectra of a $MoO_3/SiO_2/Si(100)$ model catalyst after sulfidation in $H_2S + H_2$ at the indicated temperatures show that considerable uptake of sulfur occurs already at room temperature, accompanied by reduction of Mo^{6+} to Mo^{4+}. The S 2p spectra show the presence of at least two sulfur species after low-temperature sulfidation. (From [64]).

temperatures, the catalyst continues to take up sulfur until sulfidation is complete around 900 K. The total sulfur consumption then corresponds to 1.9 sulfur per molybdenum, which is close to the stoichiometry of MoS_2.

A combination of spectroscopies has revealed the states of molybdenum and sulfur present during sulfidation. The XPS spectra in Figure 9.15 show how MoO_3 particles supported on a thin-film oxide support convert to the sulfidic state in a H_2S/H_2 mixture [64]. Exposure of MoO_3 to H_2S at temperatures between 20 and 100 °C converts almost all Mo^{6+} to Mo^{5+}, which is thought to be present in the form of oxysulfides [64, 65]. TPS (Fig. 2.6) reveals the evolution of water from the surface, indicating that sulfur from H_2S exchanges for oxygen from the catalyst particles [63]. According to Rutherford backscattering (RBS; see Fig. 4.16), the S/Mo atomic ratio after sulfidation at low temperature is around 1, confirming that sulfur uptake is indeed significant at low temperatures [66]. This sulfur is present in two forms. In addition to the normal S^{2-} ion ($E_b = 161$ eV), the disulfide, S_2^{2-}, ($E_b = 162$ eV) also appears [64]. The latter is only observed after sulfidation at relatively low temperatures, and forms a likely candidate for the excess sulfur that gives rise to the H_2S evolution peaks seen in TPS patterns [63]. The SIMS spectra (see Fig. 4.8) reveal that molybdenum ions in catalysts after low-temperature sulfidation still have oxygen in the first coordination shell, which points to the presence of oxysulfides in the surface region [66].

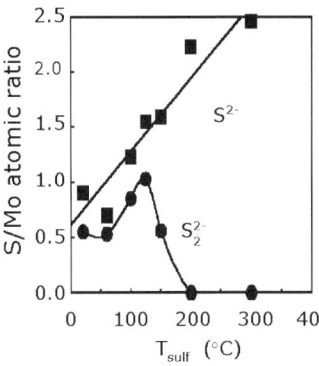

Fig. 9.16 Relative concentrations of sulfide, S^{2-}, and the intermediate disulfide species, S_2^{2-}, obtained from combining XPS and Rutherford backscattering after sulfidation at different temperatures. (From [66]).

Hence, the first steps in the sulfidation of MoO_3 include an exchange of lattice oxygen for sulfur. Weber et al. [65] have shown that terminal O-atoms of the MoO_3 lattice are the most reactive entities, and therefore are the most likely candidates to exchange with sulfur from H_2S. However, the XPS spectra tell us that a configuration with Mo^{6+} and S^{2-} is unstable, and that molybdenum ions with direct sulfur neighbors rearrange in an internal redox reaction to a pair of Mo^{5+} ions bridged by a S_2^{2-} group. This reaction pulls the molybdenum ions somewhat towards each other, which causes the lattice to break up and new terminal Mo=O entities to be formed (as confirmed by infrared spectroscopy [65]), whereupon the next H_2S molecules can react.

At sulfidation temperatures above 125 °C, the Mo 3d XPS spectra shift to those characteristic for Mo^{4+}, whereas at temperatures of 250 °C and higher the Mo and S XPS spectra become equal to those of MoS_2. The S_2^{2-} groups gradually disappear to leave S^{2-} as the only state of sulfur; the S/Mo atomic ratio as determined by RBS reaches values between 2 and 2.5, as expected for MoS_2 (see Fig. 9.16).

Quick EXAFS (QEXAFS), carried out in temperature-programmed fashion, reveals the average surroundings of Mo ions during sulfidation of a high-surface area silica-supported NiMo catalyst [67]. Figure 9.17 shows the Fourier transform of the Mo K-edge EXAFS spectra, which were measured at a heating rate of 6 °C per minute. In brief, the measurements indicate four different temperature regimes. In the first, below 100 °C, molybdenum is predominantly surrounded by oxygen neighbors, as seen by contributions from Mo–O shells between 0.5 and 1.9 Å, and a Mo–Mo shell at 3 Å (not phase-corrected), in the Fourier transform of the EXAFS signal. In the second region, between 100 and 225 °C, the longest Mo–O and the Mo–Mo shell disappear, while a new contribution due to sulfur neighbors at about 2 Å (uncorrected) appears, indicative of oxysulfides. In the third region, between 225 and 280 °C, the spectrum reveals only Mo–S and Mo–Mo contributions, attributed to sulfur-rich phases such as MoS_3, which has also been observed in other EXAFS studies [68]. The spectrum characteristic of

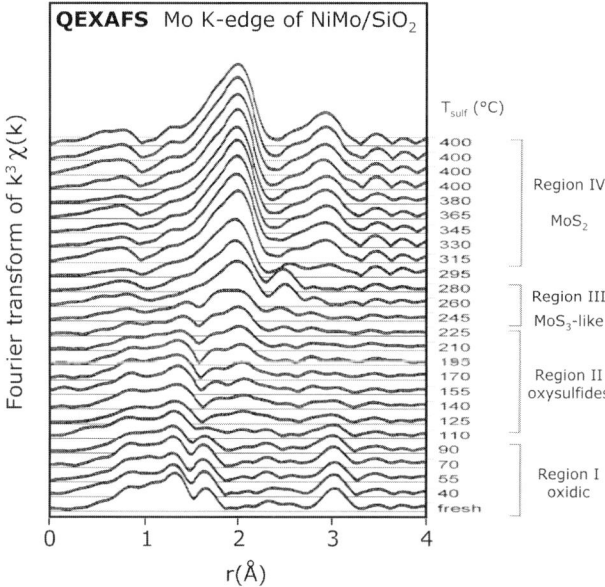

Fig. 9.17 Quick EXAFS measurements show the effect of temperature-programmed sulfidation on the Mo K-edge in Ni–Mo/SiO$_2$ catalysts. (Adapted from [67]).

MoS$_2$ is observed after sulfidation at temperatures above 300 °C. These results are easily reconciled with those from XPS and RBS described before, be it that the temperatures at which the changes occur are higher in the QEXAFS study, which is in fact expected for a temperature-programmed measurement. In addition, the QEXAFS were made on high-surface area catalysts, where diffusion limitations are likely, whereas the XPS studies concern planar, thin films, in which virtually all material is in contact with the gas phase.

The sulfidation mechanism presented by Weber et al. [65] describes the conversion from MoO$_3$ to MoS$_2$ in molecular detail. We have discussed already how sulfur exchanges with oxygen to form bridging disulfides. These, and also the terminal S$_2^{2-}$ groups which may form at defects of the MoO$_3$ lattice, are responsible for the high sulfur content of catalysts in the intermediate stages of the sulfidation process. Locally, the catalyst contains MoS$_3$-like structures [69]. At higher temperatures, however, sulfur is removed by hydrogen according to the reaction S$_2^{2-}$ + H$_2$ → 2 (S–H)$^-$ → S^{2-} + H$_2$S. The intermediate S–H groups are rather unstable and react to H$_2$S, which is indeed observed as a product in TPS measurements (see Fig. 2.6) [63].

The sulfidation mechanisms of cobalt- or nickel-promoted molybdenum catalysts are not yet known in the same detail as that of MoO$_3$, but they are not expected to be much different, as TPS patterns of Co–Mo/Al$_2$O$_3$ and Mo/Al$_2$O$_3$ are rather similar [63]. However, interactions of the promoter elements with the alumina support play an important role in the ease with which Ni and Co convert to the sulfidic state. We will return to this point after having discussed the active phase for the hydrodesulfurization reaction in more detail.

9.4.2
Structure of Sulfided Catalysts

Abundant evidence exists that sulfided Co–Mo catalysts contain molybdenum in the form of MoS$_2$. X-ray diffraction of used catalysts clearly shows the characteristic reflections of MoS$_2$ [70]. EXAFS measurements have also revealed the characteristic Mo–S and Mo–Mo distances of MoS$_2$ [71–74]. As mentioned above, both TPS and RBS measurements indicate a stoichiometry of sulfur to molybdenum close to that of MoS$_2$. Moreover, XPS spectra also yield binding energies of Mo and S which are both in agreement with assignment to MoS$_2$ [66, 75].

Molybdenum disulfide has a layered structure which resembles a sandwich consisting of a layer of Mo^{4+} between two layers of S^{2-} ions (Fig. 9.18). The sulfur ions form trigonal prisms, and half of these prisms contain a molybdenum ion in the middle. The chemical reactivity of MoS$_2$ is associated with the edges of the sandwich, whereas the basal planes are much less reactive. The edges form the sites where gases are adsorbed [76], and it seems logical to expect that the edges are also the seat of catalytic activity.

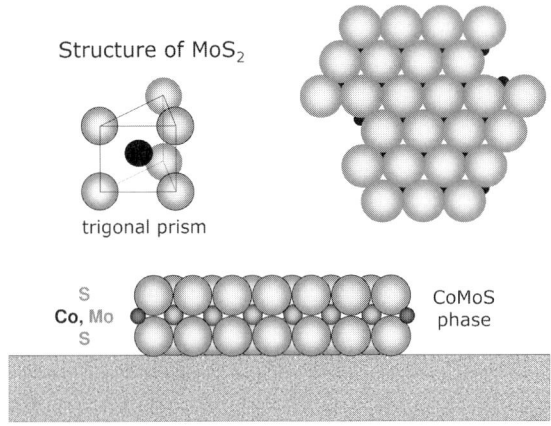

Fig. 9.18 MoS$_2$ has a layer structure; each slab of MoS$_2$ is a sandwich of Mo^{4+} ions between two layers of S^{2-} ions. Two defect sites, exposing one and two Mo ions are indicated. The bottom structure shows the Co–Mo–S phase, with cobalt atoms at the edges of a MoS$_2$ particle.

For example, a simple (possibly too naïve) picture of the HDS reaction would be that a sulfur vacancy at the edge of a MoS_2 slab is the active site. A sulfur-containing hydrocarbon such as thiophene adsorbs with its sulfur atom towards the exposed molybdenum. Next, the molecule becomes hydrogenated, the two C–S bonds in thiophene break, the sulfur-free hydrocarbon desorbs, and the catalytic site is regenerated by the removal of sulfur by hydrogen. The reaction mechanisms involved have been discussed by Prins et al. [57].

Cobalt–molybdenum catalysts are in general much more active for HDS than single molybdenum catalysts. Thus, it is essential to investigate the state of cobalt in the sulfided Co–Mo/Al_2O_3 catalyst.

The most direct information on the state of cobalt has been obtained with Mössbauer spectroscopy, applied in the emission mode. As explained in Chapter 5, such experiments are conducted with catalysts that contain the radioactive isotope ^{57}Co as the source, and a moving single-line absorber. One major advantage of this method is that the Co–Mo catalyst can be investigated under *in-situ* conditions, and that the spectrum of cobalt can be correlated to the activity of the catalyst. Care must be taken, however, because the Mössbauer spectrum obtained is, strictly speaking, not that of cobalt but rather that of its decay product, iron. The safest route to take, therefore, is to compare the spectra of the Co–Mo catalysts with those of model compounds for which the state of cobalt is known. This was the approach taken by Topsøe and co-workers [77, 78], whose results provided a breakthrough in our ideas relating to sulfided Co–Mo/Al_2O_3 catalysts.

Figure 9.19 shows the Mössbauer spectra of sulfided Co–Mo catalysts. The Mössbauer spectra of the sulfided Co–Mo/Al_2O_3 catalysts contain essentially three contributions, as indicated by bar diagrams in Figure 9.19:

- catalytically insignificant Co ions, located in the Al_2O_3 lattice;
- the bulk sulfide Co_9S_8, dominant in catalysts with a high cobalt content; and
- a formerly unknown state (labeled Co–Mo–S in Fig. 9.19), which is most evident in sulfided Co–Mo catalysts of low cobalt content.

The interesting point about this Co–Mo–S phase is that its presence correlates with the catalytic activity for the desulfurization reaction, as shown in Figure 9.20. Thus, this Co–Mo–S phase is either active itself, or is at least closely associated with the active site. Figure 9.20 represents one of the few examples in the literature where a spectroscopic signal can clearly be associated with a catalytic property.

Yet, the structure of this Co–Mo–S phase remains to be determined. A model system, prepared by impregnating a MoS_2 crystal with a dilute solution of cobalt ions, such that the model contains only ppm-levels of cobalt, appears to have the same Mössbauer spectrum as the Co–Mo–S phase. It has the same isomer shift (characteristic of the oxidation state), recoil-free fraction (characteristic of lattice vibrations), and almost the same quadrupole splitting (characteristic of symmetry) at all temperatures between 4 and 600 K [78]. Thus, the cobalt species in the

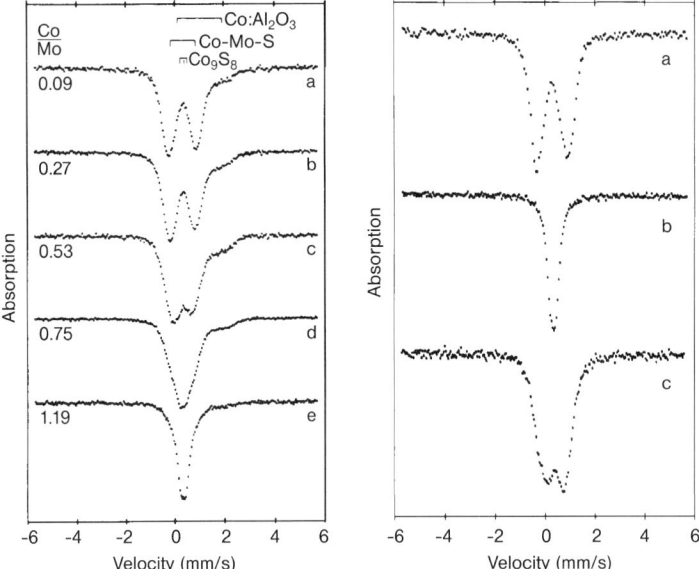

Fig. 9.19 *In-situ* Mössbauer emission spectra of ^{57}Co in a series of sulfided Co–Mo/Al$_2$O$_3$ catalysts (left) and MoS$_2$ particles doped with different amounts of cobalt (right), corresponding to Co/Mo ratios of: (a) about 3; (b) 0.05; and (c) 0.25 parts per million. The Co–Mo–S phase, active in the HDS reaction, has a spectrum unlike that of any bulk cobalt sulfide, and is most clearly observed in the spectra of Co–Mo/Al$_2$O$_3$ catalysts of low Co content, and in the MoS$_2$ particles doped with ppms of cobalt. (From [77, 78]).

ppm Co/MoS$_2$ system provides a convenient model for the active site in a Co–Mo HDS catalyst.

As the chemical reactivity of MoS$_2$ is associated with edges, we may already expect that this Co–Mo–S phase is to be found at, or near, the edges, and indeed scanning Auger spectroscopy and electron microscopy have each provided evidence that this is the case.

By taking the same approach as described above, Chianelli et al. [79] doped a single crystal of MoS$_2$ with cobalt and equilibrated it at high temperature. Chemical maps made with scanning Auger electron spectroscopy at a spatial resolution of 2 μm, reveal that the basal planes of the MoS$_2$ crystal contain Mo and S only, but that the edge regions contain also cobalt, carbon, and oxygen (Fig. 9.21). Similar experiments with electron microscopy [80] reached the same conclusion: under conditions that the model catalyst contains cobalt in the Co–Mo–S phase, the cobalt atoms are located on or near the edges.

In order to investigate the precise structure of the Co–Mo–S phase, techniques are required which provide information on the atomic scale. The infrared spectra

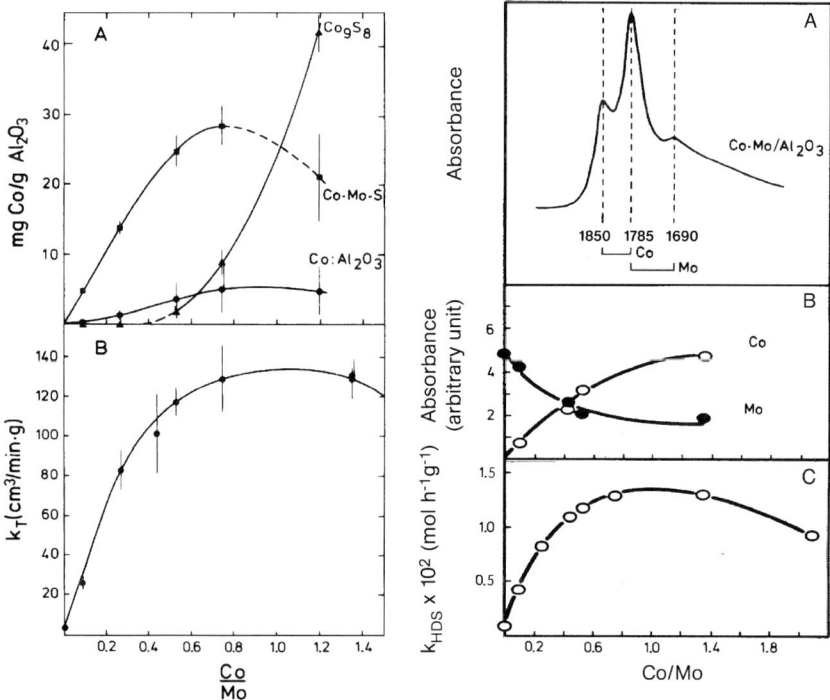

Fig. 9.20 Correlation between the activity of a series of Co–Mo/Al$_2$O$_3$ catalysts for the HDS reaction, expressed in the reaction rate constant k_T, and the cobalt phases observed in Mössbauer spectra (left) as well as the NO adsorption sites probed with infrared spectra of adsorbed NO (right). (Left figure from [77]; right figure adapted from [56, 81]).

of NO adsorbed onto sulfided Co–Mo/Al$_2$O$_3$ catalysts consist of separate signals for the stretch vibration of NO on molybdenum and NO on cobalt sites (see Fig. 8.7) [81]. This method can be used to titrate the number of Co and Mo sites in sulfided Co–Mo/Al$_2$O$_3$ catalysts. Figure 9.20 illustrates how the intensities of the NO/Mo and NO/Co signals compare with the catalytic activity. A clear correlation exists between the number of cobalt atoms accessible to NO and the HDS activity of the catalyst. At the same time, the number of molybdenum atoms which are accessible to NO decreases, indicating that cobalt inhibits the adsorption of NO onto the molybdenum. As NO is known to adsorb on the edges of MoS$_2$, the data in Figure 9.20 strongly suggest that cobalt is also located on the edges, in locations where it blocks the molybdenum sites. Hence, the Co–Mo–S phase consists of MoS$_2$ particles in which cobalt atoms decorate the edge positions. Similar results were obtained for Ni–Mo/Al$_2$O$_3$ catalysts [82].

In order to draw a crystallographic picture of the Co–Mo–S phase, precise data are required regarding the location of the cobalt atom in relation to the molybdenum and sulfur atoms. For this, EXAFS is the indicated technique.

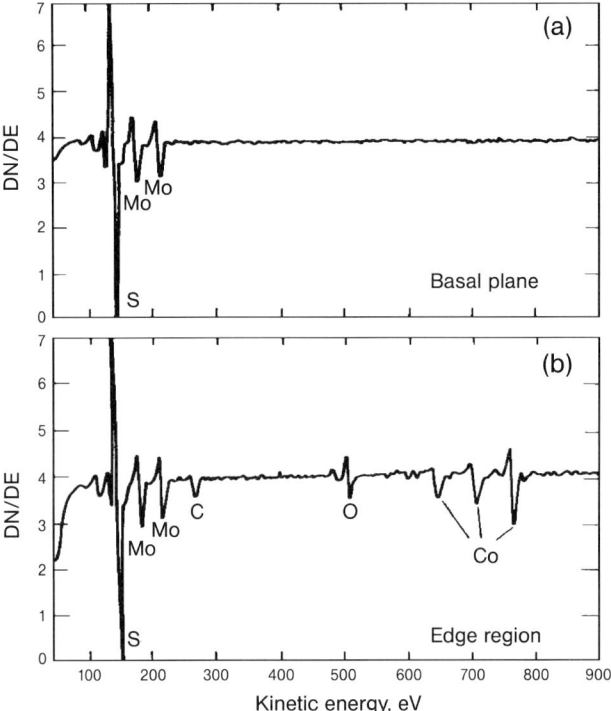

Fig. 9.21 Auger spectra of cobalt-doped MoS$_2$ crystals taken at: (a) the basal plane and (b) the edge regions. (From [79]).

Several groups [71–74, 83] have reported EXAFS studies on sulfided cobalt–molybdenum catalysts. Figure 9.22 shows the Fourier transforms of MoS$_2$ and of sulfided molybdenum and cobalt–molybdenum catalysts supported on carbon, as reported by Bouwens et al. [74, 83]. In bulk MoS$_2$, each molybdenum ion has six sulfur neighbors at a distance of 0.241 nm, and six molybdenum neighbors at 0.316 nm. However, the Mo K-edge EXAFS of sulfided catalysts reveals Mo–Mo coordination numbers significantly smaller than 6 [71–74], while some authors even report Mo–S coordination numbers below 6 [72, 74]. In order to attribute coordination numbers to a certain particle size, a molecular modeling kit is required to build MoS$_2$ structures of different sizes. Next, the neighbors are counted and average coordination numbers calculated. Over all of these studies, average MoS$_2$ crystallite sizes of 1.5 to 3 nm have been found, although of course such small crystallites will expose a relatively large number of catalytically active edge sites.

The EXAFS signal from the Co K-edge provides information on the surroundings of cobalt. As an active sulfided Co–Mo/Al$_2$O$_3$ catalyst contains at least two cobalt species – namely, ions inside the Al$_2$O$_3$ lattice and in the Co–Mo–S phase

9.4 Cobalt–Molybdenum Sulfide Hydrodesulfurization Catalysts

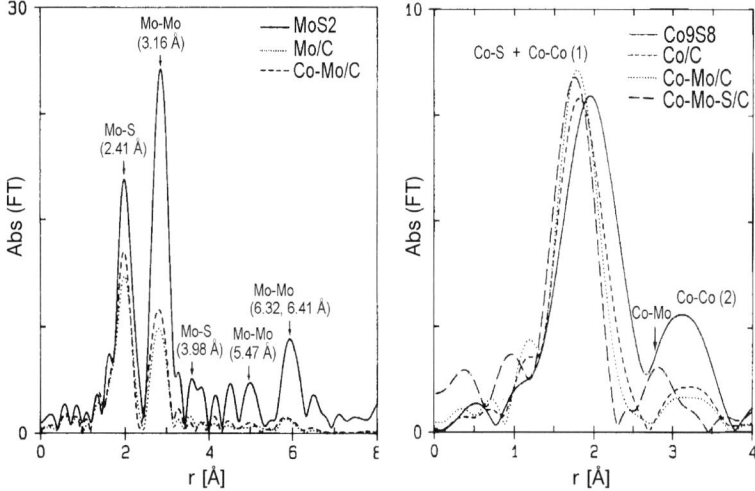

Fig. 9.22 Left: Mo K-edge EXAFS Fourier transforms of MoS$_2$ and sulfided, carbon-supported Mo and Co–Mo catalysts, showing the reduced S and Mo coordination in the first shells around molybdenum in the catalyst. (From Ref. [74].) Right: Co K-edge Fourier transforms of the same catalysts and of a Co$_9$S$_8$ reference. Note the presence of a contribution from Mo neighbors in the Fourier transform of the Co–Mo–S phase. (From [83]).

– it is better to investigate the Co–Mo–S phase in carbon-supported catalysts. The latter can be prepared such that Mössbauer spectroscopy reveals Co–Mo–S as the only cobalt phase present [84]. Figure 9.22 shows the Fourier transforms of the optimized Co–Mo–S/C catalyst and a few other samples. The peaks due to Co and S nearest neighbors overlap in the magnitude of the Fourier transform, but can clearly be distinguished in the imaginary region [83]. A clear scattering contribution from the molybdenum neighbors is also observed.

The data analysis in Table 9.3 summarizes the crystallographic information of the Co–Mo–S phase active for HDS. The Co–S distance in Co–Mo–S is 0.22 nm, with a high sulfur coordination of 6.2 ± 1.3. Each cobalt has on aver-

Table 9.3 Structural parameters from EXAFS of a carbon-supported Co–Mo–S phase [74, 83].

Mo K-edge EXAFS			Co K-edge EXAFS		
	R [Å]	N		R [Å]	N
Mo–S	2.40 ± 0.01	5.5 ± 0.5	Co–S	2.21 ± 0.02	6.2 ± 1.30
Mo–Mo	3.13 ± 0.01	3.2 ± 0.3	Co–Mo	2.80 ± 0.03	1.7 ± 0.35

age 1.7 ± 0.35 molybdenum neighbors at a distance of 0.28 nm. Based on these distances and coordination numbers, one can test structure models for the Co–Mo–S phase. The data obtained are in full agreement with a structure in which cobalt is on the edge of a MoS_2 particle, in the same plane as molybdenum.

The exact positions of cobalt and sulfur on the edge remain the subject of debate, however. Bouwens et al. [74, 83] proposed that cobalt was located at a site where it is coordinated to four sulfur atoms on the edge, which are slightly displaced from their bulk positions. The cobalt is not exactly on the site where a molybdenum atom would be if the MoS_2 lattice were continued (in that case the Co–Mo distance should be 0.316 nm), but rather is somewhat displaced towards the MoS_2 particle (a Co–Mo distance of 0.28 nm). The total sulfur coordination of six is achieved when two additional sulfur atoms are placed on the cobalt atom such that they form a distorted trigonal prism together with the four sulfur atoms of the MoS_2 edge. Nørskov and co-workers [85] have proposed a slightly different structure, based on a calculation indicating that reconstruction at the edges comes into play. Interestingly, this structure would explain the EXAFS data equally well.

Irrespective of the exact configuration around the promoter atom, we now have a detailed picture of the Co–Mo–S phase on the atomic scale, and the appearance of a working Co–Mo/Al_2O_3 HDS catalyst is summarized schematically in Figure 9.23. The catalyst contains MoS_2 particles with dimensions of a few nanometers, decorated with cobalt to form the catalytically highly active Co–Mo–S phase. The scanning tunneling microscopy (STM) images of Figure 7.24 provide a beautiful visualization of the structure. The catalyst also contains cobalt ions that are firmly bound to the lattice of the alumina support, and it may also contain crystallites of the stable bulk sulfide Co_9S_8, which has a low activity for the HDS reaction [56].

The remaining question is how exactly the Co–Mo–S phase is formed in an industrial catalyst. The preparation of a HDS catalyst involves impregnation of the

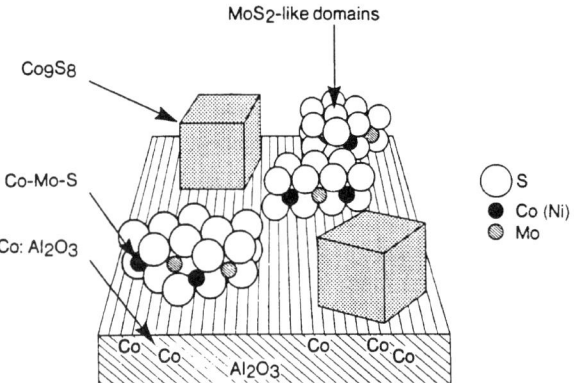

Fig. 9.23 Schematic representation of the different phases present in a typical sulfided, alumina-supported Co–Mo HDS catalyst. (From [56]).

9.4 Cobalt–Molybdenum Sulfide Hydrodesulfurization Catalysts

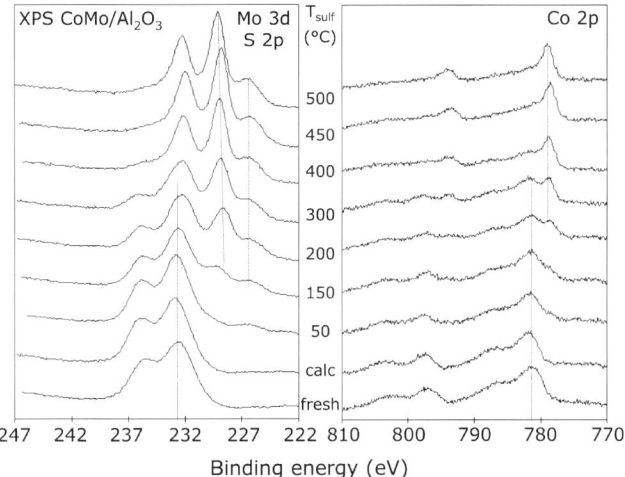

Fig. 9.24 Mo 3d (left) and Co 2p (right) XPS spectra of a calcined Co–Mo/Al$_2$O$_3$/Si(100) catalyst during sulfidation, in which Co and Mo have been impregnated from cobalt nitrate and ammonium heptamolybdate. The spectra indicate that calcination establishes firm bonding between the support and cobalt and molybdenum. As a result, the sulfidation of both is retarded to roughly the same temperature range. (From [87]).

alumina support with molybdenum and cobalt compounds, followed by drying and calcination to obtain well-dispersed oxides of these elements. The preparation is completed with a conversion of the oxides into the catalytically active phase by sulfidation. As cobalt oxide is known to form sulfides at relatively low temperatures (i.e., well below 100 °C [86]), whereas molybdenum oxide in high-surface area supports requires significantly higher temperatures (>200 °C), it is not immediately apparent how a phase in which cobalt decorates the edges of MoS$_2$ slabs can form instead of the thermodynamically expected mixture of Co$_9$S$_8$ and MoS$_2$.

Sanders et al. [87] impregnated a thin-film of alumina on Si(100) with a solution of ammonium heptamolybdate and cobalt nitrate to obtain nominal loadings of about 5 Mo and 2 Co per nm^2. After drying, the catalyst was calcined at 500 °C in air for 6 hours. This procedure is very similar to the industrial preparation of a Co–Mo/Al$_2$O$_3$ catalyst. Next, the model was sulfided at different temperatures; the corresponding XPS spectra are shown in Figure 9.24. Conversion of molybdenum into sulfides is seen to begin at about 100 °C, and is completed at 300–400 °C. Remarkably, the sulfidation of cobalt proceeds at a similar rate to that of molybdenum; in fact, the XPS spectra suggest that the sulfidation of cobalt even lags behind that of molybdenum. In any case, cobalt sulfidation is considerably retarded as compared to that of cobalt on other supports. In other words, the alumina support – and perhaps also the molybdenum oxide phase – have a stabilizing effect on the cobalt oxide which retards the sulfidation of the cobalt to temperatures where molybdenum sulfides are formed or just have been formed.

Interestingly, the retardation of either nickel or cobalt sulfidation to temperatures where MoS_2 has already been formed, can also be achieved by applying chelating agents (e.g., nitrilotriacetic acid) which bind cobalt or nickel and then release these elements at higher temperatures. In this way, one can prepare the Co–Mo–S phase on any support, including the highly inert carbon [84] and silica [88–90]. These catalysts have also been investigated extensively with EXAFS [88] and XPS [89, 90].

The characterization of Co–Mo HDS catalysts can be regarded as a success story, and the progress made since Massoth's review of 1978 [58] has been truly impressive. In hindsight, however, the breakthrough came with the application of an (at that time) exotic technique in catalysis – Mössbauer spectroscopy – which pointed towards a unique environment for cobalt in active HDS catalysts. Conclusive evidence for the location of cobalt species at the edges of the MoS_2 slabs was obtained with a well-established technique, infrared spectroscopy. However, this technique was applied in a creative manner by using NO as an indicator for its adsorption site. Structural information was obtained using a modern tool, EXAFS, and later from STM investigations. The application of surface science models of supported catalysts has provided detailed insights into the preparation chemistry, mainly by XPS, RBS and infrared spectroscopy. Owing to the combination of these *in-situ* techniques, we now have well-founded ideas of the site that is active in the HDS of molecules such as thiophene, which mimics the types of molecule present in crude oil. The challenge for the coming years is to identify how to improve the current catalysts such that they are sufficiently active to desulfurize more difficult molecules than thiophene (i.e., substituted dibenzothiophenes), as these will be required for the deep HDS of diesel fuels [91, 92].

9.5
Chromium Polymerization Catalysts

High-density polyethylene (HDPE) is a commodity chemical that is produced on a very large scale, using either the Ziegler–Natta or Phillips catalytic process. The latter procedure, which accounts for about one-third of all polyethylene production, utilizes a catalyst consisting of small amounts of chromium (0.2–1.0 wt%) on a silica support, as developed by Hogan and Banks at the Phillips Petroleum Company during the early 1950s [93, 94].

The original recipe involved the aqueous impregnation of chromic acid on silica, although nowadays less-poisonous chromium(III) salts are used. Over the years, a family of Phillips-type catalysts has emerged producing no less than 50 different types of polyethylene, and this versatility is the reason for the commercial success of the Phillips ethylene polymerization process. The properties of the desired polymer product can be tailored by varying parameters such as calcination temperature, polymerization temperature and pressure, by adding titania as

9.5 Chromium Polymerization Catalysts

a promoter to the support, or by varying the pore size of the silica support. During the reaction, the polymer molecules fill the pores, which places considerable pressure upon the support. As a consequence, the catalyst is pulverized and remains in a finely dispersed form in the end product [94].

Due to its small chromium content, the Phillips catalyst presents a major challenge to the spectroscopist. As a result, spectroscopic studies are either scarce [95] or deal with samples of higher loading [96], in the hope that chromium behaves similarly as in the real Phillips catalyst. The introduction of surface science approaches to the field of polymerization catalysis has significantly improved the opportunities to study catalysts and polymer products, both in Ziegler–Natta [97] and chromium catalysts [98–100]. The use of planar model supports consisting of a thin SiO_2 layer on a silicon disk offers the advantage that all of the chromium is optimally visible to surface-sensitive techniques such as XPS and SIMS, while as an additional benefit the polymer product can be studied with imaging techniques such as atomic force microscopy (AFM) after reaction. In the following section we will follow the state of chromium in a model catalyst during preparation, and then examine the structure of the polyethylene that grows on this model system.

The Cr/SiO_2 catalyst is prepared by spin-coat impregnation of the flat silica support with a solution of chromic acid in water. The Cr XPS spectrum of the catalyst after drying is shown in Figure 9.25. Chromium is present as a hydrated Cr(VI) oxide, with a Cr 2p binding energy characteristic of chromate species, as the reference spectra in the lower half of the figure show. An essential step in the preparation is the calcination in O_2 at relatively high temperatures of at least 550 °C. According to the corresponding XPS spectrum in Figure 9.25, chromium is still present in the Cr(VI) state, with a high binding energy of 581.3 eV. The interpretation of spectra is best carried out by making comparisons with data measured from well-defined reference materials, and in this case the authors used the Cr(VI)–siloxane complex (see Fig. 9.25) in which a Cr^{6+} ion is bound to two silicon ions via oxygen bridges. The fact that chromium in this compound and in the calcined catalyst have the same binding energy in XPS indicates that both chromium ions possess a similar coordination. Hence, the XPS spectrum shows that calcination at high temperatures leads to an anchoring of chromium ions to the SiO_2 support. However, XPS does not indicate immediately whether chromium ions are present in isolated species, or in clusters with two or more metal ions [99].

Convincing evidence to show that chromium in the active catalyst occurs in isolated species has been obtained from SIMS. Figure 9.26 shows XPS and SIMS spectra from catalysts of high and normal loadings. At relatively high loading, the XPS spectrum indicates the presence of Cr_2O_3 particles in addition to the anchored Cr(VI) ions discussed above. The negative SIMS spectrum of this catalyst (which is also shown in Fig. 9.26) clearly shows the presence of secondary cluster ions containing two chromium ions, and confirming that if the catalyst contains extended chromium phases, SIMS recognizes these by yielding secondary ions of the type $Cr_2O_x^-$. The catalyst of normal chromium content shows the

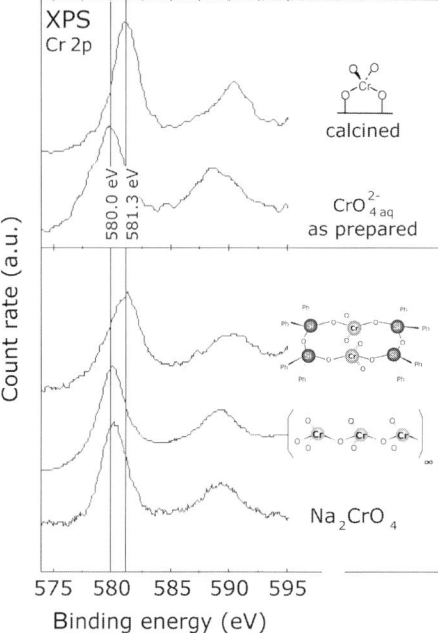

Fig. 9.25 Cr 2p spectra of the $CrO_x/SiO_2/Si(100)$ model catalyst and chromium (VI) reference compounds. Impregnated chromate features the same binding energy as alkali chromates/dichromates or bulk CrO_3. Upon calcination the binding energy of chromium in the catalyst increases by a full 1.3 eV. This unusually high binding energy is typical for chromate(VI) forming ester bonds to silica as in the cluster compound $[CrO_2(OSi(C_6H_5)_2OSi(C_6H_5)2O)]_2$. (Adapted from [99]).

expected XPS doublet of anchored Cr(VI) ions. The corresponding SIMS spectrum, however, exhibits only secondary ions with one chromium ion, providing evidence that the catalyst contains exclusively monochromate species.

Irrespective of how much chromium the catalyst contains after impregnation, the Cr content after high-temperature calcination (at 750 °C) is never higher than 1 Cr ion nm^{-2}. This is shown by the XPS intensity of the Cr signal (Fig. 9.27). If, for example, the freshly impregnated catalyst contains a high metal loading of 4 Cr nm^{-2}, calcination at 450 °C already reduces this to less than 2.5 Cr nm^{-2}, while calcination at higher temperatures decreases the chromium content further, to 1 Cr nm^{-2} at 750 °C. Consequently, chromium evaporates from the catalyst during calcination, as has been confirmed by analyzing an initially blank support placed opposite a chromium-loaded model catalyst during the calcination, by means of RBS (see Fig. 9.28). In Figure 9.28, the top spectrum is that from a freshly impregnated catalyst, the middle spectrum is that of the calcined catalyst, and the bottom spectrum has been measured from the initially blank support.

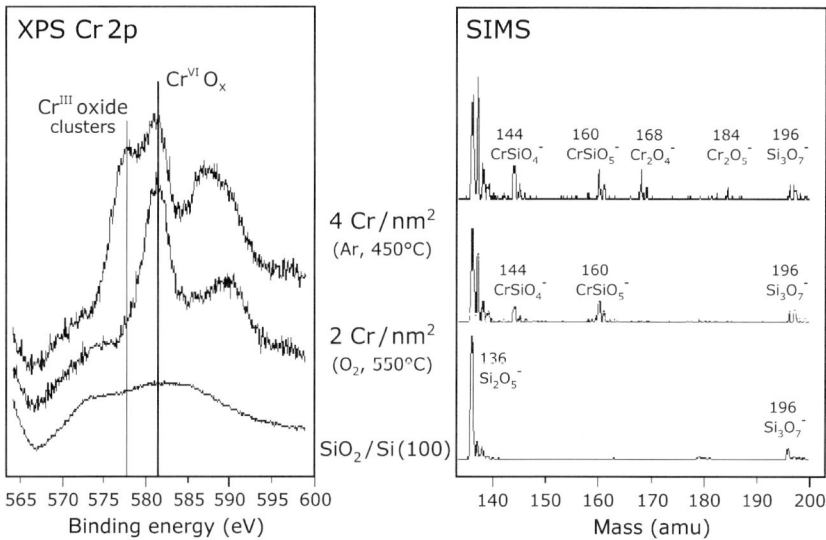

Fig. 9.26 Cr 2p spectra (left) and negative SIMS spectra (right) of two model catalysts and a blank reference. The blank (bottom) shows only Si_xO_y fragments; on a chromium-loaded catalyst Cr_1O_x fragments appear after thermal activation (Ar/O_2). If desorption of chromium is made impossible (in oxygen-free argon), Cr_2O_x clusters can also be detected. In combination, this is strong evidence that chromate anchors to the silica surface as a monomer. (Adapted from [99]).

Hence, above a certain threshold loading of about 1 atom nm^{-2}, chromium evaporates from the catalyst during calcination. If this occurs in a porous catalyst, the airborne species may deposit somewhere else and anchor subsequently to the support. Hence, gas-phase transport may be expected to play an important role in the redispersion of chromium species through the calcination of a high-surface area polymerization catalyst.

In order to produce polymer on a planar $CrO_x/SiO_2/Si(100)$ model catalyst in a flow of ethylene, one needs to consider that a sample of 1 cm^2 exposes no more than 10^{14} highly reactive chromium ions, which makes the model extremely sensitive to deactivation by impurities, such as water or acetylene. Hence, working with ultra-pure gases, cleaned by filters close to the position of the model catalyst, is crucial [100].

Polymerization on heterogeneous catalysts differs from other catalytic reactions in the sense that the product remains on the catalyst. Several techniques can be used to study the polymer product after reaction. Figures 9.29 and 9.30 show several examples of polymer that was formed at 160 °C (i.e., above its melting point), and subsequently cooled to room temperature. During cooling, polyethylene crystallizes and is expected to develop its well-known spherulite morphology [101].

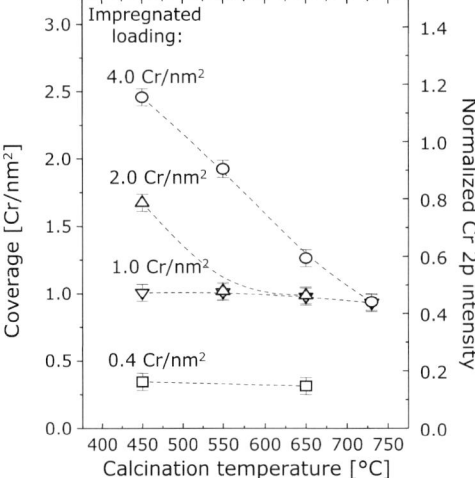

Fig. 9.27 The apparent chromium coverage decreases with increasing calcination temperature. However, at low initial loadings the Cr coverage is much more stable against calcination. We assign the decrease in apparent Cr coverage to desorption of chromate which happens most easily from highly covered surfaces. (Adapted from [99]).

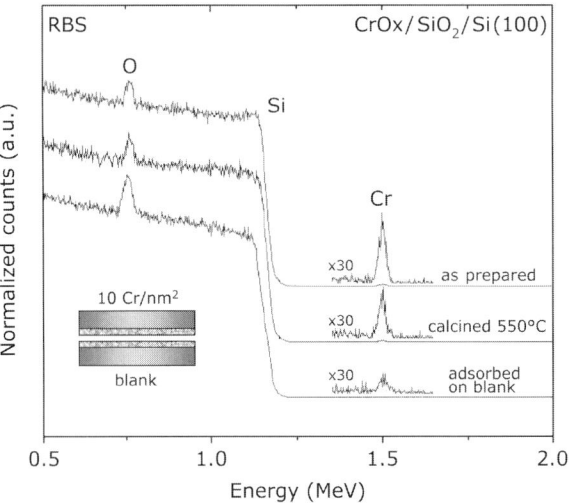

Fig. 9.28 Rutherford backscattering (RBS) proves that chromate can indeed desorb from the model catalyst during thermal activation. A wafer with 10 Cr nm^{-2} loading features 7.0 Cr nm^{-2} after calcination at 550 °C. The desorbing chromate readily re-adsorbs on an empty silica surface placed opposite to the loaded wafer. (Adapted from [99]).

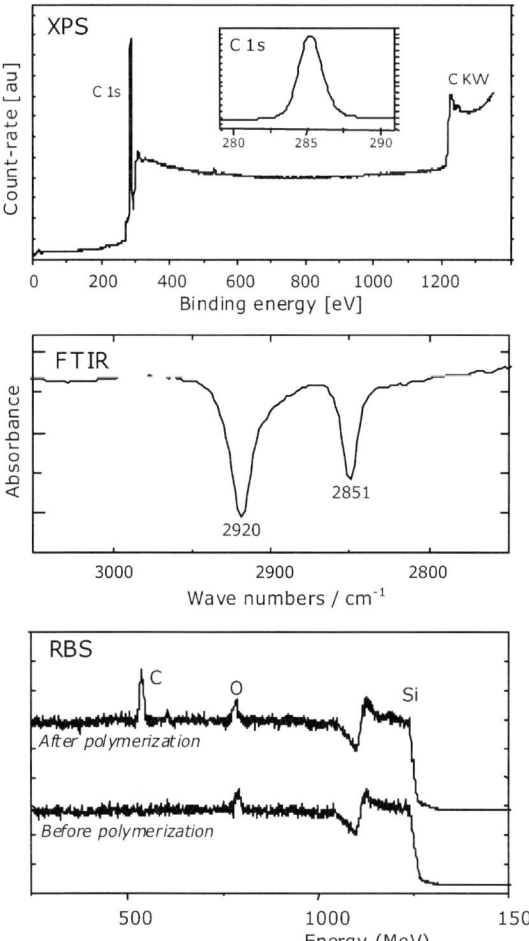

Fig. 9.29 XPS, FTIR and RBS of a CrOx/SiO$_2$/Si(100) after polymerization of ethylene at 160 °C all reveal the presence of a significant amount of polyethylene, visible in the C 1s peak in XPS, the symmetric and asymmetric C–H stretch vibrations of CH$_2$ groups in transmission IR, and a C-peak in RBS. (Adapted from [102]).

The XPS spectrum in Figure 9.29 reveals a clear and sharp C 1s peak with a binding energy characteristic of polyethylene, whereas contributions from chromium or even the silica support are no longer visible. The transmission infrared spectrum, which is measurable because the silicon disk is largely transparent in the C–H stretch region, shows the asymmetric and symmetric C–H stretch vibrations of CH$_2$ groups, whereas the corresponding vibrations of CH$_3$ groups cannot be discerned, indicating that the (–CH$_2$–)$_n$ chains, which are terminated by methyl groups, must have considerable length. Rutherford backscattering (Fig.

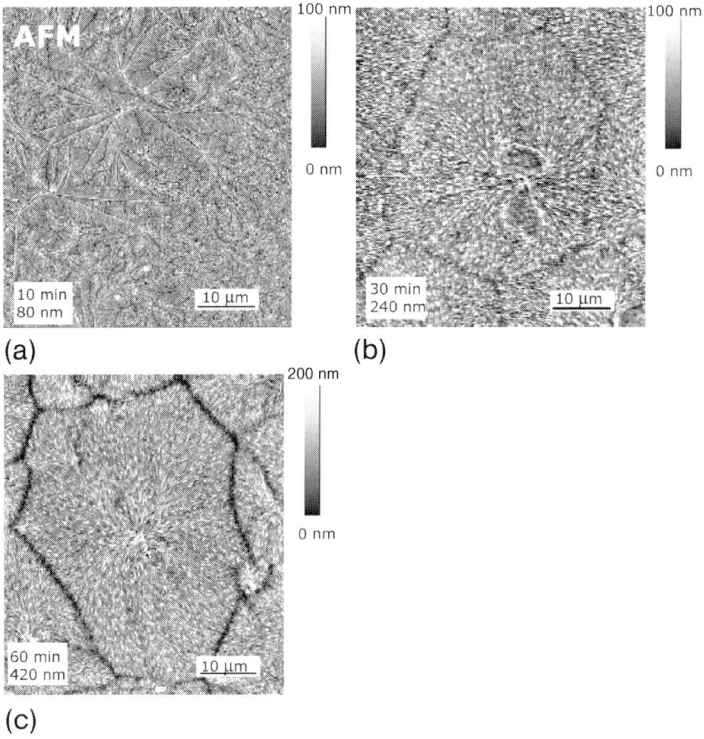

Fig. 9.30 (a–c) AFM images of polyethylene films formed on the planar CrOx/SiO$_2$/Si(100) model catalyst. The small white stripes are lamellar crystals. These form the well-known spherulite superstructure upon crystallization from the melt. Depending on the layer thickness, spherulite growth stops at different stages of development. (Adapted from [100]).

9.29) detects a thick carbon-containing layer on the model catalyst, of a few hundred nanometers thickness [102]. Hence, there is no doubt that the model catalyst has indeed produced polymer.

Scanning force microscopy offers the best information on the polyethylene product (Fig. 9.30). These images show the different stages in the formation of polyethylene spherulites when the polymer crystallizes from the melt. After a relatively short reaction time, a closed layer of polyethylene lamellae of 80 nm thickness has formed (Fig. 9.30a). In this layer, shorter paraffins coexist with polymers. Besides stacked, plate-like parts characteristic of paraffins and edge-on grown polyethylene lamellae, some crystal bundles of polyethylene grown in sheaf-like fashion are also present. After longer reaction times, the layer becomes considerably thicker, and the morphology changes drastically. Figure 9.30b shows ordering of the polyethylene into domains of roughly 50 μm in diameter, and

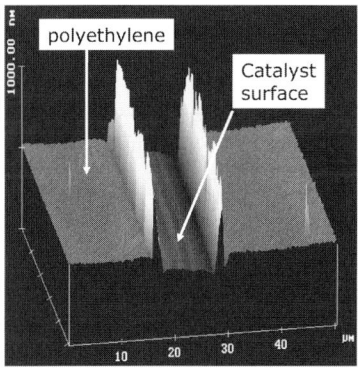 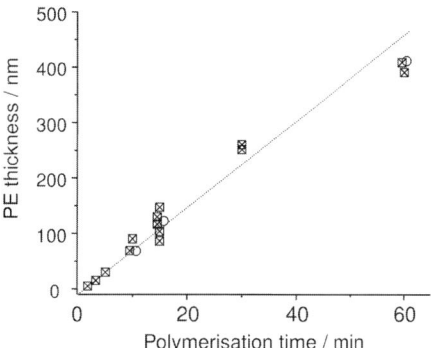

Fig. 9.31 AFM image of a scalpel scratch used to determine the thickness of the polyethylene films grown on a CrOx/SiO$_2$/Si(100) model catalyst. The polymer thickness correlates linearly with the reaction time. (Adapted from [100]).

the film is on average 240 nm thick. In the center, polyethylene lamellae grow in sheaf-like manner from a nucleation center. In fact, the lamellae attempt to order in parallel fashion, but succeed only partially because the lamellae are imperfect, due to for example loops of polymer molecules and loose ends that protrude. Figure 9.30c shows the final stage of spherulite formation, with a layer thickness of 420 nm. The nucleation center can still be recognized, but has been covered by a dome-like structure, as is usually observed for crystalline polyethylene in thick layers [101]. Owing to the fact that model catalyst produces polymer films of limited thickness, the intermediate stages in polyethylene crystallization could be observed and analyzed for the first time [100].

The thickness of the polyethylene films can be measured conveniently with AFM, by making a scratch with a scalpel, and determining the difference in height between the substrate and the surface of the polymers (Fig. 9.31). A plot of polymer thickness versus reaction time shows a straight line, indicating that there are no appreciable transport limitations for ethylene molecules in their diffusion through the progressively thicker layer of molten polymer that forms during reaction.

Due to the thick film of polymer that grows on the catalyst, it is very difficult to study the state of Cr ions of the active catalyst. Groppo et al. used XANES and EXAFS for this purpose, and found that the initial Cr^{6+} ions of the oxygen-pretreated catalyst were indeed reduced under polymerization conditions, as had been expected [103, 104].

Future challenges for polymerization model catalysts include studying the structure of polymers below their melting point, in what is termed the "nascent" morphology. Such studies can either be undertaken on silica-supported chromium catalysts (as discussed above), or on so-called single-site catalysts, such as metallocenes, applied on flat silica supports.

9.6
Concluding Remarks

The case studies on supported rhodium, sulfide, and chromium catalysts nicely illustrate the principles of the ideal strategy for research on catalysts that we introduced in Chapter 1:

- An integrated approach from catalysis and spectroscopy, in which experts in both fields cooperate closely. This ensures that the spectroscopist investigates catalysts that have undergone the correct treatments, and that the catalytic chemist disposes over correctly analyzed and interpreted spectra.

- The application of a combination of techniques. Only rarely will one single spectroscopy be capable of solving a problem in its entirety. These case studies demonstrate the advantages of using different techniques on the same catalysts.

- *In-situ* characterization. Catalysts should preferably be investigated at the conditions under which they are active in the reaction, although a variety of reasons exist why this may not be possible. For example, lattice vibrations often impede the use of EXAFS, XRD and Mössbauer spectroscopy at reaction temperatures, the mean free path of electrons and ions dictates that XPS, SIMS and LEIS are carried out in vacuum, etc. Nevertheless, one should strive to choose the conditions which are as close as possible to those of the catalytic reaction. This means that the catalyst is kept under reaction gases or inert atmosphere at low temperature to be studied by EXAFS and Mössbauer spectroscopy, or that it is transferred to the vacuum spectrometers under conditions which preserve the chemical state of the surface.

- Optimized catalysts. Samples should be suitable for investigating the particular aspect of the catalyst one in which interest is centered. For example, meaningful information on the metal support interface is only obtained if the supported particles are small, and all of the same size. In the case study on Co–Mo catalysts, the structure of the active Co–Mo–S phase could only be studied by EXAFS because it was the only phase present in the carbon-supported Co–Mo catalyst.

- Realistic model systems. Some techniques become much more informative if suitable model systems are used. Examples are the thin-film oxides used as conducting model supports, which offer much better opportunities for surface analysis than technical catalysts do. Another example is provided by the non-porous, spherical supports that have been employed successfully in electron microscopy. It is also important is that the model systems exhibit the same chemistry as the catalyst they represent.

- Suitable reference compounds, preferably measured together with the catalysts, to confirm spectral assignments.

These are the ingredients for successful research in catalysis!

References

1. R.M. Heck and R.J. Farrauto, *Catalytic Air Pollution Control*. Wiley, New York, 2002.
2. S.J. Tauster, S.C. Fung, and R.L. Garten, *J. Am. Chem. Soc.* **100** (1978) 170.
3. H.F.J. van't Blik, J.B.A.D. van Zon, T. Huizinga, J.C. Vis, D.C. Koningsberger, and R. Prins, *J. Am. Chem. Soc.* **107** (1985) 3139.
4. H.J. Borg, L.C.A. van den Oetelaar, and J.W. Niemantsverdriet, *Catal. Lett.* **17** (1993) 81.
5. P.L.J. Gunter, J.W. Niemantsverdriet, F.H. Ribeiro, and G.A. Somorjai, *Catal. Rev.-Sci. Eng.* **39** (1997) 77.
6. J.G. Chen, M.L. Colaianni, P.J. Chen, J.T. Yates, and G.B. Fisher, *J. Phys. Chem.* **94** (1990) 5059.
7. B.G. Frederick, G. Apai, and T.N. Rhodin, *J. Am. Chem. Soc.* **109** (1987) 4797.
8. D.N. Belton and S.J. Schmieg, *Surface Sci.* **199** (1988) 518.
9. J.W. Niemantsverdriet, A.F.G. Engelen, A.M. de Jong, W. Wieldraaijer, and G.J. Kramer, *Appl. Surface Sci.* **144–145** (1999) 366.
10. H.J. Borg, L.C.A. van den Oetelaar, L.J. van IJzendoorn, and J.W. Niemantsverdriet, *J. Vac. Sci. Technol.* **A10** (1992) 2737.
11. R.J. Fenoglio, W. Alvarez, G.M. Nunez, and D.E. Resasco, in: *Preparation of Catalysts V*, G. Poncelet, P.A. Jacobs, and B. Delmon (Eds.). Elsevier, Amsterdam, 1991, p. 77.
12. M.G. Mason, *Phys. Rev.* **B27** (1983) 748.
13. G.K. Wertheim, D. Cenzo, and S.E. Youngquist, *Phys. Rev. Lett.* **51** (1983) 2310.
14. T. Huizinga, H.F.J. van't Blik, J.C. Vis, and R. Prins, *Surface Sci.* **135** (1983) 580.
15. B.J. Kip, F.B.M. Duivenvoorden, D.C. Koningsberger, and R. Prins, *J. Catal.* **105** (1987) 26.
16. P. Johnston, R.W. Joyner, P.D.A. Pudney, E.S. Shpiro, and P.B. Williams, *Faraday Discuss. Chem. Soc.* **89** (1990) 91.
17. E.G. Derouane, A.J. Simoens, and J.C. Vedrine, *Chem. Phys. Lett.* **52** (1977) 549.
18. A. Bossi, F. Garbassi, G. Petrini, and L. Zanderighi, *J. Chem. Soc. Faraday Trans. I* **78** (1982) 1029.
19. D.C. Koningsberger, J.B.A.D. van Zon, H.F.J. van't Blik, G.J. Visser, R. Prins, A.N. Mansour, D.E. Sayers, D.R. Short, and J.R. Katzer, *J. Phys. Chem.* **89** (1985) 4075.
20. J.B.A.D. van Zon, D.C. Koningsberger, H.F.J. van't Blik, and D.E. Sayers, *J. Chem. Phys.* **82** (1985) 5742.
21. J.A. Horsley, *J. Am. Chem. Soc.* **101** (1979) 2870.
22. M.K. Bahl, S.C. Tsai, and Y.W. Chung, *Phys. Rev. B* **21** (1980) 1344.
23. D.N. Belton, Y.-M. Sun, and J.M. White, *J. Phys. Chem.* **88** (1984) 1690; *J. Phys. Chem.* **88** (1984) 5172.
24. J. Santos, J. Phillips, and J.A. Dumesic, *J. Catal.* **81** (1983) 147.
25. A.D. Logan, E.J. Braunschweig, A.K. Datye, and D.J. Smith, *Langmuir* **4** (1988) 827.
26. H. Knözinger, in: *Cluster Models for Surface and Bulk Phenomena*, G. Pacchioni, P.S. Bagus, and F. Parmigiani (Eds.). NATO ASI Series B: Physics Vol. 283, Plenum, New York, 1992, p. 131.
27. P. Basu, D. Panayotov, and J.T. Yates, *J. Phys. Chem.* **91** (1987) 3133.
28. B.C. Gates, *Catalytic Chemistry*. Wiley, New York, 1992.
29. J.-D. Grunwaldt, L. Basini, and B.S. Clausen, *J. Catal.* **200** (2001) 321.
30. H.-J. Freund, M. Baumer, and H. Kuhlenbeck, *Adv. Catal.* **45** (2000) 334.
31. C.R. Henry, *Surface Sci. Rep.* **31** (1998) 231.
32. S.H. Overbury, D.R. Mullins, and L. Kundakovic, *Surface Sci.* **470** (2001) 243.
33. S.H. Oh and C.C. Eickel, *J. Catal.* **128** (1991) 526.
34. C. Wong and R.W. McCabe, *J. Catal.* **107** (1987) 535.
35. J. Burkhardt and L.D. Schmidt, *J. Catal.* **116** (1989) 240.
36. J.R. Jennings (Ed.), *Catalytic Ammonia Synthesis: Fundamentals and Practice*. Plenum, New York, 1991.
37. M.E. Dry, in: *Catalysis, Science and Technology*, Vol. 1, J.R. Anderson and

M. Boudart (Eds.). Springer, Berlin, 1981, p. 159.

38 M.P. Kiskinova, in: *Studies in Surface Science and Catalysis, Vol. 70.* Elsevier, Amsterdam, 1992.

39 H.P. Bonzel, *Surface Sci. Rep.* **8** (1987) 43.

40 N.D. Lang, in: *Physics and Chemistry of Alkali Metal Adsorption*, H.P. Bonzel, A.M. Bradshaw, and G. Ertl (Eds.). Elsevier, Amsterdam, 1989, p. 11.

41 N. Rösch, in: *Cluster Models for Surface and Bulk Phenomena*, G. Pacchioni, P.S. Bagus, and F. Parmigiani (Eds.). NATO ASI Series B, Physics Vol. 283, Plenum, New York, 1992, p. 251.

42 K. Wandelt, in: *Thin Metal Films and Gas Chemisorption*, P. Wissmann (Ed.). Elsevier, Amsterdam, 1987, p. 268.

43 T.V.W. Janssens, G.R. Castro, K. Wandelt, and J.W. Niemantsverdriet, *Phys. Rev. B* **49** (1994) 14599.

44 N.D. Lang, S. Holloway, and J.K. Nørskov, *Surface Sci.* **150** (1985) 24.

45 T.V.W. Janssens, K. Wandelt, and J.W. Niemantsverdriet, *Catal. Lett.* **19** (1993) 263.

46 H. Over, H. Bludau, R. Kose, and G. Ertl, *Phys. Rev. B* **51** (1995) 4661.

47 F. Strisland, A. Beutler, A.J. Jaworowski, R. Nyholm, B. Setlik, D. Heskett, and J.N. Andersen, *Surface Sci.* **410** (1998) 330.

48 K.J. Uram, L. Ng, and J.T. Yates, *Surface Sci.* **177** (1986) 253.

49 R.A. dePaola, J. Hrbek, and F.M. Hoffmann, *J. Chem. Phys.* **82** (1985) 2484.

50 J.E. Crowell and G.A. Somorjai, *Appl. Surface Sci.* **19** (1984) 73.

51 G. Pirug and H.P. Bonzel, *Surface Sci.* **199** (1988) 371.

52 V. Dose, J. Rogozik, A.M. Bradshaw, and K.C. Prince, *Surface Sci.* **179** (1987) 90.

53 L.J. Whitman and W. Ho, *J. Chem. Phys.* **83** (1985) 4808.

54 D. Heskett, *Surface Sci.* **199** (1988) 67.

55 B.C. Gates, J.R. Katzer, and G.C.A. Schuit, *Chemistry of Catalytic Processes.* McGraw-Hill, New York, 1979.

56 H. Topsoe, B.S. Clausen, and F.E. Massoth, *Hydrotreating Catalysis.* Springer-Verlag, Berlin, 1996.

57 R. Prins, V.H.J. de Beer, and G.A. Somorjai, *Catal. Rev. – Sci. Eng.* **31** (1989) 1.

58 F.E. Massoth, *Adv. Catal.* **27** (1978) 265.

59 J.M.J.G. Lipsch and G.C.A. Schuit, *J. Catal.* **15** (1969) 179.

60 R.J.H. Voorhoeve and J.C.M. Stuiver, *J. Catal.* **23** (1971) 243.

61 A.L. Farragher and P. Cossee, in: *Proceedings 5th International Congress on Catalysis*, Palm Beach, 1972, J.W. Hightower (Ed.). North-Holland, Amsterdam, 1973, p. 1301.

62 B. Delmon, *Am. Chem. Soc., Div. Pet. Chem., Prep.* **22**(2) (1977) 503.

63 P. Arnoldy, J.A.M. van den Heijkant, G.D. de Bok, and J.A. Moulijn, *J. Catal.* **92** (1985) 35.

64 J.C. Muijsers, Th. Weber, R.M. van Hardeveld, H.W. Zandbergen, and J.W. Niemantsverdriet, *J. Catal.* **157** (1995) 698.

65 Th. Weber, J.C. Muijsers, J.H.M.C. van Wolput, C.P.J. Verhagen, and J.W. Niemantsverdriet, *J. Phys. Chem.* **100** (1996) 14144.

66 A.M. de Jong, H.J. Borg, L.J. van IJzendoorn, V.G.F.M. Soudant, V.H.J. de Beer, J.A.R. van Veen, and J.W. Niemantsverdriet, *J. Phys. Chem.* **97** (1993) 6477.

67 R. Cattaneo, Th. Weber, T. Shido, and R. Prins, *J. Catal.* **191** (2000) 225.

68 R.G. Leliveld, A.J. van Dillen, J.W. Geus, and D.C. Koningsberger, *J. Catal.* **171** (1997) 115.

69 Th. Weber, J.C. Muijsers, and J.W. Niemantsverdriet, *J. Phys. Chem.* **99** (1995) 9144.

70 S.S. Pollack, L.E. Makovsky, and F.R. Brown, *J. Catal.* **59** (1979) 452.

71 B.S. Clausen, H. Topsøe, R. Candia, J. Villadsen, B. Lengeler, J. Als-Nielsen, and F. Christensen, *J. Phys. Chem.* **85** (1981) 3868.

72 T.G. Parham and R.P. Merrill, *J. Catal.* **85** (1984) 295.

73 G. Sankar, S. Vasudevan, and C.N.R. Rao, *J. Phys. Chem.* **91** (1987) 2011.

74 S.M.A.M. Bouwens, R. Prins, V.H.J. de Beer, and D.C. Koningsberger, *J. Phys. Chem.* **94** (1990) 3711.

75 I. Alstrup, I. Chorkendorff, R. Candia, B.S. Clausen, and H. Topsøe, *J. Catal.* **77** (1982) 397.

76 K. Suzuki, M. Soma, T. Onishi, and K. Tamaru, *J. Electron. Spectrosc. Rel. Phenom.* **24** (1981) 283.

77 C. Wivel, R. Candia, B.S. Clausen, S. Mørup, and H. Topsøe, *J. Catal.* **68** (1981) 453.

78 H. Topsøe, B.S. Clausen, R. Candia, C. Wivel, and S. Mørup, *J. Catal.* **68** (1981) 433.

79 R.R. Chianelli, A.F. Ruppert, S.K. Behal, B.H. Kear, A. Wold, and R. Kershaw, *J. Catal.* **92** (1985) 56.

80 N.Y. Topsøe, H. Topsøe, O. Sørensen, B.S. Clausen, and R. Candia, *Bull. Soc. Chim. Belg.* **93** (1984) 727.

81 N.Y. Topsøe and H. Topsøe, *J. Catal.* **84** (1983) 386; *J. Electron. Spectrosc. Rel. Phenom.* **39** (1986) 11.

82 A. Lopez Agudo, F.J. Gil Llambias, J.M.D. Tascon, and J.L.G. Fierro, *Bull. Soc. Chim. Belg.* **93** (1984) 719.

83 S.M.A.M. Bouwens, J.A.R. van Veen, D.C. Koningsberger, V.H.J. de Beer, and R. Prins, *J. Phys. Chem.* **95** (1991) 123.

84 J.A.R. van Veen, E. Gerkema, A.M. van der Kraan, and A. Knoester, *J. Chem. Soc., Chem Commun.* (1987) 1684.

85 L.S. Byskov, J.K. Nørskov, B.S. Clausen, and H. Topsøe, *J. Catal.* **187** (1999) 109.

86 M.W.J. Crajé, E. Gerkema, V.H.J. de Beer and A.M. van der Kraan, in: *Hydrotreating Catalysts, Preparation, Characterization and Performance*, M.L. Occelli and R.G. Anthony (Eds.). Elsevier, Amsterdam, 1989, p. 165.

87 A.F.H. Sanders, A.M. de Jong, V.H.J. de Beer, J.A.R. van Veen, and J.W. Niemantsverdriet, *Appl. Surface Sci.* **144–145** (1999) 380.

88 L. Medici and R. Prins, *J. Catal.* **163** (1996) 28; *J. Catal.* **163** (1998) 38.

89 L. Coulier, V.H.J. de Beer, J.A.R. van Veen, and J.W. Niemantsverdriet, *Topics Catal.* **13** (2000) 99.

90 L. Coulier, V.H.J. de Beer, J.A.R. van Veen, and J.W. Niemantsverdriet, *J. Catal.* **197** (2001) 26.

91 I.V. Babich and J.A. Moulijn, *Fuel* **82** (2003) 607.

92 M. Breysse, G. Djega-Mariadassou, S. Pessayre, C. Geantet, M. Vrinat, G. Pérot, and M. Lemaire, *Catal. Today* **84** (2003) 129.

93 J.P. Hogan and L. Banks, *U.S. Patent 2,835,721* (1958).

94 M.P. McDaniel, *Adv. Catal.* **33** (1985) 47.

95 R. Merryfield, M.P. McDaniel, and G. Parks, *J. Catal.* **77** (1982) 348.

96 B.M. Weckhuysen, I.E. Wachs, and R.A. Schoonheydt, *Chem. Rev.* **96** (1996) 3327.

97 E. Magni and G.A. Somorjai, *Surface Sci.* **345** (1996) 1.

98 P.C. Thüne, C.P.J. Verhagen, M.J.G. van den Boer, and J.W. Niemantsverdriet, *J. Phys. Chem.* **101** (1997) 8559.

99 P.C. Thüne, R. Linke, W.J.H. van Gennip, A.M. de Jong, and J.W. Niemantsverdriet, *J. Phys. Chem. B* **105** (2001) 3073.

100 P.C. Thüne, J. Loos, P.J. Lemstra, and J.W. Niemantsverdriet, *J. Catal.* **183** (1999) 1.

101 F.W. Billmeyer, *Textbook of Polymer Science*, 3rd edn. Wiley Interscience, New York, 1984.

102 P.C. Thüne and J.W. Niemantsverdriet, *Isr. J. Chem.* **38** (1998) 385.

103 E. Groppo, C. Prestipino, F. Cesano, F. Bonino, S. Bordiga, C. Lamberti, P.C. Thüne, J.W. Niemantsverdriet, and A. Zecchina, *J. Catal.* **230** (2005) 98.

104 E. Groppo, C. Lamberti, S. Bordiga, G. Spoto, and A. Zecchina, *Chem. Rev.* **105** (2005) 115.

Appendix
Metal Surfaces and Chemisorption

Keywords

Theory of metal surfaces
Chemisorption on metals

A.1
Introduction

Heterogeneous catalysis deals with reactions between species that are adsorbed onto the surface of a catalyst. The role of the catalytic surface is to provide an energetically favorable pathway for the reaction. In order to find an explanation for the catalytic activity of substances, it is essential that we examine the properties of the surface, rather than at a collective property of the bulk catalyst.

The distinctive feature of a surface atom is that it has fewer neighbors than an atom in the interior. This unsaturated coordination forms the reason why the electronic and vibrational properties – and sometimes also the crystallographic positions of surface atoms – differ from those of the bulk atoms.

This appendix begins with a brief introduction to the physics of metal surfaces. We limit ourselves to those properties of surfaces that play a role in catalysis or in catalyst characterization. The second part includes an introduction to the theory of chemisorption, and is intended to serve as a theoretical background for the chapters on vibrational spectroscopy, photoemission, and the case study on promoter effects. General textbooks on the physics and chemistry of surfaces are listed in [1–8].

A.2
Theory of Metal Surfaces

A.2.1
Surface Crystallography

The most important metals for catalysis are those of the Groups VIII and IB of the Periodic Table. Three crystal structures are important: face-centered cubic (fcc: Ni, Cu, Rh, Pd, Ag, Ir, Pt, Au); hexagonal close-packed (hcp: Co, Ru, Os); and body-centered cubic (bcc: Fe) [9, 10]. Before discussing the surfaces that these lattices expose, we should perhaps mention a few general properties.

If the lattices are viewed as close-packed spheres, the fcc and hcp lattices have the highest density, as they possess about 26% empty space. Each atom in the interior has 12 nearest neighbors or, in other words, an atom in the interior has a coordination number of 12. The bcc lattice is slightly more open and contains about 32% empty space. The coordination of a bulk atom inside the bcc lattice is 8.

The difference between the fcc and hcp structure is best seen if one considers the sequence of close-packed layers. For fcc lattices this is the (111) plane (see Fig. A.1), and for hcp lattices it is the (001) plane. The geometry of the atoms in these planes is exactly the same. Both lattices can now be built up by stacking close-packed layers on top of each other. If one places the atoms of the third layer directly above those of the first, one obtains the hcp structure. The sequence of layers in the ⟨001⟩ direction is *ababab*, etc. In the fcc structure, it is every atom of the *fourth* layer that is above an atom of the first layer. Thus, the sequence of layers in the ⟨111⟩ direction is *abcabcabc* ...

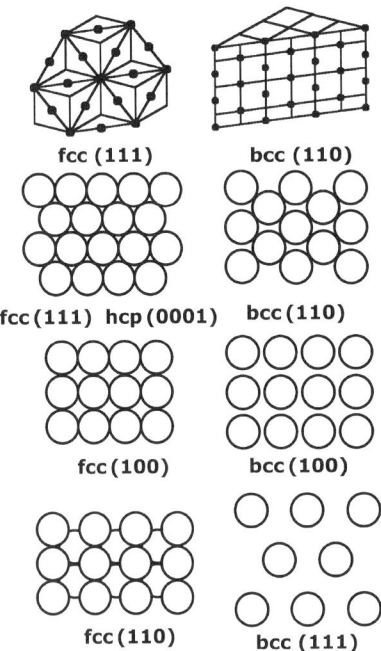

Fig. A.1 Construction of the most densely packed surfaces of fcc and bcc metals, and the outer atomic layer of the low-index surfaces. The hcp (001) surface has the same structure as the fcc (111) surface.

We can create surfaces from the fcc, hcp and bcc crystals by cutting them along a plane. There are many ways to do this, and Figure A.1 shows how one obtains the low index surfaces. Depending on the orientation of the cutting plane we obtain atomically flat surfaces with a high density of atoms per unit area or more open surfaces with steps, terraces and kinks (often referred to as corrugated or vicinal surfaces). Thus, *the* surface of a metal does not exist; one must specify its coordinates.

The surfaces are labeled according to the lattice plane that is exposed. The (111), (100), and (110) surfaces are perpendicular to the ⟨111⟩, ⟨100⟩, and ⟨110⟩ directions in the crystal. The close-packed surface of the hcp lattice, the (001) plane [or strictly speaking the (0001) plane, because four coordinates are used for hexagonal lattices], has the same structure as the fcc (111) plane [11].

The reactivity of a surface depends on the number of bonds that are unsaturated. An unsaturated bond is what is left from a former bond with a neighboring metal atom that had to be broken in order to create the surface. Thus, we want to know the number of missing neighbors, denoted by Z_s, of an atom in each surface plane. One can infer from Figure A.1 that an atom in the fcc (111) or hcp (001) surface has six neighbors in the surface and three below, but misses the three neighbors above that were present in the bulk: $Z_s = 3$. An atom in the fcc (100) surface, on the other hand, has four neighbors in the surface, four below, and misses four above the surface: $Z_s = 4$. Hence, an atom in the fcc (100) surface has one more unsaturated bond than an atom in the fcc (111) surface, and is slightly more reactive.

The fcc (110) surface, having the appearance of furrows on a plowed field, contains two types of atom. Those shown in the fcc (110) structure of Figure A.1 have two nearest neighbors in the surface, and four below, and miss as many as six neighbors as compared to an atom in the bulk. An atom in the layer just below the surface (in the furrow) is also exposed to the gas phase; it misses two neighbors.

The bcc structure contains more empty space than the fcc structure, as is clearly shown in Figure A.1. The closest packing is observed in the (110) plane. The (111) plane forms a rather open surface, which exposes atoms from three different levels. Values of Z_s are given in Table A.1.

Table A.1 Number of missing neighbors in the surface (Z_s) and neighbors in the bulk (N).

	Z_s	N		Z_s	N
fcc (111), hcp (001)	3	12	bcc (110)	2	8
fcc (100)	4	12	bcc (100)	4	8
fcc (110)	5	12	bcc (111)	4	8

Note: The fcc (110) and the bcc (111) are corrugated surfaces and possess inequivalent types of atoms.

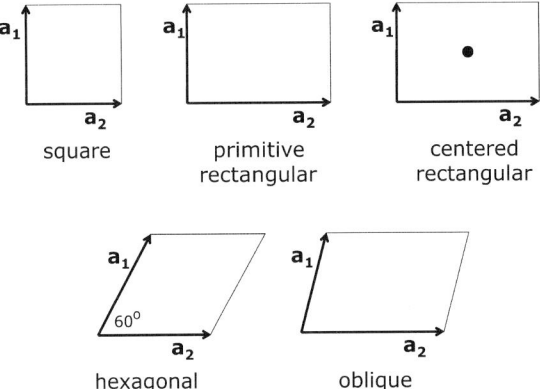

Fig. A.2 The five surface Bravais lattices: square; primitive rectangular; centered rectangular; hexagonal; and oblique.

Surfaces possess periodicity in two dimensions. Hence, two base vectors are sufficient for describing the periodic structure of a crystal surface. This does not imply that a surface must be flat, as even the unit cells of simple surfaces such as fcc (110) and bcc (111) have three-dimensional structure on the atomic scale. However, to construct a surface from this unit cell by translation, one only needs two vectors.

In two dimensions, five different lattices exist (see Fig. A.2). One recognizes the hexagonal Bravais lattice as the unit cell of the cubic (111) and hcp (001) surfaces, the centered rectangular cell as the unit cell of the bcc and fcc (110) surfaces, and the square cell as the unit cell of the cubic (100) surfaces. Translation of these unit cells over vectors $h\mathbf{a}_1 + k\mathbf{a}_2$, in which h and k are integers, produces the surface structure.

Adsorbates may form ordered overlayers, which can have their own periodicity. The adsorbate structure is given with respect to that of the substrate metal. For simple arrangements the Wood's notation is used, and some examples are given in Figure A.3. The notation Pt(110)–c(2 × 2)O means that oxygen atoms form an

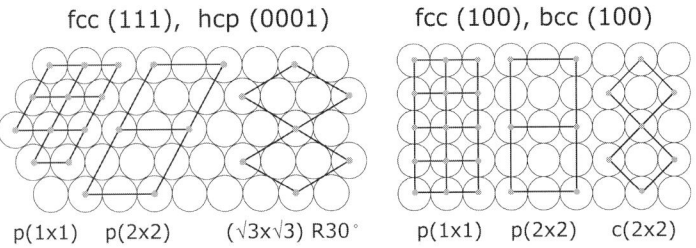

Fig. A.3 The structure of a few common ordered overlayers and the corresponding Wood's notation.

ordered overlayer with a unit cell that has twice the dimensions of the Pt(110) unit cell, and an additional O in the middle. Note that this abbreviation does not specify where the O is with respect to the Pt atoms. It may be on top of the Pt atoms but also in bridged or fourfold sites, or in principle anywhere as long as the periodic structure is c(2 × 2) with respect to the Pt(110) surface. If the unit cell of the overlayer structure becomes too complicated to be represented in the Wood's notation, one uses a 2 × 2 matrix, which expresses how the substrate vectors a_1 and a_2 transform into those of the overlayer. The same notation systems are used for indicating surface reconstructions.

For what reason do some adsorbate systems form ordered overlayers, and some surfaces such as Pt(100) or Ir(110) reconstruct? The answer is based on thermodynamics, in that the *free energy of the surface* provides the driving force behind many surface phenomena.

A.2.2
Surface Free Energy

It costs energy to create a surface, because bonds have to be broken. Thus, in going from a piece of matter to two smaller pieces, the total energy of the system increases: the surface free energy is always positive.

The surface free energy γ is related to the cohesive energy of the solid, ΔH_{coh}, and to the number of bonds between an atom and its nearest neighbors that had to be broken to create the surface:

$$\gamma = \Delta H_{coh} \frac{Z_s}{Z} N_s \tag{A-1}$$

where:
γ is the surface free energy;
ΔH_{coh} is the cohesive energy;
Z_s is the number of missing nearest neighbors of a surface atom;
Z is the coordination number of a bulk atom;
N_s is the density of atoms in the surface.

Metals possess the highest surface free energies, in the order of 1500 to 3000 ergs cm^{-2}, while values for ionic solids and oxides are much lower, roughly between 200 and 500 ergs cm^{-2}. Hydrocarbons are among the substances with the lowest surface free energies, between 15 and 30 ergs cm^{-2} [1, 2].

Expressions such as that in Eq. (A-1) form the basis of so-called "broken bond" calculations. By putting in typical numbers for an fcc metal ($\Delta H_{coh} = 500$ kJ mol^{-1}, $Z_s/Z = 0.25$, $N_s = 10^{15}$ atoms cm^{-2}), we obtain a surface free energy of approximately 2000 ergs cm^{-2}. Thus, the surface free energy of metals rises and falls with the cohesive energy, implying that transition metals with half-filled d-bands have the largest surface free energies.

Minimization of the surface free energy is the driving force behind a number of surface processes and phenomena. Here, we mention a few:

- Surfaces are always covered by a substance that lowers the surface free energy of the system. Metals are usually covered by a monolayer of hydrocarbons, oxides often also by water (OH groups).
- Clean, polycrystalline metals expose mostly their most densely packed surface, because in order to create this surface, the minimum number of bonds must be broken [see Eq. (A-1)].
- Open surfaces such as fcc (110) often reconstruct to a geometry in which the number of neighbors of a surface atom is maximized.
- In alloys, the component with the lower surface free energy segregates to the surface, making the surface composition different from that of the bulk.
- Impurities in metals, such as C, O, or S, segregate to the surface because there they lower the total energy due to their lower surface free energy.
- Small metal particles on an oxidic support sinter at elevated temperatures because loss of surface area means a lower total energy. At the same time, a higher fraction of the support oxide with its lower surface free energy is uncovered. In oxidic systems, however, the surface free energy provides a driving force for spreading over the surface if the active phase has a lower surface free energy than the support.

A.2.3
Lattice Vibrations

Atoms are not rigidly bound to the lattice, but rather vibrate around their equilibrium positions. If we were able to examine the crystal over a very brief observation time, we would see a slightly disordered lattice. Incident electrons see these deviations, and this is for example the reason that in low-energy electron diffraction (LEED) the spot intensities of diffracted beams depend on temperature. At high temperatures the atoms deviate more from their equilibrium position than at low temperatures, and a considerable number of atoms is not at the equilibrium position necessary for diffraction. Thus, spot intensities are low and the diffuse background high. Similar considerations apply in other scattering techniques, as well as in extended X-ray absorption fine structure (EXAFS) and in Mössbauer spectroscopy.

Lattice vibrations are described as follows [9, 10]. If the deviation of an atom from its equilibrium position is u, then $\langle u^2 \rangle$ is a measure for the average deviation of the atom (the symbol $\langle \ \rangle$ represents the time average; note that $\langle u \rangle = 0$). This so-called mean-squared displacement $\langle u^2 \rangle$ depends on the solid and the temperature, and is characteristic for the rigidity of a lattice. Lattice vibrations are a collective phenomenon; they can be visualized as the modes of vibration

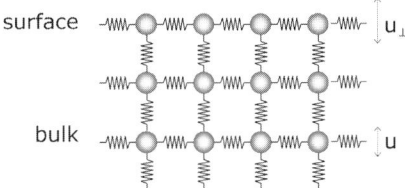

Fig. A.4 Visualization of lattice vibrations as coupled harmonic oscillations of spheres connected by springs.

one has in a system of spheres which are connected by springs (see Fig. A.4). Lattice vibrations are quantized, with the quantum being called a "phonon", and its energy distribution forming the phonon spectrum. Phonons are measured with vibrational spectroscopy, where they give characteristic absorption peaks at low frequencies, below about 500 cm^{-1}.

The simplest model to describe lattice vibrations is the Einstein model, in which all atoms vibrate as harmonic oscillators with one frequency. A more realistic model is the Debye model. In this case the atoms also act as harmonic oscillators, but now with a distribution of frequencies which is proportional to ω^2 and extends to a maximum called the Debye frequency, ω_D. It is customary to express this frequency as a temperature, the Debye temperature, defined by

$$\frac{h}{2\pi}\omega_D = k\theta_D \tag{A-2}$$

The Debye temperature characterizes the rigidity of the lattice: it is high for a rigid lattice, but low for a lattice with soft vibrational modes. The mean-squared displacement of the atom, $\langle u^2 \rangle$, can be calculated in the Debye model, and depends on the mass of the vibrating atom, the temperature, and the Debye temperature.

Atoms at the surface miss neighbors on the vacuum side. Hence, in the direction perpendicular to the surface the atoms have more freedom to vibrate than bulk atoms:

$$\langle (u_\perp^{surf})^2 \rangle > \langle (u^{bulk})^2 \rangle \rightarrow \theta_D^{surf} < \theta_D^{bulk} \tag{A-3}$$

Vibrations *in* the surface plane, however, will be rather similar to those in the bulk because the coordination in this plane is complete, at least for fcc (111) and (100), hcp (001) and bcc (110) surfaces. Thus, the Debye temperature of a surface is lower than that of the bulk, because the perpendicular lattice vibrations are softer at the surface. A rule of thumb is that the surface Debye temperature varies between about one-third and two-thirds of the bulk value (see Table A.2). Also included in Table A.2 is the displacement ratio, which is the ratio of the mean-squared displacements of surface and bulk atoms due to the lattice vibrations [1].

Table A.2 Ratio of surface and bulk displacements, and Debye temperatures (θ_D) of several metals [1].

Metal	Displacement ratio	θ_D Bulk	θ_D Surface
Ni	1.77	390	220
Rh	1.35	350	260
Pd	1.95	273	144
Ag	2.16	225	104
Pt	2.12	234	110

The Debye temperature of the bulk is a quantity that can be measured with several techniques, including X-ray diffraction (XRD), EXAFS, Mossbauer spectroscopy, and scattering of high-energy electron or neutrons. Briefly, the intensity of scattered particles is proportional to $\exp\{-k^2\langle u^2\rangle\}$, where k represents the momentum transfer in the scattering event. The Debye temperature can be calculated from $\langle u^2\rangle$. We can do the same for the surface region if we apply electron diffraction with low-energy electrons (LEED). The thickness of the layer one probes depends on the mean free path of the electrons. The best procedure is therefore to determine $\langle u^2\rangle$ as a function of electron energy and extrapolate to energies for which the mean free path is at minimum.

The fact that the surface Debye temperature is lower than that of the bulk has two consequences. First, the surface is always a weaker scatterer than the bulk. Second, the intensity of the surface signal decreases faster with increasing temperature than the intensity of the bulk signal. Sometimes, this property can be used to recognize surface behavior from measurements with bulk sensitive techniques [12].

Finally, if the temperature increases, $\langle u^2\rangle$ becomes larger until the crystal melts. The Lindemann criterion predicts that melting sets in when $\langle u^2\rangle$ becomes about 0.25 a^2, where a is the interatomic distance of the metal. Because the mean-squared displacements of surface atoms is higher, we expect that the surface melts at lower temperatures than the bulk does [2]. Indeed, evidence has been presented that the (110) surface of lead starts to melt at 560 K, whereas the bulk melting temperature is about 600 K [13].

So, how should we who are interested in catalysis investigate phonons? Lattice vibrations determine the spectral intensity in many spectroscopic techniques, and they often force us to take spectra at lower temperatures than we would prefer. Often, we cannot measure at catalytic reaction temperatures. Sometimes, however, we can use the phonons to our advantage when they enable us to associate certain spectral contributions with the surface region. Phonons also contribute to surface entropy. In fact, in special cases they may provide a driving force for segregation of species with the softer vibrations to the surface of multicomponent species [14].

A.2.4
Electronic Structure of Metal Surfaces

The purpose of this section is to highlight how the electronic properties of the surface differ from those of the bulk. In the next section, we explain how the electronic properties are involved in chemisorption, and why promoters can help molecules to chemisorb and to react on surfaces [5–8].

The theory of band structures belongs to the world of solid-state physicists, who like to think in terms of collective properties, band dispersions, Brillouin zones, and reciprocal space [9, 10]. This is not the favorite language of a chemist, who prefers to think in terms of molecular orbitals and bonds. Hoffmann provides an excellent and highly instructive comparison of the physical and chemical pictures of bonding [5]. In this appendix, we will try to use as much as possible the chemical language of molecular orbitals. However, before talking about metals we should recall a few concepts from molecular orbital theory.

The orbitals of a molecule A–B are formed from two atomic orbitals, φ_A and φ_B, by taking linear combinations. This gives two new orbitals – a bonding orbital at lower energy, and an antibonding orbital at higher energy. The difference in energy between the bonding orbital and the original atomic orbital is determined by the overlap of the two atomic orbitals (see Fig. A.5). The following rules apply:

- Each molecular orbital can accommodate two electrons. The combination of an occupied bonding and an unoccupied antibonding orbital gives optimum bonding between A and B. The combination of fully occupied bonding and antibonding orbitals does not lead to bonding between A and B (in chemisorption theory it is sometimes said that the interaction between A and B is *repulsive*).

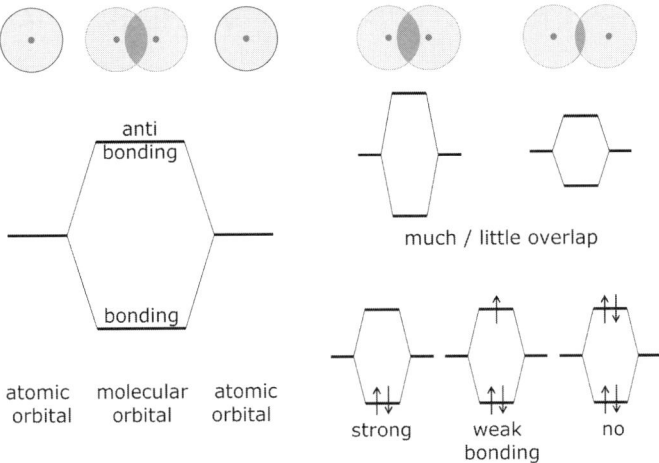

Fig. A.5 Molecular orbital energies depend on the overlap of the constituent atomic orbitals; the bond strength depends also on the occupation of the orbitals.

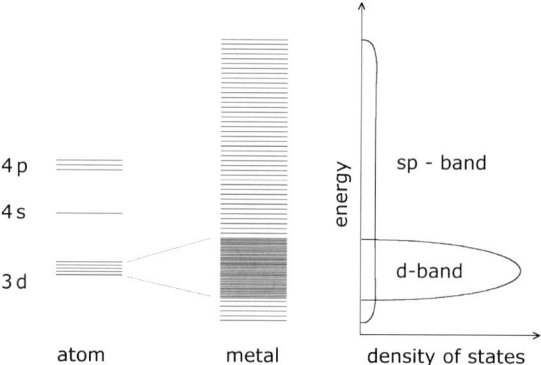

Fig. A.6 Simplified density of states of metals showing the broad band of the delocalized s- and p-electrons, and the narrower band of the more localized d-electrons.

Suppose that by means of an external cause – for example, the metal surface of a catalyst – the otherwise empty antibonding orbital of a molecule becomes partially filled. As a result, the bond is weakened and may eventually break. Before we discuss the relationship between catalysis and electronic structure, we should examine bonding in metals.

The metallic bond can be seen as a collection of molecular orbitals between a large number of atoms. As Figure A.6 illustrates, the molecular orbitals are very close and form an almost continuous "band" of levels. It is impossible to detect that the levels are actually separated from each other. Rather, the bands behave in many respects similarly to the orbitals of the molecule in Figure A.5: if there is little overlap between the electrons, the interaction is weak and the band is narrow. This is the case for the d-electrons of the metal. Atomic d-orbitals have pronounced shapes and orientations that are largely retained in the metal. This in contrast to the s electrons, which are strongly delocalized; that is, they are not restricted to well-defined regions between atoms, and form an almost free "electron gas" that spreads out over the whole metal. Hence, the atomic s-electron wave functions overlap to a great extent, and consequently the band they form is much broader.

Thus, a band is nothing but a collection of a large number of molecular orbitals. The lower levels in the band are bonding, the upper ones antibonding, and the ones in the middle non-bonding. Metals with a half-filled band have the highest cohesive energy, the highest melting points, and the highest surface free energies.

Each energy level in the band is called a *state*. However, the important quantity to consider is the density of states (DOS) – that is, the number of states at a given energy. We like to simplify and draw the DOS of transition metals as the smooth curves of Figure A.6, but in reality DOS curves show a complicated structure, due to crystal structure and symmetry [5, 15]. The bands are filled with valence electrons of the atoms up to the Fermi level. In a molecule, one would refer to this level as the Highest Occupied Molecular Orbital (HOMO).

In photoemission experiments, the Fermi level shows up as a sharp edge, and therefore it is often chosen as a convenient zero of the energy scale. It should be noted however, that electrons at the Fermi level are bound to the metal, and the energy necessary to disconnect them from the metal is equal to the work function (the molecular analogue would be the ionization potential). This property is discussed later.

For the Group VIII transition metals the d-band is partially filled and the Fermi level is in the d-band. The Group IB metals have a completely filled d-band, and here the Fermi level falls above the d-levels in the s-band. Two trends in going from left to right through the metals in the Periodic Table are that the d-band becomes narrower and that the Fermi level decreases with respect to the vacuum level.

The Fermi level can be visualized as the surface of an electron reservoir sometimes called the "Fermi sea". If for some reason an empty level exists with energy lower than E_F, it is filled immediately. Likewise, if an electron occupies temporarily a level above E_F, it will readily fall back to the Fermi level.

In order to discuss the DOS of metal surface atoms, we need to take a closer look at those orbitals that have a distinct orientation, the d-orbitals. Let us take an fcc crystal and see how the d-orbitals combine to form bands. Figure A.7 shows schematically the shape of the d-orbitals. For clarity, only the lobes in the (yz) plane are shown, while the toroidal component of the d_{z^2} orbital has been omitted. If these orbitals are placed within the fcc structure (as in Fig. A.7), one readily sees that the best overlap between nearest neighbors in the (yz) or (100) plane occurs between the d_{yz} orbitals, whereas the orbitals along the y- and z-axes of the cube overlap to a much smaller extent. This is also true in the other planes of an fcc metal, as the d_{xy}, d_{xz}, and d_{yz} orbitals overlap more than do the d_{z^2} and $d_{x^2-y^2}$. Hence, the density of states inside the metal shows a narrow band for d_{z^2} and $d_{x^2-y^2}$, and a wider band for the d_{xy}, d_{xz}, and d_{yz} orbitals [6].

The question then, is what happens to these bands when we move to the surface of the crystal? The creation of a surface implies that bonds are broken and that neighbors are missing on the outside. The orbitals affected by bond breaking no longer have any overlap with that of the removed atom, and thus the band becomes narrower. Figure A.7 illustrates this for the (yz) or (100) surface. The nearest neighbors in front of this plane are missing, which means that the overlap of the d_{xz} and d_{xy} orbitals has decreased. The d_{yz} orbital, however, lies in the plane of the surface and has remained fully coordinated, as it is in the bulk. Thus, the d_{xz} and d_{xy} bands of the surface in Figure A.7 are narrower than the d_{yz} band, which is similar to that of the bulk. The $d_{x^2-y^2}$ orbital of atoms in the (yz) surface is also partially unsaturated, as well as the s-orbital. Thus, these bands will also be narrowed at the surface [6].

Suppose that an atom adsorbs onto the surface. Which bands of the fcc (100) surface are candidates to be involved in chemisorption bonds? Most likely, these

308 | *Appendix: Metal Surfaces and Chemisorption*

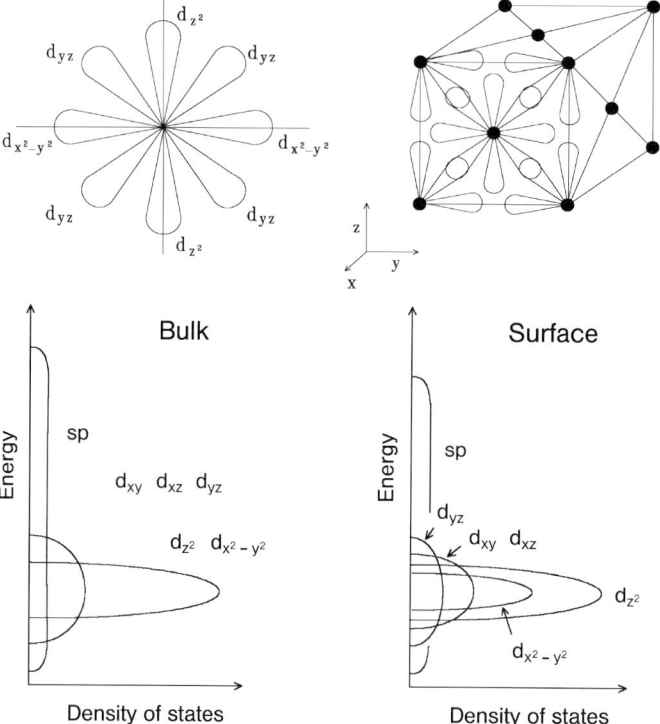

Fig. A.7 The atomic d-orbitals and the formation of d-bands in a (100) plane of an fcc metal, along with the density of states. Note that the density of states of surface atoms narrows for those bonds that are unsaturated.

will be the coordinatively unsaturated orbitals: d_{xy}, d_{xz}, $d_{x^2-y^2}$ and s, but we will return to this point later. For now, we will concentrate on the surface DOS.

In general one can say that those parts of the band that correspond to bonds which have been broken in order to create the surface, are narrower. A similar effect can be expected for small particles: the average coordination number of the atom decreases and the bands are narrower. This effect can be observed in photoemission experiments (an example is shown in Fig. 3.19).

The narrowing of d-bands at the surface may have consequences for filling of the d-band of surface atoms and the location of the Fermi level within the band. Consider a metal in which the d-band almost filled, such as palladium. Because the d-band of the surface atom narrows, the Fermi level (which is set by the entire metal) falls above the d-levels in the s-band. The implication is that the surface atoms have a slight excess negative charge. This effect is observable in the core-level binding energies of the atoms, which is known as the surface core-level shift: core-level binding energies of Pd surface atoms are slightly lower than those of bulk atoms [2]. Surface core-level shifts can only be detected if one measures XPS spectra at a synchrotron. Here, the incident X-ray energy can be chosen such that the photoelectrons have kinetic energies corresponding to the minimum inelastic mean free path (see Chapter 3).

Although our discussion of electronic structure has been limited to terms of band filling, there is of course much more to learn about band structures. The DOS is only a highly simplified representation of the actual electronic structure, and ignores the three-dimensional structure of electron states in the crystal lattice. Angle-dependent photoemission provides information on this property of the electrons, and for this the interested reader is referred to standard books on solid-state physics [9, 10] and photoemission [16, 17]. The interpretation of photoemission and X-ray absorption spectra of catalysis-oriented questions, however, is usually conducted only in terms of the electron density of states.

A.2.5
Work Function

The second electronic property that plays an important role at surfaces is the work function. The work function is a type of binding energy, and is the minimum energy needed to remove an electron from the Fermi level to a state where it is at rest, without interacting with the solid. As such it equals the energy difference between the vacuum level and the Fermi level. In "chemical language", it is similar to the ionization potential – the energy needed to remove an electron from the HOMO. If the vacuum level is taken as the potential of an electron at an infinite distance from the metal, one defines the work function as a macroscopic quantity, which is an average over the surface. It can be measured by either by UPS (as explained in Chapter 3), or by a vibrating capacitor method (the Kelvin probe) [3, 16]. First, however, we must determine the origin of the work function.

In order to bring an electron outside the solid, it must first cross a surface. If it were possible to remove the electron from the solid without going through a surface, one would have to overcome the bulk contribution to the work function (equal to the chemical potential). There is also a surface contribution to the work function, which depends on the density of atoms at the surface. Clearly, it is this contribution which makes the work function both surface- and site-specific.

In order to explain the origin of the surface contribution to the work function, we need a model for the electron distribution in the surface region of a metal; one of the simplest is the "jellium model" [18].

The hypothetical metal "jellium" consists of an ordered array of positively charged metal ions surrounded by a structure-less sea of electrons which behaves as a free electron gas (Fig. A.8). The attractive potential due to the positively charged cores is not strong enough to keep the valence electrons inside the metal, and as a result the electrons spill out into the vacuum; that is, the electron density just outside the surface is not zero. Because the charge of these electrons is not compensated by positive ions, a dipole layer exists at the surface, with the nega-

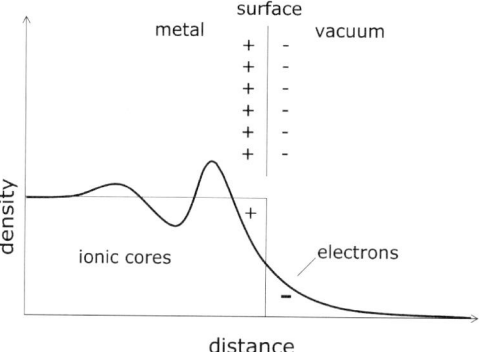

Fig. A.8 The electron distribution in the model metal jellium gives rise to an electric double layer at the surface which forms the origin of the surface contribution to the work function.

tive end to the outside. An electron traveling from the solid to the vacuum must overcome this barrier of height, $\Delta\phi$. This energy, which is necessary for surmounting the surface dipole layer, is the surface contribution to the work function. It depends very much on the structure of the surface: for the fcc metals the (111) surface is the most densely packed surface, and it has the largest work function because the dipole barrier is high. A more open surface has a smaller work function. In addition, when a surface contains many defects the work function is lower than for the perfect surface. For a given surface structure, the work function increases from left to right in the Periodic Table.

The work function plays an important role in catalysis, as it determines how easily an electron may leave the metal to perform a useful task in the activation of reacting molecules. However, strictly speaking, the work function is a *macroscopic* property, whereas chemisorption and catalysis are locally determined phenomena. The latter pair need to be described in terms of short-ranged interactions between adsorbed molecules and one or more atoms at the surface. The point to be made here is that, particularly for heterogeneous surfaces, the concept of a macroscopic work function – which is the average over the entire surface – is not very useful. Rather, it would be more meaningful to define the work function as a *local* quantity on a scale with atomic dimensions.

Let us assume that we have a facetted surface, or a surface with perfect and defect domains. The potential barrier that an electron has to surmount when leaving the solid through a defect region or through an open facet, for example of (110) orientation, is lower than that when the electrons pass through the (111) surface. The relevant quantity in chemisorption theory is not the averaged macroscopic work function, but rather is the work function of the site where a molecule adsorbs. We therefore define the local work function of a site as the difference between the potential of an electron just outside the surface dipole layer and the

Fermi level of the metal [19]. This is the value that one measures with scanning tunneling microscopy (see Chapter 7) and with photoemission of adsorbed xenon (see Chapter 3). Thus, on a heterogeneous surface we have local work functions for each type of site, and the macroscopic work function is an average over these values.

A.3
Chemisorption on Metals

In this section we provide a simple and qualitative description of chemisorption in terms of molecular orbital theory. It should provide us with a feeling for why some atoms such as potassium or chlorine acquire positive or negative charge upon adsorption, while other atoms remain more or less neutral. We explain qualitatively why a molecule adsorbs associatively or dissociatively, and we discuss the role of the work function in dissociation. This text is meant to provide some elementary background for the chapters on photoemission, thermal desorption, and vibrational spectroscopy. We avoid theoretical formulae and refer the reader to the literature for thorough treatments of chemisorption [2, 5–8].

One good starting point for discussing chemisorption is that of the resonant level model. The substrate metal is jellium, implying that we are examining metals without d-electrons, and the adsorbate is an atom. We focus on only two electron levels of the atom. Level 1 is occupied and has ionization potential I; level 2 is empty and has electron affinity ε_A. (The ionization potential is the minimum energy required to remove an electron from the atom and place it at rest just outside the atom; the electron affinity is the energy gained when an electron at rest just outside the solid falls into an empty orbital of the neutral atom.)

When the atom comes closer to the metal surface, the electron wave functions of the atom begin to feel the charge density of the metal. The result is that levels 1 and 2 broaden into so-called resonance levels, which have a Lorentzian shape. Strictly speaking, the broadened levels are no longer atomic states, but rather are states of the combined system of atom plus metal, although they retain much of their atomic character. Figure A.9 illustrates the formation of broadened adsorbate levels in a potential energy diagram. Figure A.10a provides a schematic representation of the type we will use in the following text. The UPS spectrum of this system would show the occupied part of the density of states of the metal with an additional broad peak from the occupied adsorbate level (as in Figs. 3.19 and 3.20).

In the situation as sketched in Figures A.9 and A.10a, level 1 remains occupied and level 2 empty, implying that the adsorbate atom retains the same charge as in the free atom. However, other situations can also arise. Suppose that the atom has a low ionization potential, smaller than the work function of the metal. Then, the broadened level 1 falls largely above the Fermi level of the metal, with the result that most of the electron density of level 1 ends up on the metal; hence,

Appendix: Metal Surfaces and Chemisorption

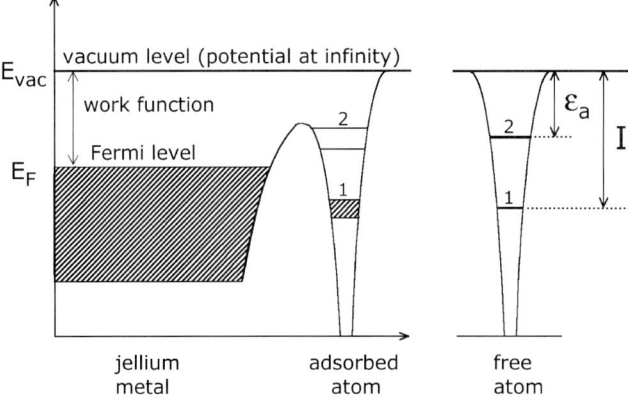

Fig. A.9 A potential energy diagram of an atom chemisorbed on the model metal jellium shows the broadening of the adsorbate orbitals in the resonant level model.

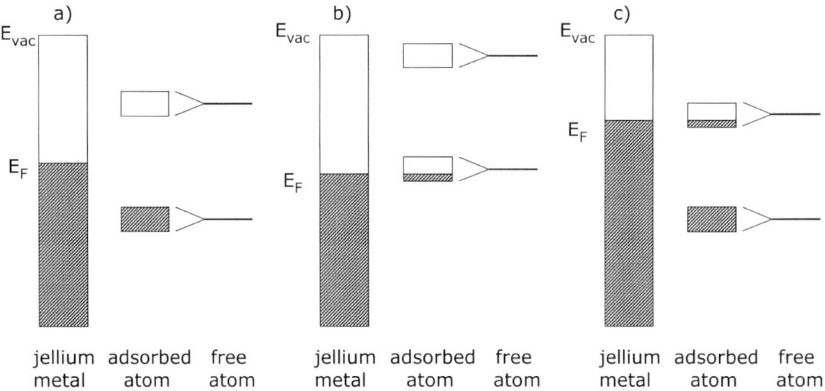

Fig. A.10 Potential energy diagrams of atoms chemisorbed on jellium in the resonant level model. The situation under (a) corresponds to Figure A.9. In (b) the adatom has a low ionization potential and consequently it donates charge to the metal (as with alkalis on metal), whereas in (c) the adatom has a high electron affinity such that it becomes negatively charged (as with fluorine on metals).

the adatom is positively charged (Fig. A.10b). This happens with alkali atoms on many metal surfaces (see for example the discussion of potassium on rhodium in Chapter 9).

The opposite occurs for atoms with a high electron affinity that is on the order of the metal work function, or higher. Here, the broadened level 2 falls partly below the Fermi level and becomes partially occupied (Fig. A.10c). In this case, the

Table A.3 Ionization potentials (I) and electron affinities (ε_A) of catalytically relevant elements.

Element	I [eV]	ε_A [eV]	Element	I [eV]	ε_A [eV]
H	13.6	0.7	Na	5.1	0.7
C	11.3	1.1	K	4.3	–
N	14.5	0	Cs	3.9	–
O	13.6	1.5	F	17.4	3.6
P	11.0	0.8	Cl	13.0	3.8
S	10.4	2.1	Br	11.8	3.5
Li	5.4	0.5	I	10.4	3.2

adatom is negatively charged. Examples are the adsorption of electronegative species such as F and Cl. The ionization potentials and electron affinities of some catalytically relevant atoms are listed in Table A.3.

This picture of chemisorbed atoms on jellium, although much too simple, illustrates a few important aspects of chemisorption. First, the electron levels of adsorbed atoms broaden due to the interaction with the s-electron band of the metal. This is generally the case in chemisorption. Second, the relative position of the broadened adsorbate levels with respect to the Fermi level of the substrate metal determines whether charge transfer between metal and adatom takes place, and in which direction.

The resonant level model readily explains the change in work function associated with chemisorption. It is well known that alkali atoms such as potassium reduce the work function of the substrate, whereas electronegative atoms such as chlorine increase the work function [2, 6, 19]. Figure A.10 indicates that potassium charges positively and chlorine negatively when adsorbed on jellium. It must be remembered that the surface contribution to the work function is caused by a dipole layer at the surface, with the negative end towards the vacuum (see Fig. A.8). The adsorption of a negatively charged particle strengthens the surface dipole at the adsorption site, and thus increases the work function. The adsorption of a positively charged particle, on the other hand, weakens the dipole moment and thus decreases the work function locally.

Although the resonant level model successfully explains a few general aspects of chemisorption, it has nevertheless many shortcomings. The model provides no information on the electronic structure of the chemisorption bond; neither does it tell us where the electrons are. Such information is obtained from a more refined model, called the "density functional method". At this point we will not explain how this works, but will simply show the results for the adsorption of Cl and Li on jellium, as reported by Lang and Williams [20].

Figure A.11 shows the change in density of states due to the chemisorption of Cl and Li. Note that the zero of energy has been chosen at the vacuum level, and that all levels below the Fermi level are filled. For lithium, we are looking at the

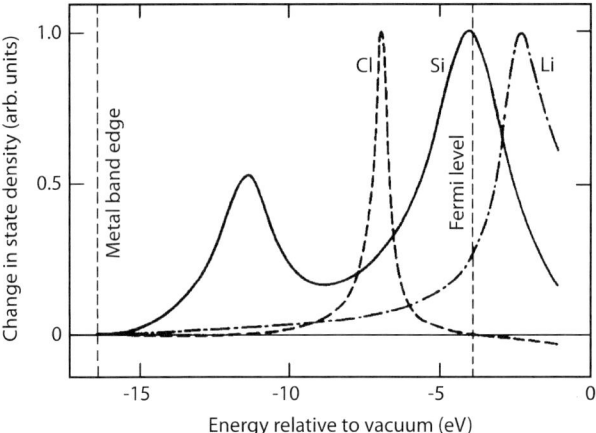

Fig. A.11 Density functional calculations show the change in the density of states induced by adsorption of Cl, Si, and Li on jellium. Lithium charges positively, and chlorine negatively. (From [20]).

broadened 2s level with an ionization potential in the free atom of 5.4 eV. The density functional calculation tells that chemisorption has shifted this level above the Fermi level so that it is largely empty. Thus, lithium atoms on jellium are present as Li$^{\delta+}$, with δ almost equal to 1. Chemisorption of chlorine involves the initially unoccupied 3p level, which has a high electron affinity of 3.8 eV. This level has shifted down in energy upon adsorption and ended up below the Fermi level, where it has become occupied. Hence, the charge on the chlorine atom is about −1.

A map of how the electron density is distributed around these atoms provides important information. It tells us to what distance from the adatom the surface is perturbed or, in catalytic terms, how many adsorption sites are promoted or poisoned by the adatom. The charge density contours in Figure A.12 are lines of constant electron density. Note that these contours follow the shape of the adsorbed atom closely, and that the electrons are very much confined to the adsorbed atom and the adsorption site.

Even more interesting are the difference plots, indicating the difference between the charge contours of the upper panel and the situation in which there would be no interaction between the adatom and the metal surface. In Figure A.12 the solid lines represent an increase in electron density (excess negative charge), and the dashed lines a decreased electron density (depletion of negative charge). These plots also make clear that Cl is charged negatively and adsorbed onto a positively charged adsorption site. This dipole increases the work function locally, most prominently on the adsorption site itself, but also marginally on the nearest neighbor site. For Li the situation is just the reverse as for Cl. The calcu-

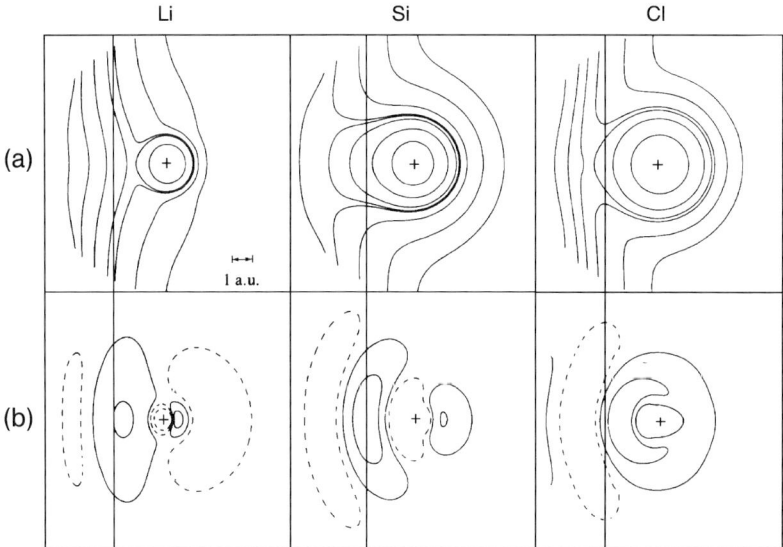

Fig. A.12 Charge density contours for the adsorption of Cl, Si, and Li on jellium. (a) Total charge and (b) induced charge. Solid lines indicate an increase in electron density, and dashed lines a decrease. Note the formation of strong dipoles for Li and Cl adsorption. (From [20]).

lations of Lang and Williams indicate that the adsorption of Li and Cl is a highly local phenomenon: sites adjacent to the adatom are perturbed, but the next nearest site is almost that of the clean metal. The latter is to some extent an artifact of the calculations, because a single adatom has been placed on a furthermore empty surface. How potassium at higher concentrations changes the potential of the entire rhodium surface was discussed in Chapter 9.

A.3.1
Adsorption of Molecules on Jellium

If we wish to understand the conditions under which a diatomic molecule such as H_2, N_2, or CO dissociates on a surface, we need to take two orbitals of the molecule into account – the highest occupied and the lowest unoccupied molecular orbital (the HOMO and LUMO of the so-called frontier orbital concept). Let us take a simple case to start with: the molecule A_2 with occupied bonding level σ and unoccupied antibonding level σ^*. We use jellium as the substrate metal and discuss the chemisorption of A_2 in the resonant level model. What happens is that the two levels broaden due to the rather weak interaction with the free electron cloud of the metal.

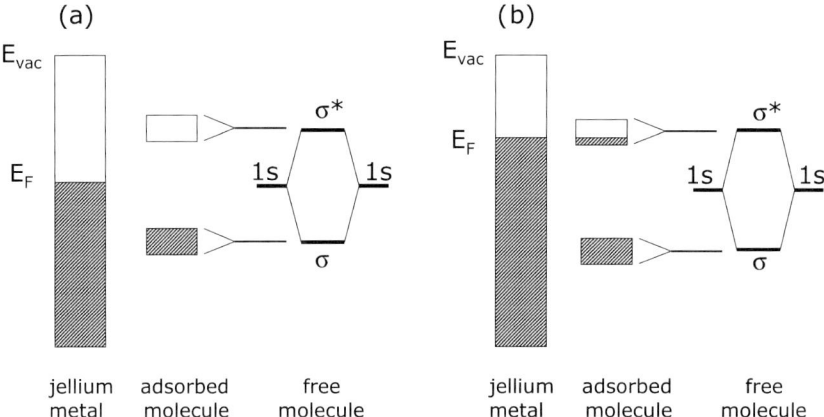

Fig. A.13 Potential energy diagram for the adsorption of a molecule on jellium. In situation (b) the antibonding orbital of the adsorbate is partially occupied, implying that the intramolecular bond is weakened.

We consider two cases (see Fig. A.13). First, the metal has a work function that is between the electron affinity (the energy of the σ^*-level) and the ionization potential (the energy of the σ-level) of the molecule. Upon adsorption, the levels broaden. However, the occupation of the adsorbate levels remains as in the free molecule. This situation represents a rather extreme case in which the intramolecular bond of the adsorbate molecule stays about as strong as in the gas phase. The other extreme occurs if both the σ-level and the σ^*-level fall below the Fermi level of the metal. Because the antibonding σ^*-level is filled with electrons from the metal, the intramolecular bond breaks. This is the case for hydrogen adsorption on many metals. Thus, a low work function of the metal and a high electron affinity of the adsorbed molecule are favorable for dissociative adsorption.

The intermediate case is that the antibonding level of the molecule broadens across the Fermi level, as shown in Figure A.13. It becomes partially filled, and consequently the intramolecular bond is weakened. The relevance for catalysis is that this type of chemisorption weakens (or "activates" in catalytic language) the A–A bond and, as a result, it may be more reactive than in the gas phase. In molecules such as CO and NO this weakening of the bond is readily observed with infrared spectroscopy. This partial filling of the antibonding orbital of a chemisorbed molecule is often referred to as "back donation".

The picture sketched in Figure A.13 is a gross oversimplification, although it comes close to the situation for H_2 chemisorption on aluminum. It also explains correctly the role that the work function plays in the dissociation of a molecule. However, a low work function is not the only reason for dissociation. The interaction of the adsorbed molecule with the d-levels of the metal is at least as important.

A.3.2
Adsorption on Metals with d-Electrons

The distinctive feature of the catalytically active metals is that they possess between six and 10 d-electrons, which are much more localized on the atoms than are the s-electrons. The d-electrons certainly do not behave as a free electron gas; instead, they spread over the crystal in well-defined bands which have retained characteristics of the atomic d-orbitals.

Clearly, chemisorption on d-metals requires a different description than chemisorption on a jellium metal, and with the d-metals we must think in terms of a "surface molecule", with new molecular orbitals made up from d-levels of the metal and the orbitals of the adsorbate. These new levels interact with the s-band of the metal, similarly as in the resonant level model. We start with the adsorption of an atom, in which only one atomic orbital is involved in chemisorption. Once the principle is clear, it is not difficult to invoke more orbitals.

The simple picture for the chemisorption of an atom on a metal with d-electrons in Figure A.14 arises as follows. First, we construct molecular orbitals from the atomic orbital of the adsorbate atom and the entire d-band. This produces a pair of bonding and antibonding chemisorption orbitals. Second, these new orbitals are broadened and perhaps shifted by the interaction with the free electron s-band of the metal.

The formation of molecular orbitals implies that there is a bonding and an antibonding orbital. The strongest bond occurs if only the bonding orbital is occupied, and several situations may arise:

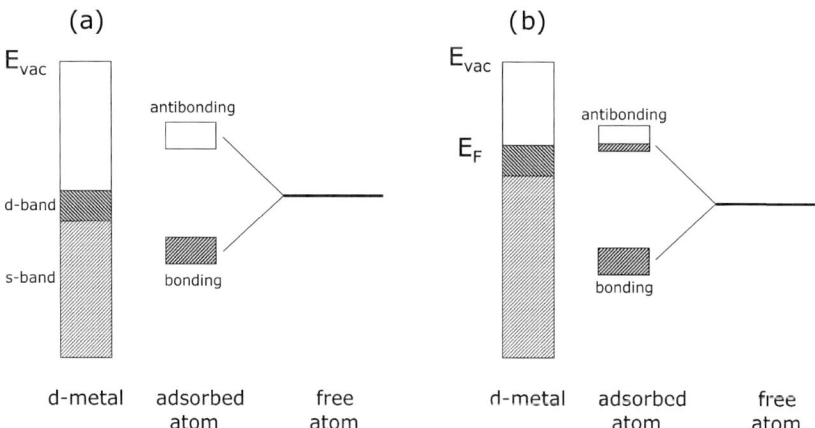

Fig. A.14 Energy diagram for the adsorption of an atom on a d-metal. Chemisorption is described with molecular orbitals constructed from the d-band of the metal and atomic orbitals of the adatom. The chemisorption bond in (b) is weaker, because the antibonding chemisorption orbital is partially filled (compare Fig. A.5).

- If the antibonding chemisorption orbital lies entirely above the Fermi level, it remains empty and a strong chemisorption bond results (Fig. A.14a).
- If the interaction between the atomic orbital and the d-band is weak, the extent of splitting between the bonding and the antibonding orbital of the chemisorption bond is small. The antibonding orbital falls below the Fermi level and is occupied; this represents a repulsive interaction and does not lead to bonding.
- Intermediate cases in which the antibonding chemisorption orbital is broadened across the Fermi level can also arise (Fig. A.14b). In such cases, the antibonding orbital is only partially filled and the atom A will be chemisorbed, albeit with a weaker chemisorption bond than in Figure A.14a.

It is not difficult to draw a qualitative picture for the chemisorption of a molecule on a d-metal (see Fig. A.15). Here, we use again the diatomic molecule A_2 with a filled bonding σ-orbital as the HOMO and an empty antibonding σ^*-orbital as the LUMO. These are the necessary steps:

(a) Construct new molecular orbitals from the HOMO, in this case the bonding orbital σ of A_2, and levels in the d-band that have appropriate orientation and symmetry.
(b) Do the same for the LUMO, here the antibonding σ^* of A_2, and d-levels of appropriate orientation and symmetry.
(c) Let all levels broaden further by the interaction with the s-band.

What are the important things to consider? First, the interaction under (a) between σ- and occupied d-levels provides in principle a repulsive interaction, be-

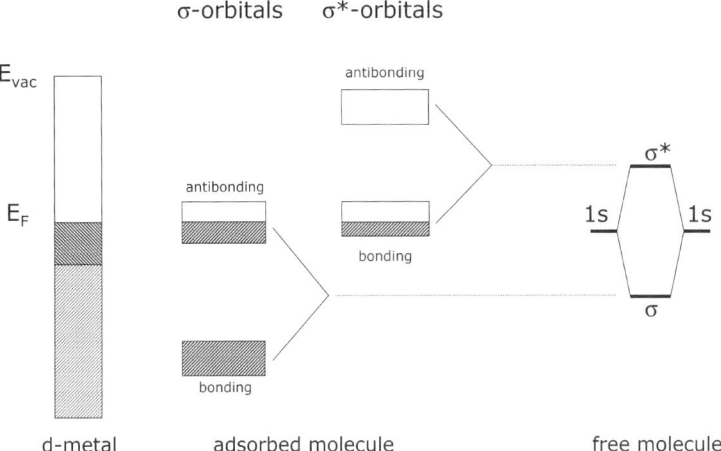

Fig. A.15 Energy diagram for the adsorption of a simple diatomic molecule on a d-metal. Chemisorption orbitals are constructed from both the bonding and the antibonding level of the molecule. As the latter becomes partially occupied, the intramolecular bond of the adsorbate has been activated.

cause both the bonding and the antibonding chemisorption orbitals will be occupied. However, the antibonding orbital may fall above the Fermi level, implying that the repulsion is partially or entirely relieved.

Second, interaction (b) gives a bonding orbital, which can either be above or below the Fermi level. If it is fully occupied, the adsorbate molecule dissociates, but if it is only partially occupied then it will contribute to bonding between A_2 and the surface, while at the same time the bond A–A in the chemisorbed molecule is weakened. This situation, which is shown in Figure A.15, very much resembles the adsorption of CO on transition metals, although the chemisorption orbitals are much more complex than the simple block-shaped bands of Figure A.15.

Density functional theory has become the standard tool to compute the geometry, energy, and vibrational frequencies of adsorbed species. Computational chemistry has developed into an almost routine ingredient of research in catalysis and surface science. Thus, the reader is referred to textbooks [21, 22] and reviews [23, 24] relating to this fascinating subject which – unfortunately – is beyond the scope of this book.

A.3.3
Concluding Remarks

Insight into chemisorption bonds is important for the interpretation of photoemission and vibrational spectra of adsorbed molecules, of temperature-programmed desorption data, and for the understanding of catalytic reactions. If we have a feeling for how electrons rearrange over orbitals when a molecule adsorbs, we may be able to understand why molecules dissociate or not, to what extent they are activated, or also how a promoter influences the reactions on the surface.

References

1 (a) G.A. Somorjai, *Chemistry in Two Dimensions: Surfaces*. Cornell University Press, Ithaca, 1981; (b) G.A. Somorjai, *Introduction to Surface Chemistry and Catalysis*. Wiley, New York, 1994.
2 A. Zangwill, *Physics at Surfaces*. Cambridge University Press, 1988.
3 G.A. Attard and C.J. Barnes, *Surfaces*. Oxford University Press, 1998.
4 R.I. Masel, *Principles of Adsorption and Reaction on Solid Surfaces*. Wiley, New York, 1996.
5 R. Hoffmann, *Solids and Surfaces*. VCH, Weinheim, 1988.
6 R.A. van Santen, *Theoretical Heterogeneous Catalysis*. World Scientific, Singapore, 1991.
7 K.W. Kolasinski, *Surface Science, Foundations of Catalysis and Nanoscience*. Wiley, Chichester, 2002.
8 I. Chorkendorff and J.W. Niemantsverdriet, *Concepts of Modern Catalysis and Kinetics*. Wiley-VCH, Weinheim, 2003.
9 N.W. Ashcroft and N.D. Mermin, *Solid State Physics*. Holt, Rinchart and Winston, New York, 1976.
10 C. Kittel, *Quantum Theory of Solids*. Wiley, New York, 1963.
11 J.M. MacLaren, J.B. Pendry, P.J. Rous, D.K. Saldin, G.A. Somorjai, M.A. van Hove, and D.D. Vvedensky, *Handbook of Surface Structures*. Reidel, Dordrecht, 1987.

12 J.W. Niemantsverdriet, A.M. van der Kraan, and W.N. Delgass, *J. Catal.* **89** (1984) 138.
13 J.W.M. Frenken and J.F. van der Veen, *Phys. Rev. Lett.* **54** (1985) 134.
14 A.D. van Langeveld and J.W. Niemantsverdriet, *J. Vac. Sci. Technol.* **A5** (1987) 558.
15 V.L. Moruzzi, J.F. Janak, and A.R. Williams, *Calculated Electronic Properties of Metals*. Pergamon, New York, 1978.
16 G. Ertl and J. Küppers, *Low Energy Electrons and Surface Chemistry*. VCH, Weinheim, 1985.
17 B. Feuerbacher, B. Fitton, and R.F. Willis (Eds.), *Photoemission and the Electronic Properties of Surfaces*. Wiley, New York, 1983.
18 N.D. Lang and W. Kohn, *Phys. Rev. B* **1** (1970) 4555.
19 K. Wandelt, in: *Thin Metal Films and Gas Chemisorption*, P. Wissmann (Ed.). Elsevier, 1987, p. 280.
20 N.D. Lang and A.R. Williams, *Phys. Rev. B* **18** (1978) 616.
21 C. Fiolhais, F. Nogueira, and N. Marques (Eds.), *A Primer in Density Functional Theory*. Springer-Verlag, Berlin 2003.
22 R.A. van Santen and M. Neurock, *Molecular Heterogeneous Catalysis: A Conceptual and Computational Approach*. Wiley-VCH, Weinheim, 2006.
23 B. Hammer and J.K. Nørskov, *Adv. Catal.* **45** (2000) 71.
24 Q. Ge, R. Kose, and D.A. King, *Adv. Catal.* **45** (2000) 207.

Index

a

activation energy 1, 17–20, 33, 25–29, 33–37, 103, 211
adsorption 2, 69
alkali promoter 70, 71, 264–271
alloy, surface segregation 208, 302
ammonia synthesis 2, 19, 97, 98, 104, 117, 121, 264
ammoxidation 2, 87, 88, 98, 117
angle-dependent XPS 59–62
annular dark field 186–189
anomalous scattering 154
area, specific 2, 3
Arrhenius equation 17, 25, 33, 35
atom probe microscopy 197
atomic force microscopy 180, 197–208, 233, 285, 290, 291
attenuated total reflection 224, 225, 233, 234
attenuation length, electron 45
Auger decay 41, 42, 74–78, 90, 113, 190
Auger electron spectroscopy 40–42, 74–81, 184, 252, 260, 280
automotive exhaust catalysis 1, 36, 37, 98, 102, 117, 252, 264

b

back donation 234, 316
backscattered electron 181, 184
backscattering (in AES) 78
backscattering amplitude (EXAFS) 162–164
band structure 305
bimetallics 26
body-centered cubic 298
bonding in metals 306
Bragg diffraction 184
Bragg's law 148–151
Bravais lattice, surface 300

Bremsstrahlung 148, 190, 192
Bremstrahlung isochromat spectroscopy 67
bridging the gap strategy 8
broken-bond model 301, 302
Brunauer–Emmett–Teller equation 2

c

catalysis, heterogeneous 1, 297
catalytic cycle 2
cathodoluminescence 182
charge-potential model 48, 49
charging 52, 53, 80, 95, 184, 185, 252
chemical potential 309
chemical promoter 4, 265
chemical shift 46
chemisorption 3, 311–319
CO hydrogenation 2, 22, 23, 100, 141–143
CO oxidation 2, 31, 35, 36, 213, 214
cohesive energy 301, 306
compensation effect 29, 30
contracting sphere model 16
coordination number 163, 164, 168, 169, 189, 258, 259, 280, 298
core hole 42, 48, 52, 69, 74–77
corrugated surface 299
Coulomb interaction 109, 112, 199
Coulomb potential 112
cross section 44, 45, 133, 239
crystallography, surface 297–301
cyclotron 110

d

Debye model 123, 124, 303
Debye temperature 124, 137, 159, 303, 304
degree of freedom 220
density functional theory 313, 319
density of states 49, 50, 66–68, 207, 265, 306–311

depth profiling 80, 81
desorption 23–35
diffraction 147–159
diffuse reflection infrared Fourier transform spectroscopy 224, 225, 230–233
dipole coupling 224, 227, 228
dipole moment 265–268
dipole scattering 243, 244
dispersion 3, 54–59, 167–170, 255
dissociative adsorption 2, 101–103, 316–319
Doppler effect 126
dual anode source 42

e

Einstein model 303
electric quadrupole splitting 129–131, 134
electromagnetic spectrum 6, 7
electron affinity 92, 311–319
electron energy loss spectroscopy 217, 224, 227, 243–247, 268–271
electron microprobe analysis 190–193
electron microscopy 179–211
electron spectroscopy for chemical analysis 39
electronegativity 47
electronic structure, of surface 305–311
electrostatic potential 266–269
elementary reaction 103, 104
ellipsometry 213, 214
ellipsometry microscopy for surface imaging 212–214
emission, kinetic and potential 90, 95
energy dispersive X-ray analysis 190, 193
ensemble effect 227–229
environmental scanning electron microscopy 185
ethylbenzene dehydrogenation 213
ethylene epoxidation 62, 98
extended X-ray absorption fine structure 147, 148, 159–175, 232, 254, 257–264, 274, 280–284, 291

f

face-centered cubic 298
fast atom bombardment 90
Fermi level 49, 50, 66–70, 173, 195, 207, 265, 269, 307–319
Fermi's Golden Rule 67
field emission 194
field emission microscopy 180, 193–197
field ion microscopy 180, 193–197
field ionization 195
final state effect 48, 51, 63, 67, 69, 164

Fischer–Tropsch synthesis 2, 15, 19, 117, 121, 135–137, 174, 175, 180, 264
fluorescence 239
Fourier Transform 162–168
Fourier Transform infra red 226
Fowler Nordheim equation 194
free electron gas 306, 309, 317
frontier orbital concept 315
fundamental vibrations 221

g

group frequency 222

h

harmonic oscillator 218–220
heat of adsorption 25
heat of vaporization 27
Heisenberg's uncertainty relation 52
heterogeneous catalysis 1, 297
hexagonally closed packed 298
high resolution electron energy loss spectroscopy *see* EELS
high-angle annular dark field 186–189
highest occupied molecular orbital 307, 309
hydrodenitrogenation 272
hydrodesulfurization 21, 22, 99, 117, 121, 208–210, 229, 230, 272–284
hyperfine interactions 122, 126–136

i

image potential 194
impact scattering 243, 244
impregnation 3
in situ 4, 9, 121, 135, 139–142, 152, 153, 171–175, 190, 211, 224, 225, 235–239, 277
inelastic mean free path *see* mean free path
inelastic neutron scattering 217
Infrared emission spectroscopy 224, 225, 235
infrared imaging 214
infrared radiation 218
Infrared reflection absorption spectroscopy 226
infrared regions 222
infrared spectroscopy 217–220, 224–235, 239, 261, 262, 279, 289
initial state effect 48, 51, 63, 67
inverse photoemission 270
ion scattering 106–110
ion scattering spectroscopy 85, 106–117
ionization potential 92–94, 195, 265, 307–319
ionization probability (in AES) 77, 88, 92

isoelectric point 241, 254
isomer shift 128–131, 134, 277
isotopes in infrared spectroscopy 220, 221, 227
isotopes in mass spectrometry 36, 37
I–V plots 158

j
jellium model 265, 266, 309, 311–313

k
Kelvin probe 309
kinematic factor 106–110, 114
kinetic parameters 103
Kubelka-Munk function 225

l
Langevin equation 139
lateral interactions 26, 28
lattice vibrations 123, 124, 164, 218, 302–304
Lennard-Jones potential 199
Lindemann criterion 304
line width 51, 52
local work function 71–74, 268, 310, 311
low energy electron diffraction 147, 155–159, 212, 245–247, 302, 304
low energy ion scattering 85, 112–117
low-index surface 298, 299

m
Madelung sum 49
magnetic hyperfine splitting 131, 134
magnetization 137
Mars-van Krevelen mechanism 98
mass spectrometer 12, 24, 87
mass transfer limitation 35
matrix effect 87, 95, 101, 105
mean free path 6, 40, 41, 44–46, 50, 59, 60, 63, 67, 78, 79, 155, 163, 304, 309
mean squared displacement 302–304
metal support interaction 259–261
metal surfaces, theory of 297–310
metallic bond 306–311
metal-support interaction 4
metastable atom excitation spectroscopy 73, 74
metastable ion excitation spectroscopy 73, 74
metathesis 95
methanol synthesis 79, 80, 98, 142, 171
microscopy 179–211
model catalysts 8, 252
model support 252

model systems 292
molecular orbital theory 305, 311–319
momentum, total 43
monochromator, infrared 226
monochromator, X-rays 52, 53
Morse potential 219, 220
Mössbauer effect 122–125
Mössbauer parameters 127, 128, 134
Mössbauer spectroscopy 121–145, 277–281, 302
multiplett splitting 50, 51

n
neutralization 90, 93, 106, 112–115
NEXAFS see X-ray absorption near edge spectroscopy
nuclear reaction in RBS 110
nucleation and growth 16, 20

o
orbital momentum 43
oscillating reaction 197, 211–214
overtone 219, 220
oxidation of CO 2

p
particle size 151, 152, 169, 170, 187–189, 192, 193
partition function 32, 33
phase shift 162–166
Phillips catalyst 185, 186, 284–291
phonon 123, 243, 303, 304
photoelectric effect 39–41
photoelectron 39–41, 159–161, 191
photoelectron diffraction 62
photoelectron emission microscopy 180, 212
photoemission 39, 42, 48, 49, 267, 268, 307–309
photoemission of adsorbed xenon 71–74, 266, 311
photon sources 7
physisorption 2, 71
plasmon 50, 75, 164
poison 4, 97, 98
polarizability 239
polyethylene 284–291
polymerization catalysis 185, 186, 205, 284–291
pore volume 3
preexponential factor 25, 27, 33–36, 211
probing depth 44
promoter 4, 70, 97, 98, 234, 264–271
pulsed field desorption mass spectrometry 197

q

quadrupole moment   129
quadrupole splitting   *see* electric quadrupole splitting
quantum numbers of electrons   43, 223
Quick EXAFS   170–172, 274–276

r

Raman scattering   224
Raman spectroscopy   217, 218, 220, 235–243
rate of desorption   25
rate of reaction   35
rate-determining step   2
Rayleigh scattering   238, 239
reciprocal lattice   156, 157
recoil free fraction   122–124, 132–134, 144
reduced mass   219
reduction   3, 13–21, 171, 172, 255–257
reference compounds   292
reflection absorption infrared spectroscopy   224–227, 234, 235, 244–247, 271
reflection anisotropy microscopy   214
reionization   112, 114
relaxation   48, 76, 164
resonance neutralization   113
rotational energy   218, 222, 234
Rutherford backscattering spectroscopy   85, 108–111, 273, 274, 286–289

s

scanning Auger microscopy   80, 278–280
scanning electron microscopy   182, 184–186
scanning force microscopy   180, 197–208, 290, 291
scanning probe microscopy   197–211
scanning transmission electron microscopy   80, 182, 186–189
scanning tunneling microscopy   180, 197, 198, 205–211, 282, 311
scanning tunneling spectroscopy   208
scattering cross section   108–110, 113
Scherrer equation   151
second order Doppler shift   127–129
secondary electrons   66, 75, 184
secondary ion mass spectrometry   85–105, 252, 257, 260, 273, 285–287
secondary neutral mass spectrometry   86, 105, 106
selection rule   217, 220, 226, 239, 244
selective chemisorption   59, 167–170
selective oxidation   2, 87, 88, 98, 117, 143

shake up, shake off   50, 51, 164
shrinking core model   16
single crystal   8
sintering   4, 302
small angle X-ray scattering   154
solid state reaction   152, 153
spectroscopic notation   43
spin-orbit splitting   43, 44
sputter yield   88–90, 93, 105
sputtering   86–90, 93, 105, 155, 117
static disorder   164
static secondary ion mass spectrometry   94, 101–104
sticking coefficient   65
Stokes band   238, 239
strong metal support interaction   252, 260, 261
structural promoter   4, 265
subsurface species   62
sulfidation   21–23, 110, 111, 241, 272–276, 283
sum frequency generation   224, 235–238
superparamagnetism   137, 139, 142
support   2, 3
supported catalyst   3, 4
surface core-level shift   308, 309
surface free energy   46, 166, 301, 302, 306
surface sensitivity   6, 41, 50, 54, 59, 79, 86, 112, 117
synchrotron   7, 40, 63, 64, 67, 148, 153, 154, 160, 171, 172

t

take-off angle   59–62
temperature programmed desorption   23–35, 29–31, 102, 270, 271
temperature programmed oxidation   11, 12, 18
temperature programmed reaction spectroscopy   22, 23, 35
temperature programmed reduction   11–13, 16–18, 254
temperature programmed secondary ion mass spectrometry   102
temperature programmed sulfidation   21, 22, 272–276
temperature programmed techniques   11–38
thermal conductivity detector   12
thermal desorption spectroscopy   23
thermodynamics, of reduction   13, 14
threeway catalyst   *see* automotive exhaust catalysis

trajectory, ion   112, 113, 115
transition state theory   26, 29, 31, 32, 34
transmission electron microscopy   180, 182–186, 260, 261
transmission infrared spectroscopy   227–230

u

ultra violet photoelectron spectroscopy   40, 41, 65–74, 266, 269–271, 309, 311
unit cell   300
UV Raman spectroscopy   239, 240
UV source   65

v

vacuum level   50, 71, 72, 307–319
vacuum, ultrahigh   8
valence band   49, 50, 66, 67, 113, 194
van de Graaf accelerator   110
van der Waals forces   200
vibrational energy   218
vibrational spectroscopy   217–247
vibrations, types of   221
vicinal surface   299

w

wave number   161
white line   173, 174
Wood's notation   156, 158, 246, 300
work function   41, 48, 50, 66–71, 76, 93, 113, 194–197, 205, 206, 265–268, 307–319

x

X-ray absorption near edge spectroscopy   147, 159–161, 172–175, 232, 254, 291
X-ray absorption spectroscopy   147, 148, 159
X-ray diffraction   147–154, 171, 172, 276
X-ray emission   190–193
X-ray fluorescence   52, 75, 77, 148, 190, 191
X-ray notation   43
X-ray photoelectron spectroscopy   39–65, 45, 51, 52, 54–58, 63–65, 148, 247, 252–257, 273, 274, 283, 284, 285–287, 289, 309
X-ray source   41, 51

z

ZAF correction   192
Zeeman effect   131
Ziegler-Natta catalyst   284, 285

Related Titles

G. Centi, R.A. van Santen (Eds.)

Catalysis for Renewables

From Feedstock to Energy Production

2007
ISBN 978-3-527-31788-2

R.A. van Santen, M. Neurock

Molecular Heterogeneous Catalysis

A Conceptual and Computational Approach

2006
ISBN 978-3-527-29662-0

B. Heaton (Ed.)

Mechanisms in Homogeneous Catalysis

A Spectroscopic Approach

2005
ISBN 978-3-527-31025-8

I. Chorkendorff, J.W. Niemantsverdriet

Concepts of Modern Catalysis and Kinetics

2003
ISBN 978-3-527-30574-2

J.F. Haw (Ed.)

In-Situ Spectroscopy in Heterogeneous Catalysis

2002
ISBN 978-3-527-30248-2